Essential Oils as Antimicrobial Agents in Food Preservation

Perishable products such as fruits and vegetables account for the largest proportion of food loss due to their short shelf life, especially in the absence of proper storage facilities, which requires sustainable, universal, and convenient preservation technology.

The existing methods to prolong the shelf life of food mainly include adding preservatives, irradiation, cold storage, heat treatment, and controlled atmosphere storage. But with disadvantages in irradiation, cold storage, heat treatment, and controlled atmosphere storage, chemical synthetic preservatives . are still the main means to control food corruption.

As the food industry responds to the increasing consumer demand for green, safe, and sustainable products, it is reformulating new products to replace chemical synthetic food additives. *Essential Oils as Antimicrobial Agents in Food Preservation* provides a comprehensive introduction to the antimicrobial activity of plant essential oils and their application strategies in food preservation. It is aimed at food microbiology experts, food preservation experts, food safety experts, food technicians, and students.

Features:

- Summarizes the application strategy and safety of essential oil in the field of food preservation
- Describes the synergistic antibacterial effect of essential oil and antimicrobial agents
- Explains the action mechanism of essential oil as antimicrobial agent against foodborne fungi, foodborne bacteria, viruses, and insects
- Analyzes the antimicrobial activity of essential oil in gas phase

The book discusses how as a natural antimicrobial and antioxidant, essential oil has great potential to be used in the food industry to combat the growth of foodborne pathogens and spoilage microorganisms. But because the essential oil itself has obvious smell and is sensitive to light and heat, it cannot be directly added to the food matrix and thus the application strategies presented in this book explain how to alleviate those issues.

Essential Oils as Antimicrobial Agents in Food Preservation

Jian Ju

CRC Press
Taylor & Francis Group
Boca Raton London

CRC Press is an imprint of the
Taylor & Francis Group, an **informa** business

First edition published 2023
by CRC Press
6000 Broken Sound Parkway NW, Suite 300, Boca Raton, FL 33487–2742

and by CRC Press
4 Park Square, Milton Park, Abingdon, Oxon, OX14 4RN

CRC Press is an imprint of Taylor & Francis Group, LLC

ISBN: 978-1-032-34874-2 (hbk)
ISBN: 978-1-032-35904-5 (pbk)
ISBN: 978-1-003-32926-8 (ebk)

DOI: 10.1201/9781003329268

Typeset in Times
by Apex CoVantage, LLC

Contents

Preface....................xiii
About the Author....................xv

Chapter 1 Application of Plant Essential Oils in Different Fields....................1

1.1 Application of Plant Essential Oil in Food Industry....................1
1.2 Application of Plant Essential Oil in Fish Culture....................4
1.3 Application of Plant Essential Oil in Livestock and Poultry Breeding....................5
1.4 Application of Plant Essential Oil in Pesticide Industry....................5
1.5 Application of Plant Essential Oil in Cosmetic Industry....................6
1.6 Application of Plant Essential Oils in Medicine....................7
1.7 Concluding Remarks....................8
References....................8

Chapter 2 The Inhibitory Activity and Mechanism of Essential Oils on Foodborne Bacteria....................13

2.1 Introduction....................13
2.2 Sources and Extraction Methods of Common Essential Oils....................14
2.3 The Main Chemical Components of Essential Oils....................14
2.4 Relationship Between Chemical Constituents and Antibacterial Activities of Essential Oils....................17
2.5 Main Evaluation Methods of Essential Oils Antibacterial Effect....................17
2.5.1 Disc Diffusion Method....................18
2.5.2 Dilution Method....................18
2.5.3 Time Sterilization....................19
2.5.4 Checkerboard Test....................19
2.6 Inhibitory Effect of Essential Oils on Foodborne Pathogenic Bacteria....................20
2.7 Antibacterial Mechanism....................20
2.7.1 Damage Effect of Essential Oil on Cell Wall....................20
2.7.2 Damage Effect of Essential Oil on Cell Membrane....................23
2.7.2.1 Effect of Essential Oil on Protein in Cell Membrane....................23
2.7.2.2 Effect of Essential Oil on Fatty Acids in Cell Membrane....................24
2.7.3 Effect of Essential Oil on Energy Metabolism....................25
2.7.4 Effect of Essential Oil on Genetic Material....................25
2.7.5 Effect of Essential Oil on Cell Morphology....................26
2.7.6 Effect on Exercise Ability and Biofilm Formation....................27

2.7.7 Effect on Quorum Sensing 27
2.7.8 Effect of Essential Oil on Bacterial Toxins 29
2.8 Concluding Remarks 29
References 31

Chapter 3 The Inhibitory Activity and Mechanism of Essential Oils on Foodborne Fungi 39

3.1 Introduction 39
3.2 Main Factors Affecting the Yield and Chemical Composition of Essential Oil 40
3.2.1 Harvest Time of Plants 40
3.2.2 Geographical Position 41
3.2.3 Weather Conditions 41
3.2.4 Extraction Method 41
3.2.5 Drying Conditions 42
3.2.6 Distillation Time 42
3.3 Common Experimental Methods for Evaluating Antimicrobials Mechanism 43
3.3.1 Permeability of Cell Membrane 43
3.3.2 Integrity of Cell Membrane 45
3.3.3 Release of Cellular Components 45
3.4 Antifungal Activity of Essential Oils 46
3.4.1 Effect on the Cell Wall and Cell Membrane 46
3.4.2 Effect on Nucleic Acids and Proteins 50
3.4.3 Effect on Mitochondria 51
3.4.4 Effect on Membrane Pumps 52
3.4.5 Effect on ROS Production 53
3.4.6 Effect of Mycotoxins 54
3.4.7 Effect on Biofilms 54
3.5 Concluding Remarks 56
References 57

Chapter 4 Essential Oils in Vapor Phase as Antimicrobial Agents 69

4.1 Introduction 69
4.2 Chemical Composition of Essential Oils 69
4.2.1 Phenols 71
4.2.2 Terpenoids 71
4.3 Detection Methods of Volatile Compounds in Essential Oils 72
4.4 Interaction Between Essential Oils and Their Components 72
4.5 Antimicrobial Activity of Essential Oils in Vapor Phase 74
4.6 Evaluation of the Antimicrobial Activity of Essential Oils in Vapor Phase 74
4.6.1 Inverted Petri Dish Method 75
4.6.2 Boxes of Different Materials to Create the Volatile Space of Essential Oils 75

4.6.3 Agar Plug Method 75
4.6.4 Other Methods 75
4.7 Potential Applications of Essential Oils in Vapor Phase 76
4.8 Concluding Remarks 78
References 78

Chapter 5 Synergistic Antibacterial Effect of Essential Oils and Antibacterial Agent 87

5.1 Introduction 87
5.2 Factors Affecting the Antimicrobial Activity of Essential Oils 88
5.2.1 Microorganisms 88
5.2.1.1 Species of Microorganism 88
5.2.1.2 Number of Microorganisms 88
5.2.1.3 Stress Response of Microorganisms 89
5.2.2 Dispersants and Emulsifiers 89
5.2.3 Cultivation Conditions 89
5.2.3.1 Time 89
5.2.3.2 Temperature 90
5.2.3.3 Oxygen Content 90
5.2.3.4 Acidity and Alkalinity 90
5.3 Methods for Evaluating the Synergistic Effect of Drugs 90
5.4 The Interaction Between Essential Oils and Its Active Components 92
5.5 Combined Application of Essential Oils and Antimicrobial Agents 93
5.5.1 The Main Causes of Bacterial Drug Resistance 93
5.5.2 The Potential of Combined Essential Oils and Food Additives 93
5.5.3 The Potential of Combined Essential Oils and Antibiotics 94
5.6 Concluding Remarks 95
References 96

Chapter 6 Antioxidant Activity and Mechanism of Essential Oils 103

6.1 Introduction 103
6.2 Oxidative Stress 104
6.2.1 Effect of Oxidative Stress on Protein 104
6.2.2 Effect of Oxidative Stress on Lipids 104
6.2.3 Effect of Oxidative Stress on Genetic Material 104
6.3 Free Radicals 105
6.4 Antioxidant System 106
6.5 Antioxidative Components in Essential Oils 106
6.6 Antioxidant Mechanism of Essential Oils 108
6.6.1 Direct Antioxidant Effect 108
6.6.1.1 Free Radical Scavenging 108

6.6.1.2 Chelation With Metal Ions 109
6.6.2 Indirect Antioxidant Effect 110
6.6.2.1 Inhibit Lipid Peroxidation 110
6.6.2.2 Regulate the Level of Antioxidant Enzymes 111
6.7 Factors Affecting the Antioxidant Activity of Essential Oils 111
6.7.1 Solubility 112
6.7.2 Extraction Method 112
6.7.3 Initiator of Oxidation 113
6.7.4 Receptor Type 113
6.7.5 Storage Condition 113
6.8 Concluding Remarks 113
References 114

Chapter 7 Antiviral Activity and Mechanism of Essential Oils 121

7.1 Introduction 121
7.2 An Overview of Viruses 122
7.3 Plant Essential Oils With Antiviral Activity and Their Active Components 123
7.4 SARS-CoV-2 Entering the Airway 124
7.5 The Main Antiviral Pathway and Mechanism of Essential Oil and Its Active Components 126
7.5.1 The Main Antiviral Ways of Essential Oils and Their Active Components 126
7.5.2 Antiviral Mechanism of Essential Oil and Its Active Components 127
7.6 Concluding Remarks 129
References 129

Chapter 8 The Repellent or Insecticidal Effect of Essential Oils Against Insects and Its Mechanism 135

8.1 Introduction 135
8.2 Active Components in Plant Essential Oils That Have Insecticidal or Insecticidal Effects 136
8.2.1 Terpene Compound 137
8.2.2 Phenolic Compounds 138
8.2.3 Alkaloid 139
8.3 Inhibitory Effect of Plant Essential Oils on the Reproductive Ability of Insect Pests 139
8.4 Repellent or Insecticidal Activity of Plant Essential Oils 139
8.5 Insecticidal Mechanism of Plant Essential Oils 140
8.6 Development of Resistance to Plant Essential Oils or Phytochemicals 141

8.7 Concluding Remarks 142
References 143

Chapter 9 Preparation Strategy and Application of Porous Starch Microcapsule and Cyclodextrin Containing Essential Oils 147

9.1 Introduction 147
9.2 The Concept and Development of Microencapsulation 147
9.3 Wall and Core Materials of Microcapsules 148
9.3.1 Wall Materials 148
9.3.2 Core Materials 149
9.4 Application of Microencapsulation Technology in Food Industry 149
9.5 Modified Starch 150
9.6 Structural Characteristics and Properties of Porous Starch 150
9.7 Preparation of Microencapsulated Porous Starch 152
9.8 Binding Characteristics of Essential Oils–Starch Microcapsules 152
9.9 Preparation of Essential Oils–Starch Microcapsules 153
9.10 Action Mechanism 158
9.11 Application of Essential Oils–Starch Microcapsules in Food Preservation 162
9.12 Application of the Combination of Cyclodextrin and Essential Oils in Food Preservation 163
9.13 Concluding Remarks 165
References 165

Chapter 10 Preparation Strategy and Application of Edible Coating Containing Essential Oils 171

10.1 Introduction 171
10.2 The Concept and Development of Edible Coatings 171
10.3 Requirements for the Use of Edible Coating Materials 172
10.4 Materials Used for Essential Oils–Edible Coatings 172
10.4.1 Polysaccharide-Based Coatings 176
10.4.2 Protein-Based Coatings 177
10.4.3 Lipid-Based Coatings 178
10.4.4 Composite Coatings 178
10.5 Definition and Production Method of Coating 179
10.5.1 Definition of Essential Oils–Edible Coating 179
10.5.2 Production Method of Coating 179
10.5.2.1 Dipping Method 179
10.5.2.2 Spraying Method 180
10.5.2.3 Spreading Method 181
10.5.2.4 Thin Film Hydration Method 181
10.6 Mechanisms of Food Protection 182

10.7 Antibacterial Activity of Essential Oils–Edible Coating....... 183
10.8 Food Sensory Properties Related to Edible Coatings............ 191
10.8.1 Appearance of the Product.................................... 192
10.8.2 Texture of the Product.. 193
10.8.3 Taste, Smell, and Other Sensory Properties of the Product.. 194
10.9 Concluding Remarks 194
References 195

Chapter 11 Preparation Strategy and Application of Nanoemulsion Containing Essential Oils 201

11.1 Introduction 201
11.2 Development of Nanoemulsion 201
11.3 Physicochemical Properties and Advantages of Nanoemulsion 202
11.4 Preparation Methods and Formation Mechanism................. 204
11.5 Nanocarriers as Essential Oil Transporters.......................... 205
11.6 Effect of Embedding on Antimicrobial Activity of Essential Oil 207
11.7 Quantification of Essential Oil in Nanosystem 212
11.7.1 Spectrophotometry 212
11.7.2 Chromatographic Method 214
11.7.2.1 Gas Chromatography............................ 214
11.7.2.2 High-Performance Liquid Chromatography.................................. 214
11.7.3 Indirect Quantization.. 215
11.8 Antimicrobial Mechanisms 215
11.9 Factors Affecting Antimicrobial Mechanisms...................... 217
11.10 Applications in Food Preservation....................................... 218
11.11 Application Prospects and Challenges 219
11.12 Concluding Remarks 220
References 221

Chapter 12 Preparation Strategy and Application of Electrospinning Active Packaging Containing Essential Oils.................................. 227

12.1 Introduction 227
12.2 The Principles of Electrospinning 228
12.3 The Main Factors Affecting the Quality of Electrospinning 230
12.3.1 Solution Parameters 230
12.3.2 Process Parameters... 233
12.3.3 Environmental Parameters 234
12.4 Electrospun Entrapment of Bioactive Substances................ 234
12.5 Applications of Electrospinning to Food Preservation and Packaging... 235

12.6 Optimization of Active-Component Loading and Simulation of Release Behavior 236
12.7 Application of Electrospun Mats Containing Essential Oil in Food Preservation 238
12.8 Concluding Remarks 239
References 240

Chapter 13 Preparation Strategy and Application of Active Packaging Containing Essential Oils 243

13.1 Introduction 243
13.2 Research Status of Antimicrobial Packaging Theory 244
13.3 Essential Oil–Active Packaging System 244
13.4 Essential Oil–Active Packaging 246
13.4.1 Active Packaging 246
13.4.2 Realization of Essential Oil–Active Packaging 253
13.4.2.1 Be Directly Mixed Into the Packaging Matrix Material 253
13.4.2.2 Be Coated on Packaging Materials or Adsorbed by Packaging Materials 254
13.4.2.3 Be Made Into Small Antibacterial Package 255
13.4.2.4 Be Added to Packaging in Gaseous Form 255
13.4.2.5 Be Microencapsulated and Then Packaged 255
13.5 Release Mechanism of Active Compounds 256
13.6 Application of EOs–Active Packaging in Food Preservation 256
13.7 Concluding Remarks 257
References 258

Chapter 14 In Vivo Experiment, Stability, and Safety of Essential Oils 263

14.1 Introduction 263
14.2 The Relationship Between Essential Oil and Food 267
14.3 How to Ensure the Safety of Plant Essential Oil Itself 267
14.4 Stability of Plant Essential Oils 268
14.5 Toxicological Studies 269
14.5.1 Genotoxicity and Mutagenicity 270
14.5.2 Cytotoxicity 270
14.5.3 Phototoxicity 275
14.6 Antimutagenicity of Essential Oil 276
14.7 Safety of Some Common Plant Essential Oils 276
14.8 Concluding Remarks 278
References 279

Index 285

Preface

Current global challenges, such as famine, public health emergencies, and climate change, are closely related to serious food waste worldwide. Perishable products such as fruits and vegetables account for the largest proportion of food loss due to their short shelf life, especially in the absence of proper storage facilities, which requires sustainable, universal, and convenient preservation technology.

The existing methods to prolong the shelf life of food mainly include adding preservatives, irradiation, cold storage, heat treatment, and controlled atmosphere storage. However, there are some disadvantages associated with irradiation, cold storage, heat treatment, and controlled atmosphere storage. For example, the use cost is high, the scope of application is narrow, the technology is difficult, and so on. Therefore, at present, chemical synthetic preservatives are still the main means to control food corruption. However, the use of chemical synthetic preservatives is being restricted because long-term use of chemical synthetic preservatives may pose a threat to human health, including carcinogenicity, teratogenicity, and acute toxicity. In addition, the slow degradation of chemical synthetic preservatives in the environment may lead to environmental pollution.

At present, the development of clean label food is increasing as the food industry responds to the increasing consumer demand for green, safe, and sustainable products. As a result, the food industry is reformulating new products to replace chemical synthetic food additives. As a natural antimicrobial and antioxidant, essential oil has great potential to be used in the food industry to combat the growth of foodborne pathogens and spoilage microorganisms. However, because the essential oil itself has obvious smell and is sensitive to light and heat, it cannot be directly added to the food matrix. This greatly limits the promotion and application of essential oil in the food industry.

Therefore, in order to solve these problems effectively, I wrote *Essential Oils as Antimicrobial Agents in Food Preservation.* This book gives a comprehensive introduction to the antimicrobial activity of plant essential oils and their application strategies in food preservation. It is an effective reference book for food microbiology experts, food preservation experts, food safety experts, food technicians, and students.

Preface

About the Author

Prof. Ph.D Jian Ju, Special Food Research Institute, Qingdao Agricultural University, China

Jian Ju has a Ph.D. At present, he is a professor and academic leader of Special Food Research Institute of Qingdao Agricultural University. He is also a Taishan Scholar Young Expert in Shandong Province. His main responsibility at Qingdao Agricultural University is to lead team members to explore biodegradable antibacterial packaging containing plant active components and apply them to extend the shelf life of food.

Dr. Ju Jian graduated from the School of Food, Jiangnan University. His main research direction is food preservation. He presided over and participated in the National Natural Science Foundation of China, the Special fund for Taishan Scholars Project, the "Top Talent" Scientific Research Initiation Fund of the Special Support Program of Qingdao Agricultural University, the Jiangsu Postgraduate Innovation Project, and the Jiangnan University Excellent Doctoral Student Cultivation Fund. At present, he is the guest editor of *Frontiers in Nutrition* and *Journal of Visualized Experiences*. In addition, He is also a reviewer of international academic journals such as *Journal of Ethnopharmacology, Colloids and Surfaces B: Biointerfaces, Phytomedicine, International Food Research*, and *Phytomedicine*. In 2020, he was invited by the Swiss Federal Institute of Technology in Lausanne. As a visiting scholar, he was sent to the School of Bioengineering of the Swiss Federal Institute of Technology in Lausanne for academic exchanges.

Over the years, Dr. Ju Jian has devoted himself to the research of the inhibitory effects of plant extracts on food-borne bacteria and fungi, especially the antibacterial mechanism of plant essential oils against microorganisms. He has successfully combined essential oil preservation technology with food processing and food packaging technology to reduce the potential harm of microorganisms and food packaging pollution to the human body. At present, more than 40 related research papers have been published in this research field. He has published 2 academic monographs. Furthermore, he has provided technical support services for several enterprises.

About the Author

[illegible]

1 Application of Plant Essential Oils in Different Fields

With the improvement of the quality of life, people are more and more interested in all kinds of natural products. Therefore, "green, safe, and degradable" botanical products are highly recognized. Plant essential oil (EO), also known as volatile oil, is known as liquid gold. Essential oils usually have fragrant odors and exist in flowers, seeds, stems, leaves, roots, and other parts of plants. They can be obtained by steam distillation extraction, chemical solvent extraction, and carbon dioxide extraction. Plant essential oils are safe, natural, and pollution-free plant products, which have many functions such as sterilization, antioxidation, anticancer, and insecticidal. At present, plant essential oils have been widely used in food industry, fish breeding industry, livestock and poultry breeding industry, cosmetics industry, pesticide industry, and pharmaceutical industry (Figure 1.1). This chapter analyzes and summarizes the application of plant essential oil in different industries in order to provide reference for promoting the deep processing and application of plant essential oil.

1.1 APPLICATION OF PLANT ESSENTIAL OIL IN FOOD INDUSTRY

Fruits and vegetables are extremely vulnerable to microbial contamination during transportation, storage, and sale. At present, the commonly used fungicides in the market are mainly chemical syntheses. Although chemical synthetic fungicides have the advantages of low cost and strong bactericidal ability, long-term use may pose a threat to human health. As a natural antibacterial agent, plant essential oil can effectively inhibit microbial pollution and prolong the shelf life of fruits and vegetables.

Therefore, in recent years, more and more researchers have paid attention to the effect of different kinds of plant essential oils on fruit preservation. In addition, the preparation of edible films or antibacterial active packaging by mixing plant essential oils with good film-forming polymers for the preservation of all kinds of fruits has been explored and some gratifying results have been achieved. For example, the active film based on tea tree essential oil and sodium alginate can effectively prolong the storage time of banana and reduce the incidence of anthracnose during storage (Yang et al., 2022). Similarly, the starch film containing clove essential oil can extend the shelf life of grapes at 5°C to 15 days, and the active film does not have any adverse effect on the sensory properties of grapes (Sneh et al., 2022). In addition, silica nanocomposites loaded with turmeric essential oil (TEO) can effectively prolong the shelf life of surimi. The active material can reduce the microbial load from

DOI: 10.1201/9781003329268-1

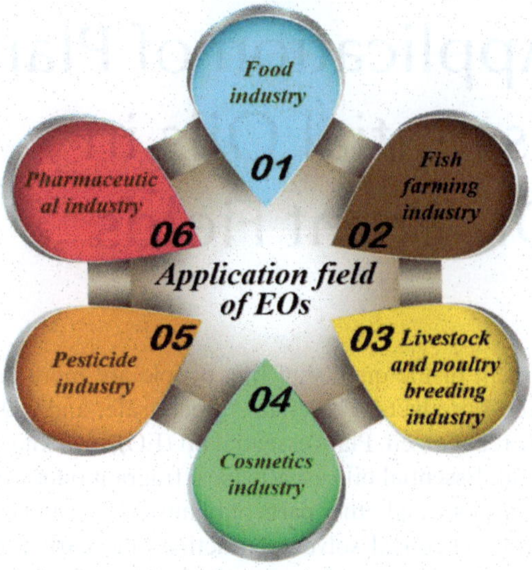

FIGURE 1.1 Application of essential oils in different fields.

4.0 log cfu/g to 2.78 log cfu/g within 14 days (Ds et al., 2022). The mixed film of starch and whey protein containing thyme essential oil and the chitosan-based film containing *Schinus molle* essential oil can prolong the shelf of tomato (Gargi and Shivani, 2022; Quetzali et al., 2022).

Livestock and poultry meat and its products are not only rich in lipids, proteins, and other nutrients, but also contain high water activity, which can be used as a natural medium for the growth and reproduction of microorganisms. Therefore, meat products are easily spoiled by microbial contamination. In recent years, a large number of reports have been carried out on the application of plant essential oils in the preservation of livestock and poultry meat. For example, the starch-based film containing oregano essential oil has significant antibacterial and antioxidant activity. The starch-based film loaded with 3% oregano essential oil could significantly ($P < 0.05$) inhibit the growth of *Bacillus subtilis*, *Escherichia coli*, and *Staphylococcus aureus*. The film has been proved to be effective in prolonging the shelf life of chicken under refrigerated conditions (Shen et al., 2022). The microemulsion containing thyme essential oil has a good fresh-keeping effect on mutton. By comparing the total number of bacteria on the surface of the sample and the control sample, it was found that the total number of bacteria in the control group was 8.97 log cfu/g during the storage period, while that in the experimental group was only 3.37 log cfu/g (Irshaan et al., 2020). Similarly, the active packaging of cellulose rice fiber/whey protein containing TiO_2-rosemary essential oil could significantly inhibit the growth of spoilage bacteria and reduce the oil oxidation rate of beef during storage. The active packaging can prolong the shelf life of beef from 6 days to 15 days (Alizadeh-Sani et al., 2020).

Aquatic products are rich in nutrition, rich in high-quality protein, and unsaturated fatty acids are important sources of protein in the diet of residents. In recent years, with the continuous improvement of people's awareness of green, safety, and environment-friendliness, biological antistaling agents including plant essential oils are considered to be the most potential fresh-keeping methods in the field of aquatic products. Therefore, there have been a large number of related reports on the application of plant essential oils in the preservation of aquatic products. For example, hemp essential oil can enhance the antibacterial properties of protein composite nanoparticles and delay the increase of PV (peroxide value), TBA (thiobarbituric acid), and TVB-N (total volatile basic nitrogen) in rainbow trout fillets. The active coating containing the essential oil can prolong the shelf life of rainbow trout fillets by 8–14 days. Therefore, the protein-based active membrane containing hemp essential oil can be considered as a natural preservative for fish coating (Majidiyan et al., 2022). The active packaging of sustained release polylactic acid/polyhydroxyalkanoate (PLA/PHA) containing oregano essential oil (OEO) can prolong the shelf life of puffer fish fillets by 2 days at 4°C ± 1°C. The amount of active substance (carvol) released from the active film to puffer fish during storage is 9.70 mg/kg (Zheng et al., 2022). In addition, related researchers have studied a nanoemulsion method to enhance the antibacterial properties of cinnamon essential oil as a natural preservative in real food systems. Compared with bulk cinnamon oil (488 and 11,429 mg/l) and disinfectant sodium hypochlorite (10 mg/l), the antibacterial activity of cinnamon essential oil nanoemulsion (1,429 and 11,429 mg/l) against artificially contaminated frozen sea bass slices of *E. coli*, *Salmonella typhimurium*, *S. aureus*, and *Vibrio parahaemolyticus* was studied. The results showed that compared with the untreated samples, the initial number of bacteria in cinnamon essential oil nanoemulsion (11,429 mg/l) decreased by about 0.5–1.5 log cfu/g. It also successfully inhibited the growth of these bacteria, especially *Vibrio parahaemolyticus*, which is a more effective drug than bulk cinnamon oil and sodium hypochlorite during storage (4°C ± 2°C). The antibacterial activity of cinnamon essential oil against pathogenic bacteria in fish fillet can be improved by using nanoemulsion preparation (Chuesiang et al., 2021). In other studies, it has been confirmed that active packaging containing essential oil can significantly ($P < 0.05$) prolong the shelf life of aquatic products (Hao et al., 2021; He et al., 2019; Hussain et al., 2021; Lan et al., 2021; Vieira et al., 2019).

At present, there are few reports about the fresh-keeping methods of plant essential oils combined with modified atmosphere packaging, ozone, and ultrahigh pressure treatment for aquatic products. Therefore, on the basis of the study on the fresh-keeping mechanism of plant essential oil, the research on the synergistic mechanism of essential oil will be the research trend in the future.

In addition to the aforementioned studies, essential oils can also be used to prolong the shelf life of dairy products. Helena et al. (2019) used oregano essential oil and rosemary essential oil as antistaling agents for refrigerated cheese and evaluated the bacteriostatic effect of essential oil by determining the total number of lactic acid bacteria and *Escherichia coli O157:H7* in the samples. The results showed that when the contents of oregano essential oil and rosemary essential oil were 0.03 and 1.32 μl/g respectively, the number of lactic acid bacteria was basically unchanged, while the total number of *E. coli* decreased 3.0 log cfu/g in the first 12 days. This shows that plant essential oils can inhibit the growth of *E. coli* without affecting cheese fermentation. In addition, in the field of baked goods preservation, Ju and his research team have done a lot of research on essential oils to extend the shelf life of baked goods.

They found that essential oils can effectively prolong the shelf life of baked goods (Ju et al., 2017, 2020). At the same time, essential oil can also be used as a preservative in the storage of cereal products such as rice and wheat (Bett et al., 2016; Kalagatur et al., 2020; Kaliamurthi et al., 2019).

1.2 APPLICATION OF PLANT ESSENTIAL OIL IN FISH CULTURE

Bacteria are the main pathogenic microorganisms that cause the disease of aquatic animals, accounting for 57.6% of biogenic diseases. Common pathogens include *Vibrio*, *Edwardsiella*, *Aeromonas*, and *Streptococcus*. At present, the prevention and treatment of bacterial diseases still mainly rely on antibiotics and synthetic antibiotics, such as neomycin sulfate powder, doxycycline hydrochloride, florfenicol, and thiamphenicol. However, due to the accumulation of antibiotics in aquatic products, more and more consumers are calling for the continued use of antibiotics in aquatic products. However, due to the accumulation of antibiotics in aquatic products, more and more consumers call for stopping the continuous use of antibiotics in aquatic products. As a result, the use of these antibiotics has been restricted in many countries. In recent years, the application of plant essential oils in fish culture has developed rapidly. A large number of studies have confirmed that essential oils can effectively control bacterial and parasitic diseases in aquatic animals. In addition, essential oil can also be used as an anesthetic and sedative for fish, as well as to enhance the nonspecific immunity of fish and improve the production performance of fish. Rodrigues et al. (2021) evaluated the anesthetic property and safety of *Nectandra grandiflora* Nees essential oil (NGN-EO) and its nanoemulsion (NEN) for Nile tilapia (*Oreochromis niloticus*). Formulations were made using the spontaneous emulsification process, and stability was tested for NEN and NGN-EO at 4, 25, and 40°C for 90 days. The NEN contained 95% of NGN-EO, the major compounds remained stable up to 60 days, and the NEN protected the active substances at all temperatures. Fish behavior was observed during a bath with NGN-EO and NEN 3–300 mg/l concentrations for 30 min. The difference between sedated and anesthetized fish was detected by their swimming behavior and response to stimuli on the caudal peduncle. Behavior score, survival, or adverse effects were accessed with 10 and 30 mg/l concentrations of NGN-EO and NEN during 24 h exposure and 72 h recovery. Only 100 mg/l of NEN induced anesthesia without side effects. Long-term exposure tests confirmed the 30 mg/l NEN safety, with a 100% survival rate for fish exposed to NEN and a 25.5% survival rate for fish exposed to surfactants. Therefore, this poorly water-soluble essential oil was converted into a novel sedative and anesthetic for fish through nanoemulsion. Thus, nanoemulsion protects the essential oil and improves the pharmacological activity, with the suggested use of 30 mg/l during 24 h, for sedation, and 100 mg/l, for anesthesia. It has no obvious antibacterial activity to cymene itself, but it has antibacterial synergistic effect. The combination of 2.5 μg/ml *p*-cymene and carvanol could significantly ($P < 0.05$) improve the antibacterial activity of carvanol against *Edwardsiella*, and the minimum inhibitory concentration (MIC) decreased from 20 to 5 μg/ml. The synergistic antibacterial effect was further confirmed by in vivo experiments. The cumulative mortality of tilapia was reduced from 50% to 30% by feeding carvanol with 200 mg/kg daily, and to 0% after feeding carvol and *p*-cymene essential oil (Chaparro-Hernández et al., 2015). Tea bark essential oil and tea leaf essential oil had significant ($P < 0.05$) inhibitory effect on water mold isolated from Atlantic salmon skin. The minimum inhibitory concentrations were 30 and 100 μg/ml, and the minimum bactericidal concentrations were 50 and 125 μg/ml, respectively. It can be seen that the essential oils extracted from different parts of the same plant have different antibacterial activities (Madrid et al., 2015).

1.3 APPLICATION OF PLANT ESSENTIAL OIL IN LIVESTOCK AND POULTRY BREEDING

Plant essential oil is a secondary metabolite produced by the plant itself, because it can promote the digestion and absorption of nutrients, inhibit the colonization of intestinal pathogenic bacteria, and enhance the immune function of animals. It is widely used in ruminant and poultry feed as a substitute for antibiotics. For monogastric animals such as chickens, the addition of oregano essential oil to chicken feed plays a significant role in improving the activity of chymotrypsin in the digestive system of broilers. Chicken intestines are also more efficient in digesting crude protein. However, adding plant essential oils to chicken feed is not the more the better. Adding plant essential oils according to the amount of 50 mg/kg is a more appropriate line (Lee et al., 2004). Plant essential oils also play an important role in promoting the development of duodenal villi in chickens. Fennel essential oil, citrus peel essential oil, and oregano essential oil are all common essential oils, which play an obvious role in improving the length of duodenal villi in broilers.

After entering the body, plant essential oil can play a role in the following two aspects: (1) The active components of plant essential oil extracts (such as thymol, eugenol, and cinnamaldehyde) can inhibit or even kill intestinal harmful microorganisms and promote the reproduction and growth of beneficial bacteria. In addition, it can increase digestion and absorption by increasing the content and activity of endogenous enzymes. (2) Plant extracts can stimulate the relevant variables of gastrointestinal tract to increase the secretion of mucus and bile acid in digestive tract and the activity of some digestive enzymes, so as to promote the digestion and absorption of nutrients in feed. For example, the addition of plant essential oil to feed is beneficial to the reproduction and growth of beneficial bacteria in the intestine. At the same time, it can inhibit the growth of harmful bacteria (*Eimeria*, *E. coli*, *Clostridium perfringens*) and stabilize intestinal flora and enhance immunity. Similarly, eugenol, the active component of essential oil, has a good inhibitory effect on the growth of intestinal *C. perfringens*. In the same way, Guo et al. (2015) reached a similar conclusion.

1.4 APPLICATION OF PLANT ESSENTIAL OIL IN PESTICIDE INDUSTRY

For a long time, people have relied on chemical pesticides to solve the problem of pest control in agricultural production. The use of chemical pesticides has not only made a great contribution to ensuring the supply and demand of agricultural products, but also brought huge economic benefits for human beings. However, a series of problems such as environmental pollution, pesticide residues in agricultural products, and drug resistance of pests/weeds urge people to actively explore efficient and environment-friendly biological pesticides. Therefore, in recent years, the research and application of essential oils in the field of pesticides are increasing. For example, the application of the essential oil from *Eupatorium buniifolium* at 3% on whitefly-infested plants can cause whitefly adult mortality without affecting the crop yield (Umpiérrez et al., 2017). Similarly, in bioassays with treated grain, *Conyza sumatrensis* and *Erigeron canadensis* essential oils exhibited excellent toxicity against adult *Tribolium castaneum* with LD50 of 3.7 and 5.6 mg per 10 g grains, whereas in a fumigation bioassay they showed LD50 of 6.6 and 10.6 mg/l, respectively. The essential oils extracted from *Chenopodium ambrosioides* and *E. canadensis* exhibited good antifungal activity against *A. flavus* (Muhammad et al., 2022). In addition, in a previous study, vanillin significantly enhanced repellent activity of selected monoterpene compounds, including carvacrol, thymol, α-terpineol, and *trans*-cinnamaldehyde 2.7–4.9-folds (Tak and Isman, 2017).

Essential oil has strong inhibitory effect on insect pests, but has low toxicity to humans and mammals, and it is easy to degrade in the environment. Essential oil is expected to be developed into a new green protective agent for grain storage. At the same time, the development and utilization of essential oil as a plant-derived acaricide and fungicide is also very promising. Therefore, as a new botanical pesticide, essential oil has its potential value and broad market. With the success of the commercialization of essential oil pesticides, it will greatly promote the research and development of essential oil botanical pesticides and the sustained and rapid development of industrialization. The use of essential oil in the prevention and control of diseases and insect pests will be the development trend in the future.

1.5 APPLICATION OF PLANT ESSENTIAL OIL IN COSMETIC INDUSTRY

The abnormal deposition of melanin can easily lead to the formation of dark yellow skin, freckles, and age spots. Tyrosinase is an important rate-limiting enzyme in the formation of melanin. At present, it is considered that the key way to inhibit melanin deposition is to inhibit tyrosinase activity (Hui-Min et al., 2014). Plant essential oil can effectively inhibit the expression of regulatory factors tryptophan (TRP-1 and TRP-2) and the signal transduction of protein kinase, and reduce the activity of tyrosinase (Huang et al., 2011). Thymol can effectively inhibit the redox reaction in the melanin synthesis pathway. At the same time, the free radical scavenging ability of thyme essential oil consumes the free electrons needed for melanin synthesis and inhibits the production of melanin (Satooka and Kubo, 2011). Skin tissue is composed of elastic fiber, collagen fiber, protein, and polysaccharide matrix. Long-term exposure to ultraviolet radiation can easily cause skin damage and skin diseases. Sunscreen is one of the effective means to prevent ultraviolet radiation (Dong et al., 2017). Studies have shown that a large number of active groups in essential oils can protect cells from deoxynucleotides and enhance the ability of cells to resist ultraviolet radiation

(Mao et al., 2018). At the same time, the essential oil has significant free radical scavenging and antioxidant effects. Thyme essential oil has stronger antioxidant activity than antioxidants butylhydroxytoluene and vitamin E. In addition, Orafidiya et al. (2005) established a photoaging mouse model, and then analyzed the tissue sections of mouse skin. It was found that peony essential oil could reduce the levels of TNF-α and IL-6 in mouse skin and increase the activity of SOD and GSH-Px in mouse skin. At the same time, essential oil also plays a good role in anti-acne and anti-inflammation. Linalool, an effective component of thyme essential oil, has the effect of inhibiting tumor necrosis and limiting proinflammatory factor interleukin. At the same time, it can also inhibit the hypersensitivity induced by acetone to reduce pain and achieve the effect of anti-inflammation and analgesia. Plant essential oil can induce skin to produce cyclophosphamide, which proves that plant essential oil can promote the normal growth of hair and the proliferation of hair follicles. Recent studies have also confirmed that carvol and eugenol chitosan nanoparticles have high antibacterial properties as environmental protection preservatives in cosmetics. Among them, terpenes showed higher antibacterial properties under the encapsulation of chitosan. The results show that the effective components of these two essential oils can be used in moisturizing cream (Mondéjar-López et al., 2022).

At present, the research on essential oil in cosmetics industry is mostly focused on antioxidation, whitening, anti-inflammation, and bacteriostasis, but there are few studies on its deep-seated mechanism. The application of oil-soluble ingredients from natural sources in cosmetics also has some limitations, such as odor, color, solubility, and so on. However, we believe that with the continuous development of related packaging technology, the application of essential oil in the cosmetics industry will create more value.

1.6 APPLICATION OF PLANT ESSENTIAL OILS IN MEDICINE

Essential oils also have extensive application value in medical and health care. At present, anxiety has become a common emotional disease, and the number of patients is increasing year by year. Many reports have shown that essential oils have antianxiety effects. For example, the anxiety level of patients who inhaled lavender essential oil was significantly ($P < 0.05$) lower, calmer, and more positive than those who did not inhale lavender essential oil (Moghadam et al., 2021). Similarly, inhaling lavender and chamomile essential oils can help reduce depression, anxiety, and stress levels in the elderly (Ebrahimi et al., 2022). Lavender essential oil also has a sedative and hypnotic effect. Gilani et al. (2000) used mice as the experimental model and found that lavender essential oil could prolong the sleep time of mice, which was similar to that of diazepam at the concentration of 5 mg/kg. In addition to the aforementioned functions, essential oils also have anticancer, anti-Alzheimer's disease, and analgesic effects. For example, studies have confirmed that *Zatariamultiflora* essential oil has an inhibitory effect on breast cancer; *Bergamot* essential oil has an inhibitory effect on human neuroblastoma cells; *Nutmeg* essential oil has an inhibitory effect on Novikoff hepatoma cell. In addition, *Tetraclinisarticulata* essential oil can reduce cognitive impairment and brain oxidative stress in Alzheimer's disease amyloidosis model (Azadi et al., 2020; Fatima et al., 2019; Fahimeh et al., 2020; Mickus et al., 2021). *Lavender* essential oil significantly ($P < 0.05$) improved the treatment of migraine and the volunteers did not find any side effects during the treatment (Rafie et al., 2016). Similarly, fennel essential oil and rose essential oil have good therapeutic effect on migraine (Mosavat et al., 2021; Niazi et al., 2017).

1.7 CONCLUDING REMARKS

Plant essential oil has been well applied in food industry, fish breeding, livestock and poultry breeding, cosmetic industry, and biomedical and pesticide industry and achieved some satisfactory results. It is believed that its role is not only in these fields. With the continuous deepening of research and development, the application of plant essential oil will be excavated at a deeper level. In the next few years, the market of plant essential oil will have a blowout surge, resulting in considerable economic benefits.

ACKNOWLEDGEMENTS

This work was supported by the National Natural Science Foundation of China (32202192), Special fund for Taishan Scholars Project.

REFERENCES

Alizadeh-Sani, M., Mohammadian, E., and Clements, D. J. 2020. Eco-friendly active packaging consisting of nanostructured biopolymer matrix reinforced with TiO_2 and essential oil: Application for preservation of refrigerated meat. *Food Chem.* 322126782.

Azadi, M., Jamali, T., Kianmehr, Z., Kavoosi, G., and Ardestani, S. K. (2020). In-vitro (2D and 3D cultures) and in-vivo cytotoxic properties of *Zataria multiflora* essential oil (ZEO) emulsion in breast and cervical cancer cells along with the investigation of immunomodulatory potential. *J. Ethnopharmacol.* 257: 112865.

Bett, P. K., Deng, A. L., Ogendo, J. O., Kariuki, S. T., Kamatenesi-Mugisha, M., Mihale, J. M., and Torto, B. 2016. Chemical composition of *Cupressus lusitanica* and *Eucalyptus saligna* leaf essential oils and bioactivity against major insect pests of stored food grains. *Ind. Crops Prod.* 51–62.

Chaparro-Hernández, S., Ruíz-Cruz, S., Márquez-Ríos, E., Ocaño-Higuera, V., Valenzuela-López, C., Ornelas-Paz, J., and Del-Toro-Sánchez, C. 2015. Effect of chitosan-carvacrol edible coatings on the quality and shelf life of tilapia (*Oreochromis niloticus*) fillets stored in ice. *Food Sci. Technol.* 35(4): 734–741.

Chuesiang, P., Sanguandeekul, R., and Siripatrawan, U. 2021. Enhancing effect of nanoemulsion on antimicrobial activity of cinnamon essential oil against foodborne pathogens in refrigerated Asian seabass (*Lates calcarifer*) fillets. *Food Control.* 122: 107782.

Dong, Y., He, H., Sheng, W., Wu, J., and Ma, H. 2017. A quantitative and non-contact technique to Characterise microstructural variations of skin tissues during photo-damaging process based on Mueller matrix polarimetry. *Sci. Rep.* 7(1): 14702.

Ds, A., Vcrb, C., Jsp, B., and Bsc, B. 2022. Fabrication of chitosan-based food packaging film impregnated with turmeric essential oil (TEO)-loaded magnetic-silica nanocomposites for surimi preservation. *Int. J. Biol. Macromol.* 203: 650–660.

Ebrahimi, H., Mardani, A., Basirinezhad, M. H., Hamidzadeh, A., and Eskandari, F. 2022. The effects of *Lavender* and *Chamomile* essential oil inhalation aromatherapy on depression, anxiety and stress in older community-dwelling people: A randomized controlled trial. *Explore*. 18(3): 272–278.

Fahimeh, S., Hossein, B., Gholamreza, K., and Sussan, K. A. 2020. Incorporation of *Zataria multiflora* essential oil into chitosan biopolymer nanoparticles: A nanoemulsion based delivery system to improve the in-vitro efficacy, stability and anticancer activity of ZEO against breast cancer cells. *Int. J. Biol. Macromol.* 143: 382–392.

Fatima, Z., S., Mostafa, El. I., Oana, C., Adriana, T., Monica, H., Lucina, H., and Paula, A. P. 2019. *Tetraclinis articulata* essential oil mitigates cognitive deficits and brain oxidative stress in an Alzheimer's disease amyloidosis model. *Phytomedicine*. 56: 57–63.

Gargi, G., and Shivani, E. 2022. Thyme essential oil nano-emulsion/Tamarind starch/Whey protein concentrate novel edible films for tomato packaging. *Food Control*. 138: 108990.

Gilani, A. H., Aziz, N., Khan, M. A., Shaheen, F., and Herzig, J. W. 2000. Ethnopharmacological evaluation of the anticonvulsant, sedative and antispasmodic activities of *Lavandula stoechas* L. *J. Ethnopharmacol*. 71(1–2): 161–167.

Guo, S. S., Sun, Q, J., Liu, D., Chen, Y. M., and Yao, X. 2015. Effects of dietary essential oil and enzyme supplementation on growth performance and gut health of broilers challenged by *Clostridium perfringens*. *Anim. Feed Sci. Technol.* 207: 234–244.

Hao, R., Roy, K., Pan, J., Shah, B. R., and Mraz, J. 2021. Critical review on the use of essential oils against spoilage in chilled stored fish: A quantitative meta-analyses. *Trends Food Sci. Technol.* 111: 175–190.

He, Q., Li, Z., Yang, Z., Zhang, Y., and Liu, J. 2019. A superchilling storage—ice glazing (SS-IG) of Atlantic salmon (*Salmo salar*) sashimi fillets using coating protective layers of *Zanthoxylum* essential oils (EOs). *Aquaculture*. 734506.

Helena, T. D., Janaína, B. S., Jessicada, S. G., Rita, C. E., Marta, S., Josean, F. T., EvandroLeite, S., and Marciane, M. 2019. A synergistic mixture of *Origanum vulgare* L. and *Rosmarinus officinalis* L. essential oils to preserve overall quality and control *Escherichia coli* O157:H7 in fresh cheese during storage. *LWT-Food Sci. Technol.* 112: 107781.

Huang, Z. H., Shi, F. J., Chen, F., Liang, F. X., Li, Q., Yu, J. L., Li, Z., and Han, X. J. 2011. In vitro and in vivo assessment of an intelligent artificial anal sphincter in rabbits. *Artif. Organs*. 35(10): 964–969.

Hui-Min, D, W., Chen, C. Y., and Wu, F. P. 2014. Isophilippinolide a arrests cell cycle progression and induces apoptosis for anticancer inhibitory agents in human melanoma cells. *J. Agric. Food Chem.* 62(5): 1057–1065.

Hussain, M. A., Sumon, T. A., Mazumder, S. K., Ali, M. M., and Hasan, M. T. 2021. Essential oils and chitosan as alternatives to chemical preservatives for fish and fisheries products: A review. *Food Control*. 129: 108244.

Irshaan, S., Pratik, B., and Preetam, S. 2020. Oil-in-water emulsions of geraniol and carvacrol improve the antibacterial activity of these compounds on raw goat meat surface during extended storage at 4°C. *Food Control*. 107: 106757–106757.

Ju, J., Xu, X., Xie, Y., Guo, Y., Cheng, Y., Qian, H., and Yao, W. 2017. Inhibitory effects of cinnamon and clove essential oils on mold growth on baked foods. *Food Chem.* 240: 850–855.

Ju, J., Xie, Y., Yu, H., Guo, Y., Cheng, Y., Zhang, R., and Yao, W. 2020. Synergistic inhibition effect of citral and eugenol against *Aspergillus niger* and their application in bread preservation. *Food Chem.* 310: 125974.125971–125974.125977.

Ju, J., Xie, Y., Yu, H., Guo, Y., and Yao, W. 2020. A novel method to prolong bread shelf life: Sachets containing essential oils components. *LWT-Food Sci. Technol.* 131: 109744.

Kalagatur, N. K., Gurunathan, S., Kamasani, J. R., Gunti, L., Kadirvelu, K., Mohan, C. D., and Siddaiah, C. 2020. Inhibitory effect of *C. zeylanicum*, *C. longa*, *O. basilicum*, *Z. officinale*, and *C. martini* essential oils on growth and ochratoxin A content of *A. ochraceous* and *P. verrucosum* in maize grains. *Biotechnol. Rep.* 27: e00490.

Kaliamurthi, S., Selvaraj, G., Gu, K., and Wei, D. Q. 2019. Synergism of essential oils with lipid based nanocarriers: emerging trends in preservation of grains and related food products. *Grain Oil Sci. Technol.* 2(1): 21–26.

Lan, W., Sun, Y., Chen, M., Li, H., Ren, Z., Lu, Z., and Xie, J. 2021. Effects of pectin combined with plant essential oils on water migration, myofibrillar proteins and muscle tissue enzyme activity of vacuum packaged large yellow croaker (*Pseudosciaena crocea*) during ice storage. *Food Packag. Shelf Life*. 30: 100699.

Lee, K. W., Everts, H., and Beynen, A. C. 2004. Essential oils in broiler nutrition. *Int. J. Poult. Sci.* 3(12): 738–752.

Madrid, A., Godoy, P., González, S., Zaror, L., Moller, A., Werner, E., and Montenegro, I. 2015. Chemical characterization and anti-oomycete activity of Laureliopsis philippianna essential oils against *Saprolegnia parasitica* and *S. australis*. *Molecules*. 20(5): 8033–8047.

Majidiyan, N., Hadidi, M., Azadikhah, D., and Moreno, A. 2022. Protein complex nanoparticles reinforced with industrial hemp essential oil: Characterization and application for shelf-life extension of Rainbow trout fillets. *Food Chem: X*. 13: 100202.

Mao, T. F., Liu, G. Q., Wu, H. B., Yen, W., Gou, Y. Z., and Lei, T. 2018. High throughput preparation of UV-Protective polymers from essential oil extracts via the Biginelli reaction. *J. Am. Chem. Soc.* 140: 6865–6872.

Mickus, R., Janiuk, G., Rakeviius, V., Mikalayeva, V., and Skeberdis, V. A. 2021. The effect of nutmeg essential oil constituents on Novikoff hepatoma cell viability and communication through Cx43 gap junctions. *Biomed. Pharmacother.* 135(3): 111229.

Moghadam, Z. E., Delmoradi, F., Aemmi, S. Z., Vaghee, S., and Vashani, H. B. 2021. Effectiveness of aromatherapy with inhaled lavender essential oil and breathing exercises on ECT-related anxiety in depressed patients. *Explore*. 30: 100985.

Mondéjar-López, M., López-Jimenez, A. J., Martínez, J. C. G., Ahrazem, O., Gómez-Gómez, L., and Niza, E. 2022. Comparative evaluation of carvacrol and eugenol chitosan nanoparticles as eco-friendly preservative agents in cosmetics. *Int. J. Biol. Macromol.* 206: 288–297.

Mosavat, S. H., Jaberi, A. R., Sobhani, Z., Mosaffa-Jahromi, M., Iraji, A., and Moayedfard, A. 2021. Corrigendum to "efficacy of anise (*Pimpinella anisum* L.) oil for migraine headache: A pilot randomized placebo-controlled clinical trial". *J. Ethnopharmacol.* 281: 113111.

Muhammad, A., Tariq, Z., Arshad, M. A., Muhammad, A., Raimondas, M., Mona S., and Mohamed, S. E. 2022. Pesticidal potential of some wild plant essential oils against grain pests *Tribolium castaneum* (Herbst, 1797) and *Aspergillus flavus*. *Int J Microbiol Res.* 1(3): 037–046.

Niazi, M., Hashempur, M. H., Taghizadeh, M., Heydari, M., and Shariat, A. 2017. Efficacy of topical Rose (*Rosa damascena* Mill.) oil for migraine headache: A randomized double-blinded placebo-controlled cross-over trial. *Complementary therapies in medicine*. 34: 35–41.

Orafidiya, L. O., Fakoya, F. A., Agbani, E. O., and Iwalewa, E. O. 2005. Vascular permeability-increasing effect of the leaf essential oil of *Ocimum gratissimum* Linn. as a mechanism for its wound healing property. *Afr. J. Tradit., Complementary Altern. Med.* 2(3): 186–193.

Quetzali, N. M. R., Wendy, A., Coyotl-Pérez, E., Rubio-Rosasc, G. S., Cortes-Ramírez, J. F., Sánchez, R., and Nemesio, V. R. 2022. Antifungal properties of hybrid films containing the essential oil of *Schinus molle*: Protective effect against postharvest rot of tomato. *Food Control*. 134: 108766.

Rafie, S., Namjoyan, F., Golfakhrabadi, F., Yousefbeyk, F., and Hassanzadeh, A. 2016. Effect of lavender essential oil as a prophylactic therapy for migraine: A randomized controlled clinical trial. *J. Herb. Med.* 6(1): 18–23.

Rodrigues, P., Ferrari, F. T., Barbosa, L. B., Righi, A., Laporta, L., Garlet, Q. I., and Heinzmann, B. M. 2021. Nanoemulsion boosts anesthetic activity and reduces the side effects of *Nectandra grandiflora* Nees essential oil in fish. *Aquaculture*. 545: 737146.

Satooka, H., and Kubo, I. 2011. Effects of thymol on mushroom tyrosinase-catalyzed melanin formation. *J. Agric. Food Chem.* 59(16): 8908–8914.

Shen, Y. P., Zhou, J. W., Yang, C. Y., Chen, Y. Y., Yang, Y., Cunsha, Z., Li, W., Wang, G. H., Xia, X. J., and Yu, H. Y. 2022. Preparation and characterization of oregano essential oil-loaded *Dioscorea zingiberensis* starch film with antioxidant and antibacterial activity and its application in chicken preservation. *Int. J. Biol. Macromol.* 212: 20–30.

Sneh, P. B., William, S., Whitesidea, F., Ozogulb, K. D., Dunnoc, G. A., and Cavendera, P. D. 2022. Development of starch-based films reinforced with cellulosic nanocrystals and essential oil to extend the shelf life of red grapes. *Food Control*. 138: 108990.

Tak, J. H., and Isman, M. B. 2017. Acaricidal and repellent activity of plant essential oil-derived terpenes and the effect of binary mixtures against *Tetranychus urticae* Koch (Acari: Tetranychidae). *Ind. Crops Prod.* 108: 786–792.

Umpiérrez, M., Paullier, J., Porrini, M., Garrido, M., Santos, E., and Rossini, C. 2017. Potential botanical pesticides from *Asteraceae* essential oils for tomato production: Activity against whiteflies, plants and bees. *Ind. Crops Prod.* 109: 686–692.

Vieira, B. B., Mafra, J. F., Bispo, A., Ferreira, M. A., and Evangelista-Barreto, N. S. 2019. Combination of chitosan coating and clove essential oil reduces lipid oxidation and microbial growth in frozen stored tambaqui (*Colossoma macropomum*) fillets. *LWT-Food Sci. Technol.* 116: 108546.

Yang, Z. K., Li, M. R., Zhai, X. D., Ling, Z., Haroon, E., Tahira, J., Shia, X. B., Zou, X. W., Huang, Z. H., and Li, J. X. 2022. Development and characterization of sodium alginate/tea tree essential oil nanoemulsion active film containing TiO_2 nanoparticles for banana packaging. *Int. J. Biol. Macromol.* 213: 145–154.

Zheng, H., Tang, H., Yang, C., Chen, J., Wang, L., Dong, Q., and Liu, Y. 2022. Evaluation of the slow-release polylactic acid/polyhydroxyalkanoates active film containing oregano essential oil on the quality and flavor of chilled pufferfish (*Takifugu obscurus*) fillets. *Food Chem.* 385: 132693.

2 The Inhibitory Activity and Mechanism of Essential Oils on Foodborne Bacteria

2.1 INTRODUCTION

Plant essential oils, which are called essential oils in botany field, volatile oils in chemistry and medicine field, and aromatic oils in commerce field, are volatile oily liquids existing in leaves, roots, barks, flowers, and fruits of aromatic plants (Ju et al., 2018b). They can be distilled with steam and have certain odor (generally fragrant or pungent). Generally, they are mostly colorless and belonging to the secondary metabolites of plant body itself, which enjoy the reputation of "liquid gold" (Nakadafreitas et al., 2022).

The EOs of plants are recognized by the US FDA as "generally recognized as safe," and 2,068 kinds of food additives are permitted. Natural EO preservatives have been the research and development focus in recent years due to their strong antibacterial, safe, nontoxic, and wide range of action. They are also the main develop direction of preservatives in the future. To date, EOs have been exploited as flavoring additives, as medicines or cosmetics, and as insecticidal, antioxidant, anti-inflammatory, antiallergic, and anticancer agents (Li et al., 2022b; Ghasemi et al., 2022; Ribeiro et al., 2022). In addition, many EOs have strong antibacterial, antiviral, and antifungal activity, which also play an antiseptic in food and beverage products. Related studies have shown that EOs can not only give special aroma to food, but also have many biological activities such as coloring, antioxidant, bacteriostatic, and antiseptic, which can be used as important natural preservatives (Drr et al., 2022). At present, EOs have been widely used in food as a food preservative.

For example, the combination of marjoram essential oil, ascorbic acid, and chitosan can significantly ($P < 0.05$) prolong the preservation time of fresh-cut lettuce (Xylia et al., 2021). The edible coating composed of whey protein concentrate and rosemary essential oil can significantly ($P < 0.05$) prolong the shelf life of fresh spinach (Abedi et al., 2021). Similarly, the combination of chitosan coating and Artemisia annua essential oil can significantly ($P < 0.05$) inhibit the lipid oxidation of fresh chicken during cold storage and prolong its shelf life (Yaghoubi et al., 2021). The corn protein coating containing ginger extract and fennel essential oil can significantly ($P < 0.05$) prolong the shelf life of packaged beef (Sayadi et al., 2021). At the same time, chitosan edible coating containing thyme essential oil can effectively maintain the quality and appearance of fresh-cut carrots (Viacava et al., 2022). In

DOI: 10.1201/9781003329268-2

addition, some researchers combined EOs with other preservatives applied to cheese, sausage, fruits, vegetables, and cereals (Artiga-Artigas et al., 2017; Catarino et al., 2017; Munhuweyi et al., 2018; Deng et al., 2017). It can be seen that EO is a natural food preservative with great development potential and application value.

2.2 SOURCES AND EXTRACTION METHODS OF COMMON ESSENTIAL OILS

In recent years, the antibacterial activity of essential oils against foodborne pathogenic bacteria has been widely studied. Many essential oils have broad-spectrum antibacterial properties. For example, cinnamon essential oil can inhibit not only *Staphylococcus aureus* and *Escherichia coli*, but also *Aspergillus flavus* and *Aspergillus niger*. Similarly, clove essential oil can also inhibit the contamination of foodborne bacteria (*E. coli*, *Salmonella*, *Listeria monocytogenes*, and *S. aureus*) and foodborne fungi in food preservation (*Candida albicans*, *Zygosaccharomyces rouxii*, *A. niger*, and *Penicillium roqueforti*). Not only that, essential oils also have antioxidant (Mollica et al., 2022; Aboutalebzadeh et al., 2022; Gargouri et al., 2021), anticancer (Nirmala et al., 2021; Khazaei et al., 2021; Shahid, 2021), antimutagenic (Cardia et al., 2021; Fahmy et al., 2020), antidiabetic (Nazir et al., 2021; Alves-Silva et al., 2021), and anti-inflammatory (Geni et al., 2021; Wei and Liu, 2021) effects. EOs not only have broad-spectrum antibacterial properties but also have a wide range of sources. Table 2.1 lists the source and main extraction methods of EOs that are currently used. From Table 2.1 we can see that the EOs are mainly distributed in composite family and lip family, and they are mainly distributed in different parts of plants such as stems, leaves, flowers, fruits, seeds, skins, and roots according to plant species. The content and composition of EOs are different from part to part even in the same plant. El-Sawi and Mohamed (2002) studied the content change of EOs in seeds, stems, and leaves of cumin under the same extraction conditions. It was found that the kinds of compounds and the content of principal components in the EOs have obvious changes. In view of this situation, in future researches appropriate plant tissues can be selected according to the relevant results of activity determination and composition analysis, so as to study the antibacterial activities of EOs.

2.3 THE MAIN CHEMICAL COMPONENTS OF ESSENTIAL OILS

The composition of EOs are very complex, and some relevant studies have shown that there are usually hundreds of components in EO (Ju et al., 2022a). EOs contain different chemical components, whose main structures are shown in Figure 2.1. The EOs may vary in quantity, quality, and composition, depending on plant organs, age, and its vegetative cycle stage, climate, and soil composition. Studies have shown that in these large amounts of EOs, small molecules of phenols, terpenes, aldehydes, and ketones are the main effective components of antibacterial. In addition, alcohols, ethers, and hydrocarbons also have some antibacterial activity (Nematollahi et al., 2020).

Recently, Eze et al. (2016) reported various important functional groups of EOs, including aldehydes (citral, citronellal, cinnamaldehyde, vanillin), alcohols (geraniol,

TABLE 2.1
The Source and Main Extraction Methods of Essential Oils

Plant family	Plant variety	Active site	Main extraction methods	References
Boswellia carterii Birdw.	*Boswellia serrata oleo-gum-resin*	Resin	Superheated steam extraction	Ayub et al. (2022)
	Laurus nobilis	Bark	Ultrasonic extraction	Alimi et al. (2021)
	Cinnamomum camphora	Fruit peel	Solvent-free microwave-assisted extraction	Liu et al. (2022)
Lauraceae	*Lavender*	Whole plant	Microwave-assisted hydrodistillation extraction	Uzkuç et al. (2021)
	Satureja montana L.	Whole plant	Supercritical extraction	Jokanović et al. (2020)
	Mentha canadensis Linnaeus	Leaf	Supercritical extraction	Šojić et al. (2020)
Osmanthus fragrans (Thunb.) Loureiro	*Forsythiae fructus*	Whole plant	Hydrodistillation	Shi et al. (2022)
Mustard family	*Iberis amara*	Seed	Ultrasonic-assisted hydrodistillation extraction	Liu et al. (2019)
Labiatae	*Sideritis raeseri*	Leaf	Microwave-assisted distillation	Drinić et al. (2021)
Asteraceae (Compositae)	*Wormwood*	Leaves or rhizomes	Steam distillation	Bao et al. (2015)
	Tarragon	Leaves or rhizomes	Steam distillation	Getahun et al. (2020)
	Perilla	Leaves	Ultrasonic extraction	Li et al. (2015)
	Ocimum basilicum	Leaves or seeds	Ultrasonic extraction	Jia et al. (2017)
	Oregano	Leaves or flowers	Steam distillation	Borgarello et al. (2015)
Lamiaceae/ Labiateae	*Lavender*	Leaves or flowers	Steam distillation	Li et al. (2013)
	Rosemary	Leaves	Steam distillation	Conde-Hernández et al. (2017)
	Lemon	Peel	Steam distillation	Zhang et al. (2015)
	Sage	Leaves or rhizomes	Steam distillation	Li et al. (2017)
	Mint	Leaves or rhizomes	Steam distillation	Verma et al. (2016)
	Cinamomum camphora	Rhizomes, leaves, or fruits	Steam distillation or Soxhlet extraction	Wang et al. (2015)
Lauraceae	*Cinnamon*	Peel	Steam distillation or ultrasonic extraction	Guo et al. (2014)
	Geranium	Peel	Steam distillation	Huang et al. (2014)
	Litsea cubeba	Fruits	Steam distillation	Peng et al. (2013)

(*Continued*)

TABLE 2.1
(Continued)

Plant family	Plant variety	Active site	Main extraction methods	References
Liliaceae	*Garlic*	Rhizomes	Steam distillation	Cheng et al. (2013)
	Onion	Leaves or rhizomes	Steam distillation	Xue et al. (2017)
	Thymus vulgaris	Leaves or rhizomes	Steam distillation or ultrasonic extraction	Gonçalves et al. (2017)
	Tea tree	Leaves	Steam distillation	Hadaś et al. (2017)
Poaceae	*Lemon grass*	Leaves	Steam distillation	Liang et al. (2017)
	Palmarosa	Leaves or rhizomes	Ultrasonic extraction	Huang et al. (2015)
	Lemon grass	Leaves or rhizomes	Steam distillation	Wang et al. (2016)
	Grape fruit	Leaves or peel	Steam distillation or Soxhlet extraction	Chen et al. (2017)
	Turmeric	Fruits	Steam distillation or Ultrasonic extraction	Wei et al. (2016)
	Cardamom	Fruits	Supercritical CO_2 extraction	Wang et al. (2015)

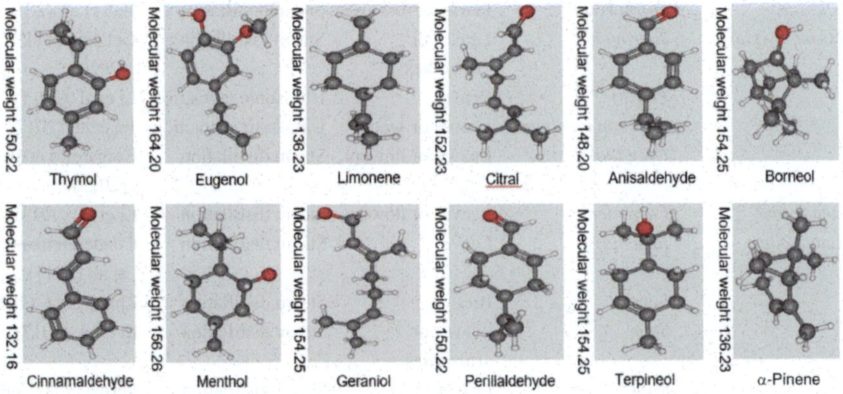

FIGURE 2.1 Structure of some important chemical components of EOs.

citronellol, menthol, linalool, terpineol, borneol), esters (benzoates, acetates, salicylates, cinnamates), ketones (camphor, carvone, menthone, pulegone, thujone), oxides (cineol), phenol ethers (anethol, safrol), phenols (eugenol, thymol, carvacrol), hydrocarbons (cymene, myrcene, sabinene, storene), terpenes (limonene, phellandrene, pinene, camphene, cedrene), and acids (benzoic, cinnamic, myristic, isovaleric). All

EOs consist of one or more of these components as key element and play vital role as antimicrobials.

Carvacrol is the main constituent of oregano (60–74%) and thyme EOs (45%). EOs rich in carvacrol have been widely reported to possess remarkable antimicrobial activity (Marchese et al., 2016). Jianu et al. (2013) isolated caryophyllene (24.1%), beta-phellandrene (16%), and eucalyptol (15.6%) from EO of *L. angustifolia*, while camphor (32.7%) and eucalyptol (26.9%) from EO of lavandin (*Lavandula* × intermedia) as major constituents which showed significant ($P < 0.05$) antibacterial activity.

2.4 RELATIONSHIP BETWEEN CHEMICAL CONSTITUENTS AND ANTIBACTERIAL ACTIVITIES OF ESSENTIAL OILS

As mentioned earlier, EOs contain a variety of chemical constituents, usually up to hundreds. The antimicrobial activity of EOs is closely related to their chemical constituents, especially those with high activity usually play a major role in their antimicrobial activity. The antimicrobial activity of EOs is mainly related to its chemical functional groups. As early as the 19th century, some researchers reported the order of antimicrobial activity of functional groups in EOs as follows: phenols (highest activity) > alcohols > aldehydes > ketones > esters > hydrocarbons (Charai et al., 1996). Then in 2003, Kalemba and Kunicka (2003) resummarized the antimicrobial activity of the functional groups of EOs based on hundreds of previous studies and made the following order: phenols > cinnamaldehyde > alcohols > aldehydes = ketones > esters > hydrocarbons. Recently, some researchers have reexamined the relationship between chemical composition and antimicrobial activity of EOs, and the results further confirm the conclusion of Kalemba et al. (Li et al., 2016). In addition to different functional groups which can cause different antimicrobial activities, the structure of different active ingredients will also have an important impact on their antimicrobial activities. For example, comparing the activity of carvacrol with its structure-related compounds, it was found that the aliphatic side chains of carvacrol could interact with cell membranes, but had little effect on the activity of carvacrol. Of course, the substitution does not affect the hydrophobicity, spatial structure, and solubility of the compounds. Therefore, it can also be seen that the hydroxyl group of carvacrol is not necessary for its bacteriostatic activity. However, the presence of hydroxyl groups did increase the specific characteristics of the antimicrobial mode of action of carvacrol (Sharifi-Rad et al., 2018). The antifungal activity of the same EO is also different from that of different microbial species. For example, the antifungal activity of garlic essential oil and citronella EOs is much higher than that of their antifungal activity, while the antifungal activity of cinnamon EOs is higher than that of their antifungal activity.

2.5 MAIN EVALUATION METHODS OF ESSENTIAL OILS ANTIBACTERIAL EFFECT

Although there is no standard method to evaluate the antimicrobial efficacy of EOs, many researchers in the industry used EUCAST (European Antimicrobial Susceptibility

Test Committee), NCCLS (National Committee of Clinical Laboratories), and CLSI (Institute of Clinical and Laboratory Standards) standards as guidelines to evaluate the antimicrobial efficacy of EOs usually (Rao et al., 2019). This section focuses on in vitro methods for evaluating the antimicrobial efficacy of EOs.

2.5.1 Disc Diffusion Method

Disc diffusion method is a mature and most widely used method for screening the antimicrobial activity of EOs or the sensitivity of microorganisms to EOs. In short, the suspension of the tested strain (10^5–10^8 cfu/ml) was inoculated into the agar-containing culture dish, and then the paper tray (6 mm) containing different concentrations of antimicrobial agents was placed on the agar plate or added to the Oxford cup. Then the culture dish was incubated under the corresponding conditions to make the strain grow. Finally, the antimicrobial effect of the antimicrobial agent was determined by measuring the diameter of the antimicrobial zone. The larger the diameter of antimicrobial zone, the better the bacteriostatic effect of essential oil. On the contrary, the smaller the diameter of antimicrobial zone, the weaker the antibacterial effect of essential oil. Besides its simple operation and low cost, the method can simultaneously screen a large number of microbial strains for EOs sensitivity; compared with other in vitro tests, the disk diffusion method has become a routine detection method. However, this method is difficult to quantify the antimicrobial efficacy of EOs. Therefore, disk diffusion method and dilution method are usually combined to further evaluate the antimicrobial efficacy of EOs.

2.5.2 Dilution Method

The dilution method can better quantify the microbial inhibition activity of EOs, because it can be used to determine the minimum antimicrobial concentration (MIC) or minimum antimicrobial concentration (MBC) of EOs to microorganisms (Arulmozhi et al., 2018). MIC and MBC have been widely used to evaluate the antimicrobial properties and efficacy of EOs. MIC is the drugs content of 1 ml dilution, the minimum concentration of which can completely inhibit the growth of bacteria using tube dilution method; that is, the sensitivity of the tested bacteria to the drug. Because of the different kinds of drugs and bacteria, it is necessary to specify the dilution range of drugs, the type of medium, the amount of bacteria added, the culture conditions, and the time for judging the results. Similarly, MBC is the lowest concentration of dilution that bacteria cannot grow (sterilize) in a culture medium. Broth or agar dilution method can be used to quantitatively measure the antimicrobial activity of EOs in vitro. The general procedure of this method is to add the strain to agar or broth medium containing a series of diluted EOs first, and then determine whether there is bacterial growth after the treatment with different concentrations of EOs. The MIC of the tested EOs was calculated after the specified incubation time. In fact, the MIC value of specific EOs for testing microbial strains is affected by many factors, including inoculation conditions such as the size of bacterial content, inhibition time, preparation method, growth medium type, preparation method of EOs suspension, etc.

2.5.3 Time Sterilization

The time sterilization method is mainly used to evaluate the changes of microbial inhibition or bactericidal activity of EOs with time. This method is usually used in conjunction with MIC and MBC. In short, the EOs were placed in broth medium containing 10^5–10^8 cfu/ml of bacterial strains suspension for time killing test. In most of the reported literature, the concentration of EOs applied in the test suspension is usually 0, 1, 2, and 3 MIC (Kong et al., 2019). After a certain incubation time (24, 36, or 48 h), the germicidal curve of EOs over time was established by monitoring the number of bacteria surviving at each time interval.

2.5.4 Checkerboard Test

Chessboard test is the most important method to identify the biological activity of different EOs or between EOs and other antimicrobial agents. The combination of different antimicrobial agents produces different results such as synergistic, additive, irrelevant, or antagonistic effects (Kong et al., 2016). In this method, two kinds of EOs (A and B) were added to 96-well plate in a certain proportion and diluted twice, vertically and horizontally (Figure 2.2), respectively. Suspensions containing 10^5–10^8 cfu/ml bacterial strains were added to each pore, then 96-well plates were incubated at the corresponding temperature, and their growth was evaluated by measuring the turbidity of the suspension. This chessboard method is usually combined with fractional inhibitory concentration (FIC) to quantify the antimicrobial effect of EO combinations. FIC = (MIC of A in combination/MIC of A) + (MIC of B in combination/ MIC of B). The interaction was defined as synergistic if the FIC index was 0.5 or less, as additive if the FIC index was between 0.5 and 1, as no interaction if the FIC

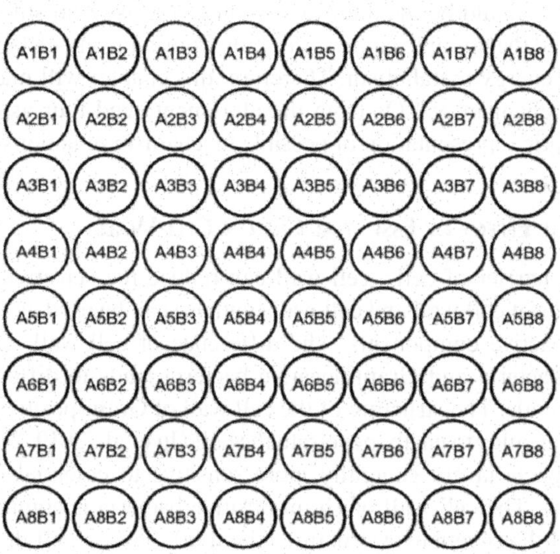

FIGURE 2.2 Chessboard diagram.

index was between 1 and 2, and as antagonistic if the FIC index was greater than 2 (Ju et al., 2022b).

2.6 INHIBITORY EFFECT OF ESSENTIAL OILS ON FOODBORNE PATHOGENIC BACTERIA

Microbial contamination is the main reason for the decrease of food quality and the shortening of shelf life. For a long time, chemical synthetic preservatives have been widely used in food industry to inhibit the growth of microorganisms and prolong the shelf life of food. At present, with the improvement of people's living standards, consumers' preference for natural food additives, and concerns about the safety of synthetic additives have prompted relevant researchers to develop natural preservatives (Ju et al., 2018a). It is hoped that natural preservatives have broad-spectrum antibacterial activity and can improve the quality and shelf life of perishable food. The essential oil is rich in a variety of bioactive components. Therefore, at present, plant essential oil is widely studied and used as a natural preservative, which can not only prolong the shelf life of food, but also prevent and assist the treatment of a variety of diseases. At present, more than 3,000 kinds of essential oils are known, of which more than 300 kinds have important commercial value in pharmaceutical, agronomic, food, health, cosmetic, and spice industries (Ju et al., 2018b).

2.7 ANTIBACTERIAL MECHANISM

The antibacterial activities of EOs are determined by the main components of EOs or synergistic effects of various components. The antibacterial mechanisms of different components may be different. Therefore, the antibacterial mechanism of EOs are usually not a single action mode, but a multiple action mechanism. Several locations in microbes should be the main action sites of EOs (Khorshidian et al., 2018). Figure 2.3 showed the possible pathways of EOs from cell wall, cell membrane, DNA, respiration, and energy metabolism. Therefore, this chapter will mainly analyze the antibacterial mechanism of EOs from the aforementioned four aspects. Figure 2.4 provides schematic representation of possible mechanism of actions of EOs.

2.7.1 Damage Effect of Essential Oil on Cell Wall

As a protective barrier of cells, cell wall can reduce the sensitivity of drugs. The main components of cell wall and cell wall synthase can be used as important targets of essential oil active components. The difference of cell wall structure between Gram-positive bacteria and Gram-negative bacteria leads to different tolerance of cells to essential oils. The structure of the cell wall of Gram-positive bacteria allows hydrophobic molecules to easily penetrate the cell (Ju et al., 2019, 2020a). Therefore, Gram-positive bacteria are generally more sensitive to essential oils than Gram-negative bacteria. The cell wall of Gram-positive bacteria is composed of peptidoglycan with multilayer reticular structure (about 90–95% of cell dry weight), and is connected by lipoteichoic acid and teichoic acid. The cell wall of Gram-negative bacteria contains monolayer peptidoglycan (about 20% of cell dry weight) and does not

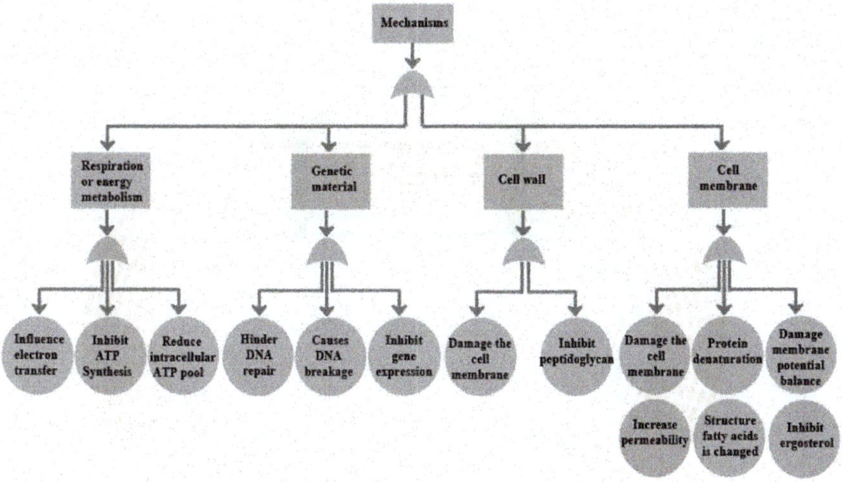

FIGURE 2.3 The possible action pathways of EOs.

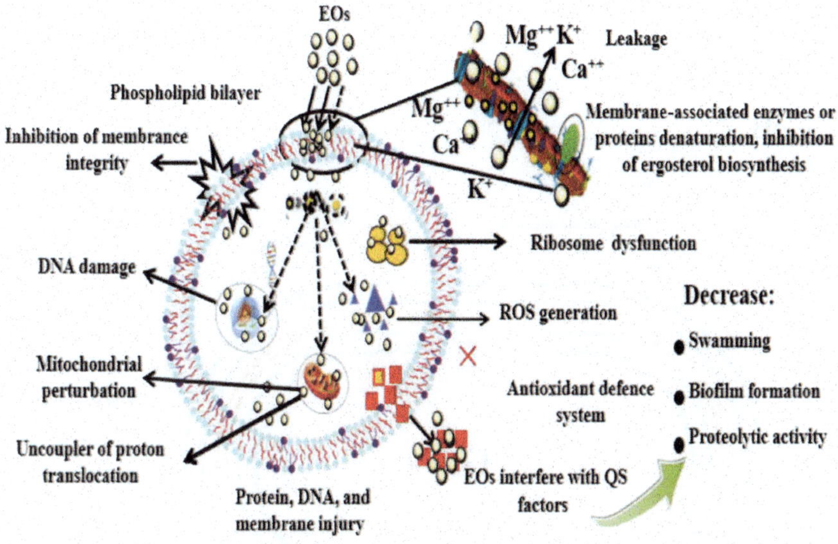

FIGURE 2.4 Schematic representation of possible mechanism of actions of EOs.

contain teichoic acid. In addition, the peptidoglycan is covered with a lipid bilayer called adventitia. It consists of lipopolysaccharide, lipoprotein, membrane porin, and phospholipid (Figure 2.5). The structure of the cell wall of Gram-positive bacteria allows hydrophobic molecules to pass through easily. However, the cell wall of Gram-negative bacteria contains lipopolysaccharide, which gives the cells a hydrophilic surface, which makes Gram-negative bacteria more resistant to essential oils and other hydrophobic natural extracts with antibacterial activity. However, phenolic

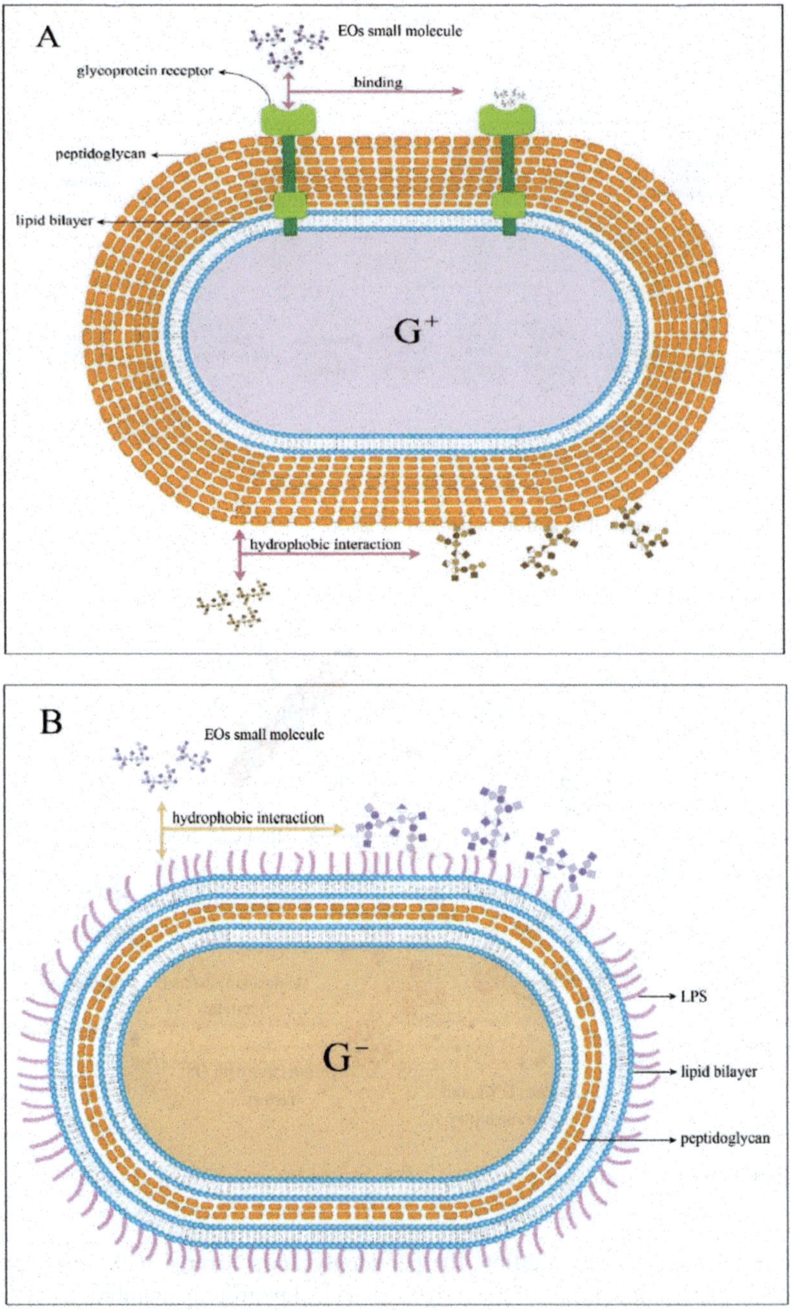

FIGURE 2.5 Absorption mechanism of small molecules of essential oils by bacteria. (a) Mechanism of absorption of small molecules of plant essential oils by Gram-positive bacteria. (b) Mechanism of absorption of small molecules in plant essential oils by Gram-negative bacteria.

Source: Zhou et al. (2022).

compounds in essential oils, such as thymol, eugenol, and carvol, can interfere with the outer membrane and release lipopolysaccharide (Serena et al., 2018). In addition, thymol can also destroy the cell wall of *L. monocytogenes*, making it thicker and rough. At the same time, terpenoids can penetrate the bacterial cell wall and destroy the lipid layer, resulting in cytoplasmic leakage, cell lysis, and eventually cell death. The components of essential oils can also interfere with proteins in the cell wall. For example, *trans*-cinnamaldehyde can reduce the expression of OmpA, OmpC, OmpR, and amino acid transporters in outer membrane proteins, and destroy the activity of bacterial transport and diffusion (Vasconcelos et al., 2018). Alkaline phosphatase is an enzyme that exists between the cell wall and the cell membrane. Under normal circumstances, its activity cannot be detected outside the cell, but when the cell wall is damaged and the permeability increases, the alkaline phosphatase will leak out of the cell (Authman, 2016).

Related research cases have also proved that the alkaline phosphatase activity of *S. aureus* and *E. coli* treated with chrysanthemum essential oil decreased by 54.02% and 47.54%, respectively (Cui et al., 2018).

2.7.2 Damage Effect of Essential Oil on Cell Membrane

The permeability barrier provided by cell membrane is essential for many cellular functions, including energy conversion, material transport, information recognition and transmission, cellular immunity, and metabolic regulation. At the same time, cell membrane is also a necessary condition to control turgor. Hydrophobicity is a typical characteristic of essential oil. The hydrophobicity of essential oil enables it to enter the bacterial cell membrane, resulting in increased membrane permeability, destruction of membrane structure, leakage of cell contents, and finally cell death. Therefore, the antibacterial activity mechanism of essential oil is usually explained by the toxic effect of essential oil on the structure and function of cell membrane. Cell membrane is usually the most critical target for essential oils to play a role. Essential oils affect the normal physiological function of plasma membrane by affecting fatty acids and proteins on the membrane (Figure 2.6).

2.7.2.1 Effect of Essential Oil on Protein in Cell Membrane

Essential oils can change the permeability and function of cell membranes. Usually, essential oils rich in phenols can be embedded in the phospholipid bilayer of bacteria and combine with proteins to prevent proteins from performing their normal functions. Eugenol in clove essential oil can denature the protein in the cell membrane, and then react with the phospholipid of the cell membrane to destroy the permeability of the cell membrane, thus inhibiting the growth of microorganisms. In addition, thyme essential oil changes the permeability of cell membrane by affecting the integrity of bilayer phospholipids, resulting in the leakage of nucleic acid, protein, and potassium ions from inside the cell, which can cause cell death in 5 min (He et al., 2022). Similarly, the cell membranes of *E. coli* and *Salmonella typhimurium* treated with carvol and thymol degrade and lead to intracellular leakage (Jossana et al., 2017).

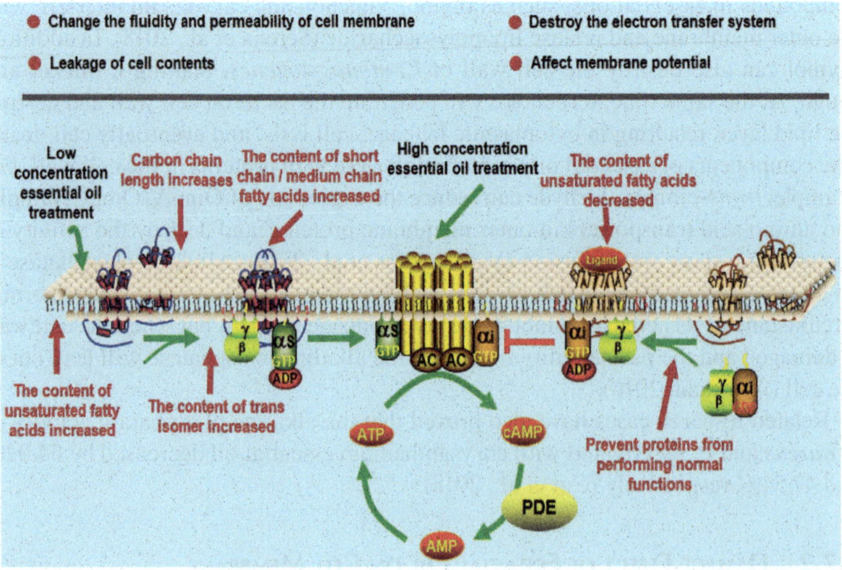

FIGURE 2.6 Effects of essential oils and their active components on microbial cell membrane.

Essential oils and their components can also change the membrane potential by affecting membrane fatty acids and membrane proteins, destroying the electron transport system, leading to leakage of intracellular substances such as ions, cytoplasm, ATP, nucleic acid, and protein, and eventually leading to cell disintegration (Ju et al., 2020a). Carvacrol can increase the fluidity of cell membrane, cause potassium ion leakage, cause membrane potential collapse, and inhibit ATP synthesis (Cherrata et al., 2016). The antibacterial mechanism of peppermint essential oil against *E. coli*, *L. monocytogenes*, and *Salmonella enteritis* in pineapple and mango juice shows that the essential oil has multitarget antibacterial mode, including destroying cell membrane, increasing membrane permeability and potential depolarization, and inhibiting efflux pump and respiratory activity (Hashemi et al., 2022).

2.7.2.2 Effect of Essential Oil on Fatty Acids in Cell Membrane

The hydrophobicity of essential oils and their components enables them to spread to lipid bilayers, affecting the proportion of unsaturated fatty acids in cell membranes and changing their structure. When bacterial cells are exposed to sublethal environmental pressure, the composition of cell membrane changes to adapt to environmental changes. The changes of membrane lipids depend on species and environmental pressure, including saturation, carbon chain length, branching position, *cis–trans* isomerization, and conversion of unsaturated fatty acids to cyclopropane. The treatment of bacteria with essential oil at sublethal concentration could increase the proportion of unsaturated fatty acids in cell membrane and increase the fluidity of cell membrane. Through this mechanism, we can maintain the structure and function of cell membrane and enhance cell adaptability. However, under the condition of higher than the minimum inhibitory concentration, the content of saturated fatty acids in the cell membrane increased, thus increasing the surface density of the cell membrane and inhibiting the entry of fat-soluble components into the cell. In addition, essential oil may also cause oxidative stress in bacterial cells and accelerate the degree of membrane lipid peroxidation. For example, cinnamaldehyde can cause oxidative stress in *E. coli* (He et al., 2018). Thymol can cause *L. monocytogenes* to produce oxidative stress. The treatment of essential oil and its components will lead to the increase of volatile components in microorganisms. For example, aldehydes, alcohols, lactones, short- and medium-chain fatty acids and their esters, and these volatile components are produced by oxidizing unsaturated fatty acids.

2.7.3 Effect of Essential Oil on Energy Metabolism

The growth and reproduction of bacteria must have the absorption and transport of substances such as glucose and calcium, and these activities are inseparable from energy metabolism. If there is a problem with the energy metabolism of microorganisms, then the growth and reproduction of microorganisms will also be affected. During respiration, the electron transport chain on the cell membrane produces a transmembrane proton gradient, which is necessary for the synthesis of ATP. This process is catalyzed by a variety of enzymes with ATP activity, including ATP-dependent transporters and F1F0-ATP enzyme complexes. Related studies have also confirmed that ginger rhizome essential oil can inhibit the respiratory metabolism and ATPase activity of *Escherichia coli O157:H7*, resulting in the death of bacteria (Zhou et al., 2021). Mutlu-Ingok et al. (2017) used cardamom, cumin, and dill essential oils to treat *Campylobacter jejuni* and *Campylobacter coli* and found that the concentration of extracellular ATP increased significantly ($P < 0.05$) after treatment, indicating that the destruction of cell membrane led to the loss of ATP. Eugenol of 5 mmol/l and cinnamaldehyde of 40 mmol/l could inhibit the production of ATP by *L. monocytogenes*. After culture in glucose-rich medium, the intracellular ATP of *L. monocytogenes* was rapidly depleted by cinnamaldehyde treatment, but the intracellular ATP of *L. monocytogenes* treated with eugenol was not consumed. This may be because eugenol limits the glucose utilization ability of *L. monocytogenes* and inhibits the activity of F1F0-ATP enzyme. When the oxidation and decomposition of sugar is inhibited, it will hinder the normal energy supply of the cell, then affect the cell function and damage it, and even lead to biological death. Bacteriostatic substances can achieve bacteriostatic effect by preventing the absorption and utilization of glucose by pathogenic microorganisms to prevent energy supply and hinder the growth and reproduction of pathogenic bacteria. Terpenoids in the active components of essential oils play an important role in inhibiting the respiratory metabolism of bacteria. This is also one of the main factors that plant essential oils have good antibacterial activity.

2.7.4 Effect of Essential Oil on Genetic Material

DNA is the main genetic material, and parents copy part of their own DNA and pass it on to their offspring to complete the transmission of traits. When DNA replication is inhibited or DNA is damaged, the biological characteristics of the organism will also be affected, thus inhibiting the growth and reproduction of the organism. Components in essential oils can destroy the structure of DNA or inhibit gene expression—just as terpenes and phenols can block DNA repair by inhibiting the oxidation of *E. coli*. In addition, cold-pressed orange essential oil can induce partial upregulation and downregulation of DNA metabolism-related genes in drug-resistant *S. aureus* and interfere with DNA replication, recombination, and repair. In addition, orange essential oil promotes cell lysis by affecting the expression of cell wall–related genes (Muthaiyan et al., 2012). The specific mechanism of DNA damage caused by essential oil is still unclear and needs to be further studied. However, some authors believe that the bactericidal ability of antimicrobials may be due to the stable interaction between drug-bound topoisomerase and lytic DNA. In the presence of antimicrobial agents, the stagnation of replication forks leads to the inhibition of DNA replication mechanism and the inhibition of DNA synthesis, resulting in cell death.

2.7.5 Effect of Essential Oil on Cell Morphology

The growth morphology of bacteria is the basic criterion for their normal physiological activity. Essential oils and their components damage the morphology of bacterial cells by destroying cell walls and cell membranes and causing leakage of cell contents. *E. coli O157:H7* and *L. monocytogenes* were damaged after treated with oregano, cinnamon, and peppermint essential oils. Among them, *E. coli* cells treated with oregano essential oil have holes or white spots on the cell wall (Li et al., 2022a). Similarly, the cells adhered to each other, deformed, sunken, shrunk, and ruptured after treated with *Alpinia guilinensis* essential oil. Transmission electron microscope observation showed that the untreated *E. coli* cells were intact, the cell wall and cell membrane were clearly discernible, and the cytoplasm was dense and uniformly distributed. However, the cells treated with essential oil were obviously damaged, the cell wall and cell membrane were destroyed, the cytoplasm was unevenly distributed, and some of the cell walls became thicker and even indistinguishable, which indicated that the cell membrane was lytic (Zhou et al., 2021). In addition, grass fruit essential oil, oregano essential oil, and sage essential oil can also destroy the cell wall of *E. coli*, resulting in morphological changes such as shrinkage, damage, and even collapse (Danilović et al., 2021).

The effect of essential oil on the cell morphology of Gram-negative bacteria was stronger than that of Gram-positive bacteria. Carvacrol is more likely to affect the cell morphology of Gram-negative bacteria, and the surface roughness of almost all Gram-negative bacteria increased significantly ($P < 0.05$) after carvacrol treatment, which may be because the outer membrane of Gram-negative bacteria is the primary target of carvacrol. However, it is generally believed that Gram-negative bacteria are more resistant to essential oils, so the changes in cell surface structure of Gram-negative bacteria can also be interpreted as adaptive responses to stress. Carvacrol treatment may increase the exposure of bacterial outer membrane components (such as proteins and lipids), resulting in increased roughness. In Gram-positive bacteria, carvanol first passes through the peptide polysaccharide layer, and then acts on the cell plasma membrane, resulting in structural changes of the cell membrane, such as changes in fluidity (Li et al., 2022a; Hasanvand et al., 2021).

2.7.6 Effect on Exercise Ability and Biofilm Formation

Exercise ability plays a key role in the reproduction, transmission, and interaction with the host of microorganisms. The ability of pathogenic bacteria to transmit, colonize, and produce toxin includes a number of programs, such as signal transduction, chemotaxis, and flagellar movement. In addition, the regulation of cell movement may also affect the formation of complex structures in the biofilm. The formation of a biofilm is not a single action process but a coordinated collective behavior.

The basic process of biofilm formation includes microbial colonization and attachment to the external (abiotic or biological) surface. Then, the biofilm changes from a planktonic state to a fixed state. An extracellular polysaccharide matrix is produced through intercellular aggregation and proliferation, and the cells are gradually bonded together to form small colonies (Chen and Wang, 2020; Park et al., 2019). With the continuous proliferation of microorganisms, a three-dimensional (3D) biofilm structure slowly forms (Figure 2.7). As the biofilm matures, it gradually degrades or separates the planktonic cells for the next cycle of biofilm formation (Vasconcelos et al., 2018). The growth of biofilms by microorganisms has many advantages, such as preventing self-drying, cooperating with each other through metabolism, and improving genetic diversity through gene transfer. Recent studies have shown that hexanal inhibits biofilm formation in *Pseudomonas fluorescens* and *E. coli* (Yu et al., 2022; Caballero-Prado et al., 2021). In addition, cinnamaldehyde also has an anti-biofilm effect on *S. aureus*, *Suppurative streptococcus*, *E. coli*, *Pseudomonas aeruginosa*, *L. monocytogenes*, and *Salmonella* (Nuţă et al., 2021; Jauro et al., 2021).

2.7.7 Effect on Quorum Sensing

Quorum sensing (QS) is an information exchange program discovered during the study of the luminous mechanism of *Vibrio fischeri* in the 1880s. QS is the behavior that bacteria perceive bacterial density and the environment outside the cell by synthesizing and secreting an autologous inducer called a signal molecule. Figure 2.8 shows a simplified Bcc bacteria QS circuit diagram. The concentration of signal molecules also increases as bacterial density increases. When the concentration reaches a certain threshold, bacteria turn on the expression of cell-dependent genes and regulate bacterial population behavior. Inhibiting QS regulates a variety of physiological and biochemical functions of bacteria, such as bioluminescence, biofilm formation, extracellular enzyme secretion, antibiotic production, and other metabolites and virulence factors. Some studies have confirmed that cinnamaldehyde downregulates the expression of bcsA and luxR in *E. coli*, and both are involved in the QS reaction. In addition, cinnamaldehyde also significantly inhibits the QS reaction in *P. aeruginosa* and *Streptococcus pyogenes* (Mangal et al., 2022; Faleye et al., 2021). Different EOs may have different inhibitory effects on QS of different cell types. Finding new ways to inhibit QS in bacteria is an interesting strategy for developing new antibiotics. An

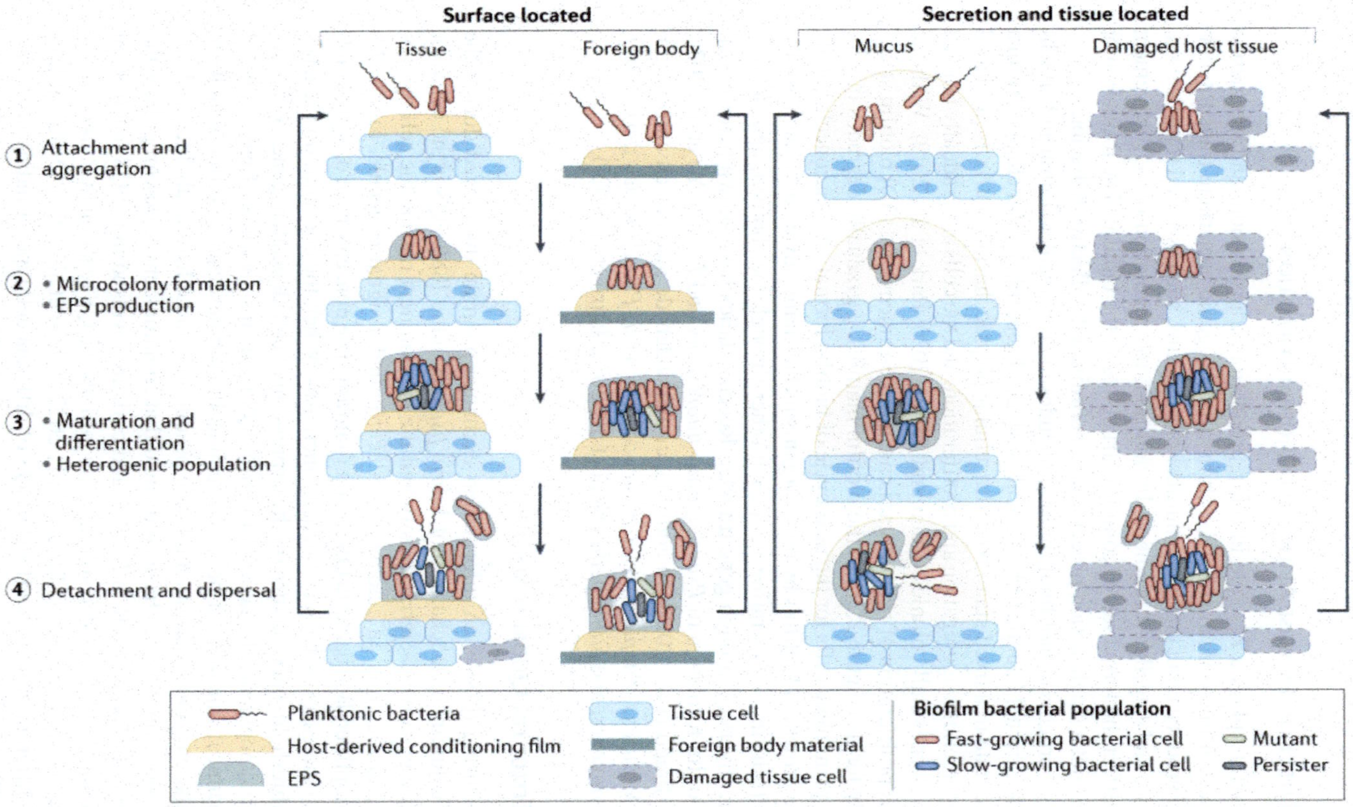

EPS, extracellular polymeric substance.

FIGURE 2.7 Biofilm formation and characteristics.

Source: Oana et al. (2022).

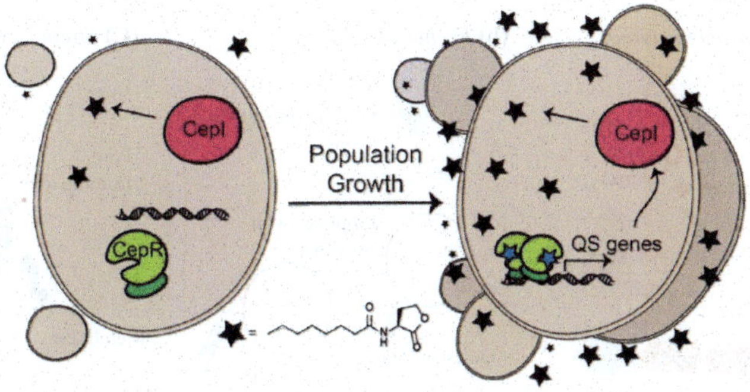

FIGURE 2.8 Schematic diagram of quorum sensing.

Source: Betty et al. (2019).

anti-QS compound could overcome the pathogenicity of bacteria without causing drug resistance.

The action mechanism of EOs against microorganisms may not coexist, so it is difficult to appear alone. Therefore, it is necessary to further investigate the action targets of different EOs or their active components on different microorganisms, and further explore the action mechanism of EOs at the molecular level to improve the specific antibacterial effect of antimicrobial agents on target microorganisms.

2.7.8 Effect of Essential Oil on Bacterial Toxins

Essential oil has inhibitory effect on bacterial toxins. Studies have shown that bacterial toxins are essential for the control and treatment of widespread bacterial diseases. More and more reports have emphasized that essential oils may inhibit the production of toxins by some bacteria, such as *S. aureus*, *E. coli*, *Streptococcus*, *Vibrio cholerae*, and *P. aeruginosa*. The mechanism of essential oil inhibiting bacterial toxins includes two pathways of disrupting toxin production and upstream gene transcriptional regulation system (Figure 2.9). Related studies have confirmed that oregano essential oil, thymol, and allicin can block regulatory factors or participate in gene transcription produced by *S. aureus* toxins. In addition, they can inhibit the expression of genes encoded by *E. coli* shiga toxin and block the cleavage cycle of phage, as well as the activities of streptococcal hemolysin (SLO) and pneumolysin (PLY) (Huang et al., 2017; Magry et al., 2021). Similarly, *Litsea cubeba* essential oil can inhibit the transcriptional level of the main virulence gene stx1, stx2, ehx Apene of *E. coli O157:H7*. The *Moringa oleifera* essential oil can inhibit the transcriptional expression of virulence islands (prf A, plc A, hly, act A, plc B) and LPI-2 virulence islands (inl A, inl B) of *L. monocytogenes*.

2.8 CONCLUDING REMARKS

Food safety is a major global public health problem. Although science and technology has developed to a considerable level, food contamination caused by foodborne

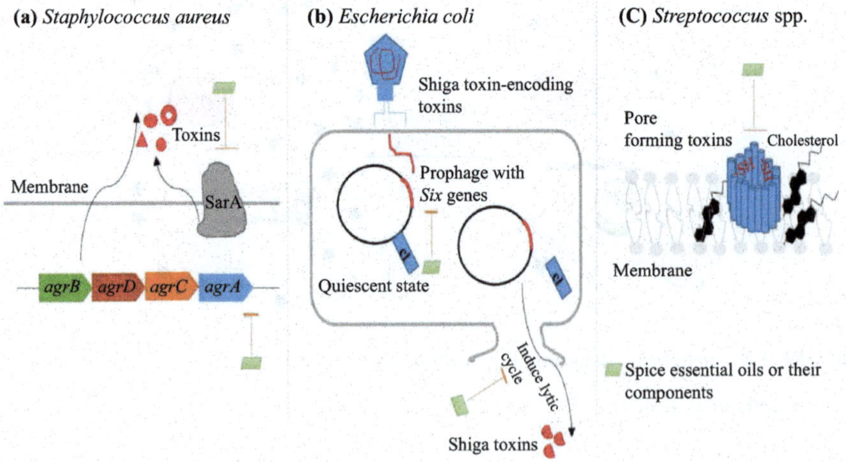

FIGURE 2.9 The schematic diagrams of the mechanisms of toxin inhibition by spice essential oils in *Staphylococcus aureus* (a), *Escherichia coli* (b), and *Streptococcus* spp. (c).

Source: Zhang et al. (2020).

diseases is still widespread in many countries around the world, and the food safety situation is still very grim. Therefore, this chapter focuses on the research of effective contribution of EOs to inhibit the growth of foodborne pathogens, systematically summarized the commonly used EOs and their main chemical components, extraction methods, the antibacterial mechanism of EOs on foodborne pathogens, and systematically analyzes the specific application case of EOs in food discussed.

In view of the extensive antibacterial, antioxidant, and antiviral activities of EOs, EOs have been used in food, medicine, and cosmetics. However, EOs have some particularity, such as low water solubility, strong sensory flavor, and poor stability. Therefore, finding new ways to solve these problems is one of the major challenges faced by researchers: thinking about the characteristics of EOs which could be used in synergy with other molecules, as active ingredient, to obtain antimicrobial or antioxidant packaging, able to extend food shelf life, without significantly impacting on the product sensory characteristics, and also preserving the product quality during time. At present, although there has been relevant reports on the study of active packaging containing EOs, there has been almost no large-scale industrial application so far; so it is necessary to make it possible to have production possibilities and industrial applications. We also should establish a relationship between the EOs and the species that it inhibits, so as to select the optimum EOs based on different spoilage microbes. In this way, it can play an even greater role in its antimicrobial activities. Besides, despite the numerous published scientific papers and the different suggested practical applications, it is important to carry on more in vivo studies to confirm the good results already obtained in vitro and also to deepen the knowledge on the toxicological aspects.

ACKNOWLEDGEMENTS

This work was supported by the National Natural Science Foundation of China (32202192), Special fund for Taishan Scholars Project.

REFERENCES

Abedi, A., Lakzadeh, L., and Amouheydari, M. 2021. Effect of an edible coating composed of whey protein concentrate and rosemary essential oil on the shelf life of fresh spinach. *J. Food Process. Preserv.* 45(4): 15284.

Aboutalebzadeh, S., Esmaeilzadeh-Kenari, R., and Jafarpour, A. 2022. Nano-encapsulation of sweet basil essential oil based on native gums and its application in controlling the oxidative stability of kilka fish oil. *Journal of Food Measurement and Characterization.* 16(3): 2386–2399.

Alimi, D., Hajri, A., Jallouli, S., and Sebai, H. 2021. In vitro acaricidal activity of essential oil and crude extracts of *Laurus nobilis*, (Lauraceae) grown in Tunisia, against arthropod ectoparasites of livestock and poultry: *Hyalomma scupense* and *Dermanyssus gallinae. Veterinary Parasitology.* 298: 109507.

Alves-Silva, J. M., Zuzarte, M., Giro, H., and Salgueiro, L. 2021. The role of essential oils and their main compounds in the management of cardiovascular disease risk factors. *Molecules.* 26(12): 3506.

Artiga-Artigas, M., Acevedo-Fani, A., and Martín-Belloso, O. 2017. Improving the shelf life of low-fat cut cheese using nanoemulsion-based edible coatings containing oregano essential oil and mandarin fiber. *Food Control.* 76: 1–12.

Arulmozhi, P., Vijayakumar, S., and Kumar, T. 2018. Phytochemical analysis and antimicrobial activity of some medicinal plants against selected pathogenic microorganisms. *Microb. Pathog.* 123: 219–226.

Authman, S. H. 2016. In vivo antibacterial activity of alkaline phosphatase isolates from *Escherichia coli* isolated from diarrhea patients against *Pseudomonas aeruginosa. The Pharma Innovation.* 77: 35–42.

Ayub, M. A., Hanif, M. A., Blanchfield, J., Zubair, M., Abid, M. A., and Saleh, M. T. 2022. Chemical composition and antimicrobial activity of *Boswellia serrata* oleo-gum-resin essential oil extracted by superheated steam. *Nat. Prod. Res.* 1–6.

Bao, Y. H., Wang, F., Deng, Q., Duan, W. L., and Sun, Y. X. 2015. Response analysis of optimization of extraction technology of p-cymene surface method and chemical composition. *Science and Technology of Food Industry.* 36(14): 287–292.

Betty, L., Slinger, J. J., Josephine, R., and Helen, B. 2019. potent modulation of the cepR quorum sensing receptor and virulence in a *Burkholderia cepacian* complex member using non-native lactone ligands. *Sci. Rep.* 9: 13449.

Borgarello, A. V., Mezza, G. N., Pramparo, M. C., and Gayol, M. F. 2015. Thymol enrichment from oregano essential oil by molecular distillation. *Sep. Purif. Technol.* 153: 60–66.

Caballero-Prado, C. J., Merino-Mascorro, J. A., Heredia, N., Dávila-Aviña, J., and García, S. 2021. Eugenol, citral, and hexanal, alone or in combination with heat, affect viability, biofilm formation, and swarming on Shiga-toxin-producing *Escherichia coli. Food Sci. Biotechnol.* 30(4): 599–607.

Cardia, G. F. E., de Souza Silva-Comar, F. M., da Rocha, E. M. T., Silva-Filho, S. E., Zagotto, M., Uchida, N. S., and Cuman, R. K. N. 2021. Pharmacological, medicinal and toxicological properties of lavender essential oil: A review. *Research, Society and Development.* 10(5): e23310514933–e23310514933.

Catarino, M. D., Alves-Silva, J. M., Fernandes, R. P., Gonçalves, M. J., Salgueiro, L. R., Henriques, M. F., and Cardoso, S. M. 2017. Development and performance of whey protein active coatings with *Origanum virens* essential oils in the quality and shelf-life improvement of processed meat products. *Food Control*. 80: 273–280.

Charai, M., Mosaddak, M., and Faid, M. 1996. Chemical composition and antimicrobial activities of two aromatic plants: *Origanum majorana* L. and *O. compactum* Benth. *J. Essent. Oil Res*. 8(6): 657–664.

Chen, H. L., Liu, K. Yang, X. A, Guo, Q. Y., and Chun, J. 2017. Grapefruit essential oil and its molecular distillation components to study the antibacterial activity. *J. Nanjing Norm. Univ*. 38(3): 81–84.

Chen, J., and Wang, Y. 2020. Genetic determinants of *Salmonella* enterica critical for attachment and biofilm formation. *Int. J. Food Microbiol*. 320: 108524.

Cheng, Y. C., Chang, M. H., Tsai, C. C., Chen, T. S., Fan, C. C., Lin, C. C., and Huang, C. Y. 2013. Garlic oil attenuates the cardiac apoptosis in hamster-fed with hypercholesterol diet. *Food Chem*. 136(3): 1296–1302.

Cherrata, L., Dumasb, E., Bakkalia, M., Degraeveb, P., and Oulahalb, N. 2016. Effect of essential oils on cell viability, membrane integrity and membrane fluidity of *Listeria innocua* and *Escherichia coli*. *J. Essent. Oil-Bear. Plants*. 19(1): 155–166.

Conde-Hernández, L. A., Espinosa-Victoria, J. R., Trejo, A., and Guerrero-Beltrán, J. Á. 2017. CO_2 -supercritical extraction, hydrodistillation and steam distillation of essential oil of rosemary (*Rosmarinus officinalis*). *J. Food Eng*. 200: 81–86.

Cui, H., Bai, M., Sun, Y., Abdel-Samie, M. A. S., and Lin, L. 2018. Antibacterial activity and mechanism of *Chuzhou chrysanthemum* essential oil. *J. Funct. Foods*. 48: 159–166.

Danilović, B., Đorđević, N., Milićević, B., Šojić, B., Pavlić, B., Tomović, V., and Savić, D. 2021. Application of sage herbal dust essential oils and supercritical fluid extract for the growth control of *Escherichia coli* in minced pork during storage. *LWT-Food Sci. Technol*. 141: 110935.

Deng, J, Li, W, Zhang, J., and Liu, M. J. 2017. The research method of ASLT compound essential oil microcapsule for rice preservation effect. *Mod. Food Sci. Technol*. 33(3): 196–202.

Drinić, Z., Pljevljakušić, D., Janković, T., Zdunić, G., Bigović, D., and Šavikin, K. 2021. Hydro-distillation and microwave-assisted distillation of *Sideritis raeseri*: Comparison of the composition of the essential oil, hydrolat and residual water extract. *Sustainable Chem. Pharm*. 24: 100538.

Drr, A., Aa, A., and Mdl, A. 2022. Encapsulated essential oils: A perspective in food preservation. *Future Foods*. 5: 100126.

El-Sawi, S. A., and Mohamed, M. A. 2002. Cumin herb as a new source of essential oils and its response to foliar spray with some micro-elements. *Food Chem*. 77(1): 75–80.

Eze, U. A. 2016. In vitro antimicrobial activity of essential oils from the Lamiaceae and Rutaceae plant families against β-Lactamse producing clinical isolates of Moraxella catarrhalis. *EC Pharm. Sci*. 2: 325–337.

Fahmy, M. A., Aly, F., Hassan, E. M., Farghaly, A. A., and Samea, N. 2020. *Cymbopogon citratus* essential oil has hepato/renal protection and anti-genotoxicity against carbon tetrachloride. *Comun. Sci*. 11: e3219.

Faleye, O. S., Sathiyamoorthi, E., Lee, J. H., and Lee, J. 2021. Inhibitory effects of cinnamaldehyde derivatives on biofilm formation and virulence factors in vibrio species. *Pharmaceutics*. 13(12): 2176.

Gargouri, B., Amor, I. B., Attia, H., Messaoud, E. B., Elaguel, A., and Kallel, I. 2021. Antioxidant capacity and antitumoral activity of citrus paradisi essential oil. *Biomedical Journal of Scientific & Technical Research*. 40: 167–173.

Geni, M. S., Aksi, J. M., Stoi, M., Randjelovi, P. J., and Radulovi, N. S. 2021. Linking the antimicrobial and anti-inflammatory effects of immortelle essential oil with its chemical composition-the interplay between the major and minor constituents. *Food Chem. Toxicol.* 158: 112666.

Getahun, T., Sharma, V., Kumar, D., and Gupta, N. 2020. Chemical composition, and antibacterial and antioxidant activities of essential oils from *Laggera tomentosa* Sch. Bip. ex Oliv. et Hiern (*Asteraceae*). *Turkish J. Chem.* 44(6): 1539–1548.

Ghasemi, B., Varidi, M. J., Varidi, M., Kazemi-Taskooh, Z., and Emami, S. A. 2022. The effect of plant essential oils on physicochemical properties of chicken nuggets. *Journal of Food Measurement and Characterization.* 16(1): 772–783.

Gonçalves, N. D., de Lima Pena, F., Sartoratto, A., Derlamelina, C., Duarte, M. C. T., Antunes, A. E. C., and Prata, A. S. 2017. encapsulated thyme (*Thymus vulgaris*) essential oil used as a natural preservative in bakery product. *Food Res. Int.* 96: 154–160.

Guo, J. Yang, R. F., Fan, X. D., and Qiu, T. Q. 2014. Comparison of different extraction methods of cinnamon oil. *Science and Technology of Food Industry.* 35(14): 95–99.

Hadaś, E., Derda, M., and Cholewiński, M. 2017. Evaluation of the effectiveness of tea tree oil in treatment of acanthamoeba, infection. *Parasitol. Res.* 116(3): 997–1001.

Hasanvand, T., Mohammadi, M., Abdollahpour, F., Kamarehie, B., Jafari, A., Ghaderpoori, A., and Karami, M. A. 2021. A comparative study on antibacterial activity of carvacrol and glutaraldehyde on *Pseudomonas aeruginosa* and *Staphylococcus aureus* isolates: An in vitro study. *J. Environ. Health Sci. Eng.* 19(1): 475–482.

Hashemi, H., Hashemi, M., Talcott, S., Castillo, A., Taylor, T. M., and Akbulut, M. 2022. Nanoimbibition of essential oils in triblock copolymeric micelles as effective nanosanitizers against food pathogens *Listeria monocytogenes* and *Escherichia coli* O157: H7. *ACS Food Sci. Technol.* 2(2): 290–301.

He, Q., Zhang, L., Yang, Z., Ding, T., Ye, X., Liu, D., and Guo, M. 2022. Antibacterial mechanisms of thyme essential oil nanoemulsions against *Escherichia coli* O157: H7 and *Staphylococcus aureus*: Alterations in membrane compositions and characteristics. *Innovative Food Sci. Emerging Technol.* 75: 102902.

He, T. F., Wang, L. H., Niu, D. B., Wen, Q. H., and Zeng, X. A. 2018. Cinnamaldehyde inhibit *Escherichia coli* associated with membrane disruption and oxidative damage. *Arch. Microbiol.* 156: 451–463.

Huang, J. W., Tang, L., and Chen, H. 2015. Process optimization of ultrasonic extraction of rose essential oil and pigment grade. *J. Food Saf. Food Qual.* 06: 2123–2130.

Huang, T. W., Wang, J. Q., Song, H. Y., Zhang, Q., and Liu, G. F. 2017. Chemical analysis and in vitro antimicrobial effects and mechanism of action of *Trachyspermum copticum* essential oil against *Escherichia coli. Asian Pac. J. Trop. Med.* 663–669.

Huang, X. Li, G. S., Liu, L. F., Pu, X. W., and Liu, Y. Y. 2014. Molecular distillation of *Geranium* essential oil and its application in cigarettes. *Food Industry.* 35(3): 202–206.

Jauro, A. H., Shuâ, I., Lawan, G., Adamu, M. T., Iliyasu, M. Y., and Umar, A. F. 2021. Inhibitory effect of sodium citrate, sodium nitrite and cinnamaldehyde on biofilm-forming *Escherichia coli* O157: H7. *Int. J. Environ. Biorem. Biodegrad.* 4(2): 11–16.

Jia, X. L., Cao, F. H., Sun, Y. H., and Dong, Y. N. 2017. Extraction and composition analysis of essential oil. *Food Sci.* 24(1): 34–40.

Jianu, C., Pop, G., Gruia, A. T., and Horhat, F. G. 2013. Chemical composition and antimicrobial activity of essential oils of lavender (*Lavandula angustifolia*) and lavandin (*Lavandula* x intermedia) grown in Western Romania. *Int. J. Agric. Biol.* 15: 772–776.

Jokanović, M., Ivić, M., Škaljac, S., Tomović, V., Pavlić, B., Šojić, B., and Ikonić, P. 2020. Essential oil and supercritical extracts of winter savory (*Satureja montana* L.) as antioxidants in precooked pork chops during chilled storage. *LWT-Food Sci. Technol.* 134: 110260.

Jossana, Pereira, de, Sousa, Guedes, Evandro, Leite, de, and Souza. 2017. Investigation of damage to *Escherichia coli, Listeria monocytogenes* and *Salmonella enteritidis* exposed to *Mentha arvensis* L. and *M. piperita* L. essential oils in pineapple and mango juice by flow cytometry. *Food Microbiology*. 76: 564–571.

Ju, J., Chen, X., Xie, Y., Yu, H., Guo, Y., Cheng, Y., and Yao, W. 2019. Application of essential oil as a sustained release preparation in food packaging. *Trends Food Sci. Technol.* 92: 22–32.

Ju, J., Guo, Y., Cheng, Y., and Yao, W. 2022. Analysis of the synergistic antifungal mechanism of small molecular combinations of essential oils at the molecular level. *Ind. Crops Prod.* 188: 115612.

Ju, J., Xie, Y., Guo, Y., Cheng, Y., Qian, H., and Yao, W. 2018a. Application of starch microcapsules containing essential oil in food preservation. *Crit. Rev. Food Sci. Nutr*. 27–45.

Ju, J., Xie, Y., Guo, Y., Cheng, Y., Qian, H., and Yao, W. 2018b. Application of edible coating with essential oil in food preservation. *Crit. Rev. Food Sci. Nutr*. 1–14.

Ju, J., Xie, Y., Yu, H., Guo, Y., Cheng, Y., Zhang, R., and Yao, W. 2020a. Major components in Lilac and *Litsea cubeba* essential oils kill *Penicillium roqueforti* through mitochondrial apoptosis pathway. *Ind. Crops Prod.* 149: 112349.

Ju, J., Xie, Y., Yu, H., Guo, Y., Cheng, Y., Qian, H., and Yao, W. 2022b. Synergistic interactions of plant essential oils with antimicrobial agents: A new antimicrobial therapy. *Crit. Rev. Food Sci. Nutr.* 62(7): 1740–1751.

Ju, J., Xu, X., Xie, Y., Guo, Y., Cheng, Y., Qian, H., and Yao, W. 2018. Inhibitory effects of cinnamon and clove essential oils on mold growth on baked foods. *Food Chem.* 240: 850–855.

Kalemba, D. A. A. K., and Kunicka, A. 2003. Antibacterial and antifungal properties of essential oils. *Curr. Med. Chem.* 10(10): 813–829.

Khazaei, M., Dastan, D., and Ebadi, A. 2021. Binding of *Foeniculum vulgare* essential oil and its major compounds to double-stranded DNA: In silico and in vitro studies. *Food Bioscience*. 41(8): 100972.

Khorshidian, N., Yousefi, M., Khanniri, E., and Mortazavian, A. M. 2018. Potential application of essential oils as antimicrobial preservatives in cheese. *Innovative Food Sci. Emerging Technol.* 45: 62–72.

Kong, J., Xie, Y. F., Guo, Y. H., Cheng, Y. L., Qian, H., and Yao, W. R. 2016. Biocontrol of postharvest fungal decay of tomatoes with a combination of thymol and salicylic acid screening from 11 natural agents. *LWT-Food Sci. Technol.* 72: 215–222.

Kong, J., Zhang, Y., Ju, J., Xie, Y., Guo, Y., Cheng, Y., and Yao, W. 2019. Antifungal effects of thymol and salicylic acid on cell membrane and mitochondria of *Rhizopus stolonifer* and their application in postharvest preservation of tomatoes. *Food Chem.* 285: 380–388.

Li, B., Zheng, K., Lu, J., Zeng, D., Xiang, Q., and Ma, Y. 2022a. Antibacterial characteristics of oregano essential oil and its mechanisms against *Escherichia coli* O157: H7. *Journal of Food Measurement and Characterization*. 1–10.

Li, H. Z., Zhang, Z. J., Hou, T. Y., Li, X. J., and Chen, T. 2015. Optimization of ultrasound-assisted hexane extraction of perilla oil using response surface methodology. *Ind. Crops Prod.* 76: 18–24.

Li, S. M., Gu, Y. L., Li, W. X., and Yu, S. S. 2013. The ultrasound assisted extraction of lavender essential oil by steam distillation. *Foods*. 2: 41–44.

Li, W. R., Shi, Q. S., Xie, X. B., and Duan, Y. S. 2016. Research progress on chemical constituents and antibacterial activity of plant essential oils. *Microbiology*. 43(6): 1339–1344.

Li, Y. H., Wang, W. J., Yang, L., and Zhang, Y. 2017. Effect of sage oil in bacon sensory quality and antioxidant properties. *China Condiment*. 42(6): 57–60.

Li, Z., Wang, H., Pan, X., Guo, Y., Gao, W., Wang, J., and Chen, F. 2022b. Enzyme-deep eutectic solvent pre-treatment for extraction of essential oil from *Mentha haplocalyx* Briq. leaves: Kinetic, chemical composition and inhibitory enzyme activity. *Ind. Crops Prod.* 177: 114429.

Liang, Z. W., Zou, X. J., Chen, M. Y., and Huan, S. Q. 2017. The effects of different extraction conditions on the yield of essential oil of lemongrass. *Anim. Husb. Feed Sci.* 38(9): 17–19.

Liu, X. Y., Ou, H., Xiang, Z. B., and Gregersen, H. 2019. Optimization, chemical constituents and bioactivity of essential oil from *Iberis amara* seeds extracted by ultrasound-assisted hydro-distillation compared to conventional techniques. *J. Appl. Res. Med. Aromat. Plants.* 13: 100204.

Liu, Z., Li, H., Zhu, Z., Huang, D., Qi, Y., Ma, C., and Ni, H. 2022. *Cinnamomum camphora* fruit peel as a source of essential oil extracted using the solvent-free microwave-assisted method compared with conventional hydrodistillation. *LWT-Food Sci. Technol.* 153: 112549.

Magry, A., Olender, A., and Tchórzewska, D. 2021. Antibacterial properties of *Allium sativum* L. against the most emerging multidrug-resistant bacteria and its synergy with antibiotics. *Arch. Microbiol.* 203: 2257–2268.

Mangal, S., Singh, V., Chhibber, S., and Harjai, K. 2022. Natural bioactives versus synthetic antibiotics for the attenuation of quorum sensing-regulated virulence factors of *Pseudomonas aeruginosa. Future Microbiol.* 17(10): 773–787.

Marchese, A., Orhan, I. E., Daglia, M., Barbieri, R., Di Lorenzo, A., Nabavi, S. F., and Nabavi, S. M. 2016. Antibacterial and antifungal activities of thymol: A brief review of the literature. *Food Chem.* 210: 402–414.

Mollica, F., Gelabert, I., and Amorati, R. 2022. Synergic antioxidant effects of the essential oil component γ-terpinene on high-temperature oil oxidation. *ACS Food Science & Technology.* 2: 180–186.

Munhuweyi, K., Caleb, O. J., van Reenen, A. J., and Opara, U. L. 2018. Physical and antifungal properties of β-cyclodextrin microcapsules and nanofibre films containing cinnamon and oregano essential oils. *LWT-Food Sci. Technol.* 87: 413–422.

Muthaiyan, A., Martin, E. M., Natesan, S., Crandall, P. G., Wilkinson, B. J., and Ricke, S. C. 2012. Antimicrobial effect and mode of action of terpeneless cold-pressed Valencia orange essential oil on methicillin-resistant *Staphylococcus aureus. J. Appl. Microbiol.* 112(5): 1020–1033.

Mutlu-Ingok, A., and Karbancioglu-Guler, F. 2017. Cardamom, cumin, and dill weed essential oils: Chemical compositions, antimicrobial activities, and mechanisms of action against campylobacter spp. *Molecules.* 22(7): 1191.

Nakadafreitas, P. G., Santos, C. A., Magalhães, T. H., Bustamonte, S. S., Santos, D. C. D., Cardoso, A. I. I., and Catão, H. C. R. 2022. Effect of thyme, lemongrass and rosemary essential oils on *Aspergillus flavus* in cauliflower seeds. *Hortic. Bras.* 40: 71–75.

Nazir, N., Zahoor, M, Uddin, F., and Nisar, M. 2021. Chemical composition, in vitro antioxidant, anticholinesterase, and antidiabetic potential of essential oil of *Elaeagnus umbellata* Thunb. *BMC Complementary Med. Ther.* 21(1): 4–11.

Nematollahi, Z., Ebrahimi, M., Raeisi, M., Shahamat, Y. D., and Shirzad, H. 2020. The antibacterial activity of cinnamon essential oil against foodborne bacteria: A mini-review. *J. Hum. Environ. Health Promot.* 6(3): 101–105.

Nirmala, M. J., Durai, L., Anusha, G. S., and Nagarajan, R. 2021. Nanoemulsion of *Mentha arvensis* essential oil as an anticancer agent in anaplastic thyroid cancer cells and as an antibacterial agent in *Staphylococcus aureus. Bio Nano Science.* 11(4): 1017–1029.

Nuţă, D. C., Limban, C., Chiriţă, C., Chifiriuc, M. C., Costea, T., Ioniţă, P., and Zarafu, I. 2021. Contribution of essential oils to the fight against microbial biofilms—A review. *Processes.* 9(3): 537.

Oana, C., Claus, M., Peter, Ø. J., and Niels, H. 2022. Tolerance and resistance of microbial biofilms. *Nat. Rev. Microbiol.* 22: 682–694.

Park, E. J., Hussain, M. S., Wei, S., Kwon, M., and Oh, D. H. 2019. Genotypic and phenotypic characteristics of biofilm formation of emetic toxin producing *Bacillus cereus* strains. *Food Control.* 96: 527–534.

Peng, B. Wang, W. Q., Liu, H. J., and Zhang, N. 2013. Study on extraction process of *Litsea cubeba* oil and citral. *Appl. Chem. Industry.* 42(10): 1786–1788.

Rao, J., Chen, B., and McClements, D. J. 2019. Improving the efficacy of essential oils as antimicrobials in foods: Mechanisms of action. *Annu. Rev. Food Sci. Technol.* 10–18.

Ribeiro, H. M., Araujo, A., Cunha, C., and Rodrigues, M. 2022. Essential oils used in dermocosmetics: Review about its biological activities. *Journal of Cosmetic Dermatology*. 21(2): 513–529.

Sayadi, M., Langroodi, A. M., and Jafarpour, D. 2021. Impact of zein coating impregnated with ginger extract and *Pimpinella anisum* essential oil on the shelf life of bovine meat packaged in modified atmosphere. *Journal of Food Measurement and Characterization*. 15: 5231–5244.

Serena, D. A., Serio, A., López, C., and Paparella, A. 2018. Hydrosols: Biological activity and potential as antimicrobials for food applications. *Food Control*. 86: 126–137.

Shahid, M. 2021. Lemongrass essential oil components with antimicrobial and anticancer activities. *Antioxidants*. 11: 20–31.

Sharifi-Rad, M., Varoni, E. M., Iriti, M., Martorell, M., Setzer, W. N., del Mar Contreras, M., . . . and Sharifi-Rad, J. 2018. Carvacrol and human health: A comprehensive review. *Phytotherapy Research*. 32(9): 1675–1687.

Shi, G., Lin, L., Liu, Y., Chen, G., Fu, S., Luo, Y., and Li, H. 2022. Multi-objective optimization and extraction mechanism understanding of ionic liquid assisted in extracting essential oil from Forsythiae fructus. *Alexandria Eng. J*. 61(9): 6897–6906.

Šojić, B., Pavlić, B., Tomović, V., Kocić-Tanackov, S., Đurović, S., Zeković, Z., and Škaljac, S. 2020. Tomato pomace extract and organic peppermint essential oil as effective sodium nitrite replacement in cooked pork sausages. *Food Chem*. 330: 127202.

Uzkuç, N. M. Ç., Uzkuç, H., Berber, M. M., Kuzu, K. T., Toğay, S. Ö., Hoşoğlu, M. İ., and Yüceer, Y. K. 2021. Stabilisation of lavender essential oil extracted by microwave-assisted hydrodistillation: Characteristics of starch and soy protein-based microemulsions. *Ind. Crops Prod*. 172: 114034.

Vasconcelos, N. G., Croda, J., and Simionatto, S. 2018. Antibacterial mechanisms of cinnamon and its constituents: A review. *Microbial Pathogenesis*. S0882401018305667.

Verma, S. K., Goswami, P., Verma, R. S., Padalia, R. C., Chauhan, A., Singh, V. R., and Darokar, M. P. 2016. Chemical composition and antimicrobial activity of bergamot-mint (*Mentha citrata* Ehrh.) essential oils isolated from the herbage and aqueous distillate using different methods. *Ind. Crops Prod*. 91: 152–160.

Viacava, G. E., Cenci, M. P., and Ansorena, M. R. 2022. Effect of chitosan edible coatings incorporated with free or microencapsulated thyme essential oil on quality characteristics of fresh-cut carrot slices. *Food Bioprocess Technol*. 15(4): 768–784.

Wang, J. Wei, D. F., and Wang, D. 2015. Study on the extraction of essential oil by supercritical carbon dioxide. *Amomun Kravanh Food Industry*. 36(12): 12–14.

Wang, W. X., Xu, J. M., and Zhou, L. G. 2016. Lemon grass and *Tithonia diversifolia* leaf oil nematicidal activity. *Research and development of natural products*. 8: 17–23.

Wei, J. Su, H. L., Zhang, X. T., and Bi, Y. 2016. Study on antioxidant activity of Curcuma oil. *Sci. Technol. Food Ind*. 10: 22–31.

Wei, Q., and Liu, Y. 2021. Composition and anti-inflammatory and antioxidant activity of essential oil from *Aucuba japonica* var. variegata. *Chem. Nat. Compd*. 57(5): 961–964.

Xue, S. J., Li, L. Yang, D. Wang, X. L., and Guan, J. 2017. A study on ultrasonic assisted extraction of onion essential oil. *Food Industry*. 2: 13–16.

Xylia, P., Chrysargyris, A., and Tzortzakis, N. 2021. The combined and single effect of marjoram essential oil, ascorbic acid, and chitosan on fresh-cut lettuce preservation. *Foods*. 10(3): 575.

Yaghoubi, M., Ayaseh, A., Alirezalu, K., Nemati, Z., and Lorenzo, J. M. 2021. Effect of chitosan coating incorporated with artemisia fragrans essential oil on fresh chicken meat during refrigerated storage. *Polymers*. 13(5): 716.

Yu, H., Liu, Y., Yang, F., Xie, Y., Guo, Y., Cheng, Y., and Yao, W. 2022. The combination of hexanal and geraniol in sublethal concentrations synergistically inhibits Quorum Sensing of *Pseudomonas fluorescens*-in vitro and in silico approaches. *Appl. Microbiol*. 10: 11446.

Zhang, B., Hou, X. Z., Q, Y. Ding, X. Deng, Q. H., and K, Y. T. 2015. The chemical composition of lemon peel essential oil, antioxidant and antibacterial activity. *Sci. Technol. Food Ind.* 36(5): 126–131.

Zhang, D., Gan, R. Y., Zhang, J. R., Farha, A. K., Li, H. B., Zhu, F., and Corke, H. 2020. Antivirulence properties and related mechanisms of spice essential oils: A comprehensive review. *Compr. Rev. Food Sci. Food Saf.* 19(3): 1018–1055.

Zhou, C., Li, C., Siva, S., Cui, H., and Lin, L. 2021. Chemical composition, antibacterial activity and study of the interaction mechanisms of the main compounds present in the *Alpinia galanga* rhizomes essential oil. *Ind. Crops Prod.* 165: 113441.

Zhou, Y., Chen, X. X., Chen, T. T., and Chen, X. Q. 2022. A review of the antibacterial activity and mechanisms of plant polysaccharides. *Trends Food Sci. Technol.* 123: 264–280.

3 The Inhibitory Activity and Mechanism of Essential Oils on Foodborne Fungi

3.1 INTRODUCTION

Due to the misuse and abuse of synthetic antimicrobial agents, microbial resistance and environmental pollution are global concerns (Wu et al., 2022; Silvestre et al., 2019; Debonne et al., 2019). Plant compounds, such as water or alcohol extracts and plant essential oils (EOs), play important roles as natural and safe antimicrobial agents (Cui et al., 2020; Zhu et al., 2020; Ju, Chen et al., 2019). Due to their chemical structures, active ingredients from plants can provide broad-spectrum antimicrobial activity (Xu et al., 2020; Kujur et al., 2020; Oueslati et al., 2020; Ahmady et al., 2019).

EOs have a long history of use, first in the Middle East followed by North Africa and Europe. The term "essential oil" was coined by the Swiss scholar Paracelsus von Hohenheim in the 16th century (Macwan et al., 2016). EOs, which comprise a complex mixture of natural polar and nonpolar compounds, have antimicrobial (antifungal and antibacterial) and medicinal properties (analgesic, anti-inflammatory, antioxidant, glucose-lowering, and anticancer properties) (Tohidi et al., 2019; Borges et al., 2019; Ju et al., 2018). Due to their antimicrobial and antioxidant activities, EOs are used as natural food additives (Falleh et al., 2020; Ahmad et al., 2020; Ju, Chen et al., 2019).

EOs can be extracted from all plants, but mainly from herbs and spices. There are approximately 3,000 EOs, of which more than 300 are commercially available (Ding et al., 2022; Ju et al., 2019). These oils are mainly used in fragrances and cosmetics. In nature, EOs protect plants against microorganisms and reduce herbivore infestation (by inducing unpleasant taste). In contrast, EOs may attract insects to promote the dispersion of pollen and seeds. Therefore, EOs play important roles in the interaction between plants and the environment. The main compounds in EOs are terpenes and terpenoids. Terpenes (e.g., *p*-cymene, limonene, terpinene, α-pinene, and β-pinene) are hydrocarbons derived from isoprene (C_5H_8), which has various chemical and biological characteristics. Monoterpenes ($C_{10}H_6$) and sesquiterpenes ($C_{15}H_{24}$) are the main terpenes. Monoterpenes (e.g., limonene, *p*-cymene, α-pinene, and oxygen-containing monoterpenes, such as carvacrol, thymol, and camphor) are composed of two isoprene units and represent 90% of EOs (Nazzaro et al., 2017).

DOI: 10.1201/9781003329268-3

Sesquiterpenes are composed of three isoprenoids. The structure and function of sesquiterpenes are similar to those of monoterpenes.

Terpenoids (e.g., linalyl acetate, 1,8-cineole, linalool, piperone, citronellal, geraniol, and menthol) are similarly derived from isoprene (Alsaraf et al., 2020; Salehi et al., 2019; Aisa et al., 2020). However, unlike terpenes, the isoprene unit in terpenoids is modified via the addition or removal of methyl groups or addition of oxygen (Lopezreyes et al., 2013). Terpenoids can be classified as monoterpenoids, sesquiterpenoids, or diterpenoids based on the number of isoprene units present. Furthermore, terpenoids can be classified according to the number of cyclic structures.

The chemical composition of EOs is dependent on the plant, geographical location, environment, maturity, and extraction method (Charfi et al., 2019; Huang et al., 2018). EOs have antimicrobial effects on microorganisms, especially their cell membrane and cytoplasm. In some cases, EOs completely change the shape of cells (Cheraif et al., 2020; Mobasseri, 2019). Humans and plants are susceptible to fungal infection. While synthetic fungicides have been used to control fungal infections, their use is limited by the emergence of resistant strains and their toxic effects (Sirgamalla et al., 2020; Khosravi et al., 2020; Mohsen et al., 2020; Sirelkhatim et al., 2019). Today, humans pay considerably more attention to the protection of the ecological environment. Therefore, there has been increasing interest in the use of EOs as alternatives to conventional synthetic fungicides (Chen et al., 2020; Foddai et al., 2019).

3.2 MAIN FACTORS AFFECTING THE YIELD AND CHEMICAL COMPOSITION OF ESSENTIAL OIL

The composition and content of chemical components in essential oils are affected by many factors, such as plant harvest time, planting area, maturity and extraction methods, etc. In fact, a great deal of research has been done on the factors affecting the yield and chemical composition of essential oils in the past few years. This section will discuss and analyze some of the main factors.

3.2.1 Harvest Time of Plants

The harvest time of plants has a great influence on the composition and content of chemical components in essential oils. Saharkhiz et al. (2005) investigated the influence of harvesting period on *Trachyspermum ammi* Sprague EO content and composition. It was found that the EO yield was highest (7.1% w/w) for fruits harvested at pasty stages (with 54% moisture) in contrast to those collected at ripening stages (with 5% moisture) with a percentage yield of 3.2% w/w accordingly. Moreover, 12 constituents were identified in the EO obtained from ripe fruits (main components: thymol 36.7%, γ-terpinene 36.5%, and *p*-cymene 21.1%), while only 10 constituents were identified in the fruits harvested at pasty stage (main components: γ-terpinene 43.2%, thymol 32.4%, and *p*-cymene 20.7%). In addition, related researchers investigated the effects of three different growth stages on the composition and yield of *Daucus Sahariensis* Murb. essential oils. Monoterpenes were found to be the most important compounds in plant extracts harvested at germination and flowering stages. However, phenylpropane compounds were found to account for the largest proportion in mature plant extracts (Flamini et al., 2013). At the same time, some researchers have found that the yield of *Coptis chinensis* in Morocco is also affected by the harvest season. The highest output of essential oil was 0.32% in May, the average quantity of essential oil was about 0.26% in June, and decreased to 0.18% in January (Haloui et al., 2015).

3.2.2 Geographical Position

The changes of geographical location and altitude have an important impact on the terrestrial plant ecosystem. Changes in geographical location or altitude may lead to changes in water sources, light, relative humidity, wind speed, and other factors. However, these environmental factors are the decisive factors affecting plant growth. Changes in environmental conditions may change the physiological response of plants. For example, altitude changes affect the biosynthesis of terpenoids and oxygen-containing monoterpenes in *Kundmannia anatolica* Hub.-Mor essential oils, which are more significant at low altitudes, while sesquiterpenes are higher at high altitudes (Sanli and Karadoğan, 2017).

3.2.3 Weather Conditions

There is also a certain correlation between the yield and chemical composition of essential oil and meteorological conditions. Based on the analysis of the yield and composition of essential oil in cumin fruit picked from March to October for three consecutive years (2001–2003), it was found that compared with 2002 and 2001, the rainfall in 2003 was higher than that in 2002 and 2001. The lower temperature of atmosphere and surface soil led to lower yield of essential oil and higher content of monoterpene. It is worth noting that monoterpenes are the chemical constituents most easily affected by climatic conditions (Aprotosoaie et al., 2010).

3.2.4 Extraction Method

With the continuous development of extraction technology, the extraction rate of essential oil is increasing. Compared with steam distillation (SD), solvent-free microwave distillation (SFMD-HE) is more efficient for essential oil extraction. Under 540 W microwave radiation power, the yield of essential oil from 40 min treated by SFMD-HE was 6.50 ± 0.31 ml/kg, while the yield of essential oil treated with SD was 5.46 ± 0.22 ml/kg (Zhao et al., 2018). The reason for this phenomenon may be that in SD technology, a small amount of essential oil components often form hydrosol with condensed water, which is difficult to separate and collect. Therefore, the formation of hydrosol may lead to the loss of some compounds. In SFMD-HE, aromatic plants are placed in a microwave reactor without adding solvents and distilled at atmospheric pressure. When aromatic plants are irradiated by microwave, the oily plant cells rupture due to the increase of internal pressure. When the essential oil is released, it is distilled in situ. Therefore, this method increases the yield of essential oil by preventing the production of hydrosol. Similarly, some researchers have explored the effects of steam distillation and microwave steam diffusion gravity (MHG) on the extraction of essential oil from rosemary. The results showed that the yield of 15 min separated by MHG was similar to that of 180 min separated by SD. In addition, the number of oxygen-containing compounds obtained by MHG method was slightly more than that obtained by SD method, which were 26.16% and 29.54%, respectively (Bousbia et al., 2009). Essential oils contain organic compounds that strongly absorb microwave energy. Microwave extraction can extract compounds with high and low dipole moments in different proportions. Therefore, compared with aromatic compounds with low dipole moment (such as monoterpene hydrocarbons), organic compounds such as oxygen compounds with high dipole moment interact more strongly with microwave, so it is easier to extract. Through the comparison of scanning electron microscope, it was found that the plant cell wall was affected and the oil-bearing glands were damaged after 15 min treated with MHG. On the contrary, there was no significant change in plant materials treated with SD for 3 h compared with untreated plant materials (Bousbia et al., 2009).

3.2.5 Drying Conditions

Drying conditions and drying time will also affect the yield of essential oils. Drying lavender before steam distillation results in a loss of about 44% of essential oils. In addition, there are differences in the quantity and quality of essential oils extracted from fresh and dried flowers (Smigielski et al., 2011). Seventy-three compounds were detected in dried flower extract, while only 65 compounds were identified in fresh flower extract. In addition, compared with fresh lavender essential oils, dried lavender essential oils contain fewer monoterpene esters (reduced by 12.9%). Drying may activate hydrolases, which may lead to specific changes in the proportion of volatile compounds. Similarly, the drying time also had an important effect on the content of essential oil in lemon leaves, and the yield of essential oil reached the highest value of 1.26% ± 0.07% on the 4th day of drying. On the 5th day of drying, the content of citronellal reached the highest value (87.58% ± 0.33%) (Wulandari et al., 2019).

3.2.6 Distillation Time

Steam distillation is a commonly used method for essential oil extraction. Usually, the distillation time has a great influence on the extraction rate of essential oil. When the distillation time was 1.5 min, the contents of anethol and cineole were the highest. However, with the extension of distillation time, its content decreased gradually. When the distillation time is 7.5–15 min, the camphor concentration reaches the maximum. The concentration of linalool acetate is the highest at the distillation time of 30 min (Chagonda et al., 2020). On the other hand, *t*-caryophyllene increases with the increase of distillation time. Therefore, it can be minimized or maximized by using shorter and longer distillation time, respectively. The essential oil composition of lemon grass also varies with the distillation time. For example, the main components vanillin and geraniene reach the maximum value at the distillation time of 10–40 min, which is consistent with the optimal yield of the distillation time. However, the minor components caryophyllene oxide and *t*-caryophyllene increased with the increase of distillation time. If these ingredients are not needed, a shorter distillation time can be used to produce the main components. The oil yield of lemon grass distillation at 20 min is the highest. Shorter distillation time may produce essential oils with purer geraniol properties. Therefore, if a higher concentration of geranyl acetate is required, distillation is required for 160 min or more. Therefore, the distillation time is of great significance for optimizing the yield of essential oil and obtaining essential oil with specific chemical characteristics.

3.3 COMMON EXPERIMENTAL METHODS FOR EVALUATING ANTIMICROBIALS MECHANISM

Various strategies have been used to elucidate the antibacterial mechanism of essential oils and their components, including (1) release of cellular components, such as OD260, OD280, and electrical conductivity (Ju et al., 2020a; Zhang et al., 2020; Huang et al., 2019); (2) intracellular and extracellular pH size and ATP concentration (Zhan et al., 2022); and (3) electron microscopic observation of cell damage (Ju et al., 2020a). According to the current research, one of the most important action sites of essential oil on microbial cells is the cell membrane (Ju et al., 2018); for example, outer cell membrane and cytoplasmic membrane. Therefore, many techniques have been used to monitor the changes of cell function at this location, including the release of (a) cell membrane permeability; (b) cell membrane integrity; and (c) cell components. With the increase of cell membrane permeability, the integrity of cell membrane is destroyed, which eventually leads to the leakage of cell contents. The commonly used experimental methods for evaluating the changes of cells after exposure to essential oils are shown in Figure 3.1.

3.3.1 Permeability of Cell Membrane

The permeability of cell membrane can be measured by fluorescence-based methods, such as fluorescence spectroscopy, laser confocal microscopy, inverted fluorescence microscopy, and flow cytometry. In general, the cell membrane is stained with fluorescent dyes that preferentially accumulate in the cytoplasm. When the permeability of the membrane increases due to exposure to essential oils, the dye is released into the culture medium, resulting in an increase in fluorescence intensity (Figure 3.1). A variety of fluorescent dyes can be used to achieve this purpose, including carboxyfluorescein (cF), ethidium bromide, gentian violet, and rhodamine. Flow cytometry based on fluorescence is one of the most commonly used methods to clarify how essential oils interact with plasma membranes. The principle of this method is that carboxyfluorescein diacetate (cFDA) is used as the fluorescence substrate, which provides the viability and rapid analysis of individual cells with high sensitivity. CFDA is a lipophilic nonfluorescent dye, which is mainly used to evaluate cell enzyme activity. When diffused through the cell membrane, the diacetate group of cFDA is hydrolyzed by nonspecific esterase in the cell to produce highly fluorescent compounds, which are retained in living cells with intact plasma membranes. According

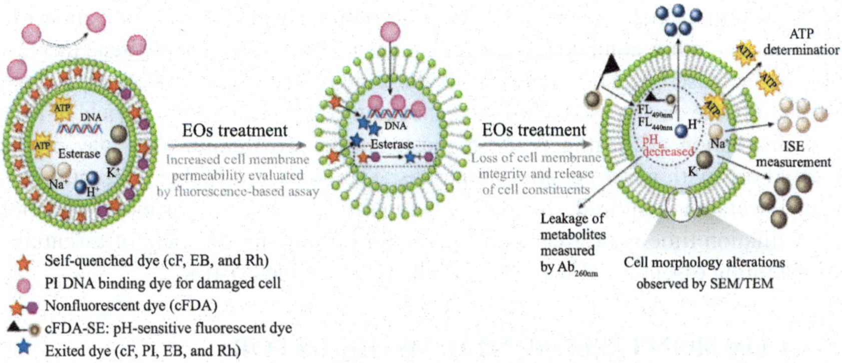

FIGURE 3.1 An overview of common experimental methods used to evaluate the antibacterial mechanism of essential oils. Based on fluorescence analysis, morphological changes of cell membrane, ATP assay, metabolite leakage, and electric field measurement, the permeability, integrity, and release of cell components of cell membrane were evaluated.

Abbreviations: cF, carboxyfluorescein; cFDA-SE, carboxyfluorescein diacetate; EB, ethidium bromide; ISE, ion selective electrode; PI, propidium iodide; Rh, rhodamine; SEM, scanning electron microscope; TEM, transmission electron microscope.

Source: Rao et al. (2019).

to this method, Ju et al. (2020b) have tested the effects of eugenol and citral on cell membrane permeability of *Aspergillus niger.* In addition, this method has also been used to study the degree of membrane damage of cinnamon, oregano, and thyme essential oils to *Salmonella typhimurium*, *Staphylococcus aureus*, *Escherichia coli*, *Rhizopus stolonifer*, and *Monilinia fructicola* (Mortazavi and Aliakbarlu, 2019; Li et al., 2022; Yan et al., 2021; Andrade-Ochoa et al., 2021; Xu et al., 2021). This method allows people to estimate the effect of essential oils on cell permeability based on fluorescence density. In investigating the mechanism of action of thymol against *E. coli*, some authors have shown that thymol treatment can significantly ($P < 0.05$) increase the permeability of *E. coli* cell membrane. Similar results were obtained when studying the effect of cinnamon essential oil on the membrane permeability of *S. aureus* (Zhang et al., 2022; Andrade-Ochoa et al., 2021).

PI is a dye that can dye DNA. It is an analogue of ethidium bromide and emits red fluorescence when intercalated into double-stranded DNA. Although PI cannot pass through the living cell membrane, it can dye the nucleus through the damaged cell membrane. PI is often used with fluorescent probes such as Calcein-AM or FDA to stain both living and dead cells. Ju et al. (2020a) used this method to investigate the damage degree of the combination of eugenol and citral on the cell membrane of *Penicillum roqueforti*. The results showed that the damage degree of cell membrane caused by eugenol was greater than that of citral. This may be mainly due to the fact that the hydroxyl group carried by eugenol can increase the permeability of cell membrane.

The lipid profile changes of several bacteria treated with essential oils (such as thymol, eugenol, genitol, and cinnamaldehyde) were analyzed by gas chromatography.

The results showed a strong decrease in unsaturated fatty acids and an increase in saturated fatty acids in the outer membrane. This indicates that the adaptation mechanism of cells to environmental stress affects membrane function and permeability (He et al., 2021; Tomaś et al., 2021; Krauze et al., 2021; Heydari et al., 2021; Bhavaniramya et al., 2019).

3.3.2 Integrity of Cell Membrane

In bacterial strains, the proton power produced across the cell membrane is essential for ATP synthesis and the transport of various solutes. The loss of membrane integrity caused by proton power collapse leads to the consumption of ATP and the decrease of transmembrane ion gradient, which may eventually lead to cell lysis. Therefore, the most common way to assess cell membrane integrity is to measure intracellular and/or extracellular ATP concentrations or ion leakage. For example, Costa et al. (2022) evaluated the antibacterial activity and mechanism of lemon essential oil against *S. aureus*. The results show that the MIC and MBC of the essential oil are 128–512 and 256–1024 μg/ml, respectively. The growth curve test showed that *S. aureus* was obviously inhibited by the essential oil. In addition, antibacterial mechanism experiments showed that the essential oil could destroy the integrity of *S. aureus* cell membrane, resulting in intracellular ion, protein, and nucleic acid leakage.

The increase of extracellular pH value means the decrease of intracellular pH value and the acidification of cells due to the accumulation of H+, which is easy to cause irreversible damage to intracellular physiological and biochemical processes (Ju et al., 2020c). Therefore, the measurement of extracellular or intracellular pH of microbial cells can also be used to study the antibacterial mechanism of essential oils. It is reported that the extracellular pH of *Penicillium* spp. in 0–30 min increased rapidly when α-terpineol with MIC concentration was added (Li et al., 2014).

Scanning electron microscopy (SEM) and transmission electron microscopy (TEM) are commonly used to observe the changes in microbial morphology or the degree of rupture of the cell membrane after EO treatment. For example, Yang et al. (2022) confirmed through SEM that nanoemulsions containing thyme essential oil can destroy the cell morphology of *E. coli O157:H7* and cause severe cytoplasmic leakage. He et al. (2022) investigated the antibacterial mechanism of nanoemulsion containing thyme essential oil against *E. coli O157:H7* and *S. aureus* by TEM. The results showed that the essential oil could change the microstructure of these two bacteria.

3.3.3 Release of Cellular Components

The most commonly used methods for measuring cell content leakage are OD260, OD280, and relative conductivity. OD260 and OD280 represent the leakage of proteins and nucleic acids in cells (Kong et al., 2019). The higher the relative conductivity, the more the leakage of electrolyte and the more serious the damage of cell membrane.

Related research cases include thyme essential oil which can cause the leakage of proteins and nucleic acids in *E. coli* cells (Yang et al., 2022). Cinnamon essential oil can significantly ($P < 0.05$) increase the concentration of protein, nucleic acid, and electrical conductivity in *E. coli* and *S. aureus* cell suspension (Zhang et al., 2016). Similarly, the combination of eugenol and citral had a significant synergistic antifungal effect on *Penicillium roqueforti* and *A. niger*, which could lead to significant ($P < 0.05$) leakage of cell contents and increased electrical conductivity of the two fungi (Ju et al., 2020c, 2002d).

3.4 ANTIFUNGAL ACTIVITY OF ESSENTIAL OILS

Compared with bacterial infections, fungal infections are more difficult to identify (Garnier et al., 2020; Pizzolitto et al., 2020). Even though chemical antifungal agents are very effective, resistant fungal strains and species emerge. The degree of fungal infections depends on the fungus and the immune status of the host. EOs are probably the most promising antifungal natural agents. Table 3.1 lists the antifungal activities of some essential oils and their active components in recent years, including minimum inhibitory concentration (MIC). Like other phytochemicals, EOs attenuate microbial growth and biofilm development through specific mechanisms. Due to their broad antimicrobial activity, EOs can be used to control microbial spoilage and extend shelf life of foods (Falleh et al., 2020). EOs are "generally recognized as safe" by the US Food and Drug Administration, and because they are natural, they are generally accepted by consumers (Ju et al., 2020a). The antifungal activity of EOs may be attributed to the presence of terpenes and terpenoids (Ju et al., 2020a; Rao et al., 2019). Due to their lipophilicity and low molecular weight, terpenes and terpenoids damage cell membranes, cause cell death, and inhibit spore formation and germination (Zhang et al., 2017). Table 3.2 lists the main attack sites of EOs and/or their active components against fungi.

According to Ju et al. (2020c), antifungal agents work by negatively impacting the structure and function of cell membrane or organelles or by inhibiting protein synthesis. The antifungal mechanisms of EOs are shown in Figure 3.2.

3.4.1 Effect on the Cell Wall and Cell Membrane

The cell walls play an important role in fungal growth (Krishnamoorthy et al., 2021). There are three main cell wall components: dextran, chitin, and mannan (Feng et al., 2021). Chitin is a homopolymer of β-1,4-linked *N*-acetylglucosamine units, which is synthesized in a reaction catalyzed by chitin synthetase. Chitin is not only an essential component of cell walls, but a necessary component for survival of fungi. When chitin polymerization is inhibited, cell division and growth are negatively impacted. Anethole, the main component of anise oil, inhibits chitin synthetase in a dose-dependent manner. Eugenol significantly ($P < 0.05$) inhibits growth of *A. niger*, resulting in changes in cell morphology and cell membrane permeability (Ju et al., 2022). Zhang et al. (2011) reported that cinnamaldehyde and citral denature the morphological structure of mycelium and sporangium of *A. niger*, and the effect of citral was significantly ($P < 0.05$) greater than that of cinnamaldehyde. Tea tree EOs alter the permeability and fluidity of *Candida albicans* by impacting cell membrane structure and function. Certain EOs may deform the mycelium cell wall (Hammer et al., 2004).

TABLE 3.1
Antifungal Activity of Essential Oils or Active Ingredients, Including Minimum Inhibitory Concentration (MIC)

Microorganism	Essential oil or active component	Antifungal activity	Reference
Malassezia restricta	*Zanthoxylum schinifolium* Siebold & Zucc. essential oil (ZSEO)	ZSEO showed antifungal activity against *M. restricta*, with minimum inhibitory concentration (MIC) and minimum fungicidal concentration (MFC) values of 2.5 and 10.0 mg/ml, respectively	Liao et al. (2022)
Candida albicans	Chrysanthemum essential oil	The minimum inhibitory concentration is 31.25 μg/ml, the minimum bactericidal concentration is 62.5 μg/ml. The antifungal mechanism of the essential oil includes the destruction of cell plasma membrane and the damage of mitochondria and DNA	Zhan et al. (2021)
Candida albicans	*Leome viscosa* essential oil	The MIC and MFC values ranged from 16.5 to 33 μl/ml with significant reduction on biofilm of *C. albicans* isolates	Krishnamoorthy et al. (2021)
Candida albicans	Thyme essential oil	The MIC 90 of the essential oil against *C. albicans* was 0.031 μg/ml	Moazeni et al. (2021)
Botrytis cinerea	*Origanum vulgare* essential oil	The lowest EC50 was for thymol (17.56 mg/l), followed by carvacrol (26.22 mg/l), the EO (52.92 mg/l), methyleugenol (112.43 mg/l), and β-caryophyllene (>250 mg/l)	Zhao et al. (2021)
Fusarium spp.	*Asarum heterotropoides* var. *mandshuricum*	The minimum inhibitory concentrations of the essential oil, which obtain from the roots and rhizomes of AHVM of five kinds of *Fusariums* spp. (*F. avenaceum*, *F. trichothecioides*, *F. sporotrioides*, *F.sambucinum*, and *F. culmorum*) were determined to be 1.25, 1.5, 1.5, >2, and >2 mg/ml, respectively	Xiao et al. (2021)
Aflatoxin	Nutmeg	MIC and MAIC are 2.75 and 1.75 mg/ml, respectively	Das et al. (2020)
Aflatoxin	*Pimenta dioica*	MIC and MAIC are 2.5 and 1 μl/ml, respectively	Chaudhari et al. (2020)
Candida albicans	*Trachyspermum ammi* essential oil	The minimum inhibitory concentration is 6.9 μg/ml	Dutta et al. (2020)
Botrytis cinerea	*Artemisia sieberi* Besser	The minimum inhibitory concentration is 100 μl/l	Ghasemi et al. (2021)
Candida spp.	*Thymus kotschanus* essential oil	The minimum inhibitory concentrations and the minimum fungicidal concentrations of EO are 4–16 and 16–64 μg/ml, respectively	Muslim and Hussin (2020)

TABLE 3.2
The Main Targets of EOs and/or Their Components against Fungi

Principal effects	Essential oils or components	Reference	Principal effects	Essential oils or components	Reference
Cell membrane/ wall	Eugenol	Yu et al. (2022)	Cell membrane/ wall	*Salvia sclarea*	Tariq et al. (2019)
	1,8-Cineole	Sharma et al. (2022)		Anethole	Lima et al. (2021)
	Citral	Wang et al. (2021)		Thymus	Tariq et al. (2019)
	Carvacrol	Kong et al. (2019)		Cinnamomum	Du et al. (2022)
	Cinnamaldehyde	Muhoza et al. (2022)		Citrus	Chen et al. (2013)
	p-Cymene	Tian et al. (2018)		*Coriaria nepalensis*	Nazzaro et al. (2017)
	Citronellal	Ouyang et al. (2020)		*Coriandrum sativum*	Nazzaro et al. (2017)
	Thymol	Ranjbar et al. (2022)		*Juniperus communis*	Haque et al. (2016)
	Terpinene-4-ol	Haque et al. (2016)		*Litsea cubeba*	Haque et al. (2016)
	α-Terpinene	Haque et al. (2016)		*Melaleuca alternifolia*	Ahmad et al. (2013)
	α-pinene	Ahmad et al. (2013)		*Mentha piperita*	Ahmad et al. (2013)
	Linalyl acetate	Haque et al. (2016)		*Ocimum basilicum*	Haque et al. (2016)
	Linalool	Haque et al. (2016)		*Origanum*	Haque et al. (2016)
	Limonene	Cai et al. (2019)			
Cell growth and morphology	Thymol	Kong et al. (2021)	Inhibition of efflux pump	Thymol	Dorman et al. (2000)
	Terpinene-4-ol	Haque et al. (2016)		Cinnamaldehyde	Lima et al. (2013)
	γ-Terpinene	Ahmad et al. (2013)		Carvacrol	Laorenza and Harnkarnsujarit (2021)
	α-Terpinene	Haque et al. (2016)		*Thymus vulgaris*	Chen et al. (2013)
	Citronellal	Stevic et al. (2014)		*Origanum vulgare*	Chen et al. (2013)
	p-Cymene	Weisany et al. (2019)		*Ocimum basilicum*	Chen et al. (2013)
	1,8-Cineole	Rodenak-Kladniew et al. (2020)		*Mentha*	Chen et al. (2013)
	α-Pinene	da Silva Bomfim et al. (2020)		*Melaleuca alternifolia*	Chen et al. (2013)
	Carvacrol	Kong et al. (2019)		Eucalyptus	Chen et al. (2013)
	Thymus	Jafri et al. (2020)		Citrus	Singh et al. (2021)
	Eucalyptus	Mieres-Castro et al. (2021)		Cinnamomum	Ivanov et al. (2021)

TABLE 3.2
(Continued)

Principal effects	Essential oils or components	Reference	Principal effects	Essential oils or components	Reference
Action on fungal mitochondria	*Anethum graveolens*	Ma et al. (2019)	ROS production antinitric oxide	Eugenol	Ju et al. (2020a)
	Artemisia herba-alba	Mohammed et al (2021), Boukhennoufa et al. (2021)		Citral	Ju et al. (2020a)
	Cananga odorata	Ghalem et al. (2016)		Carvacrol	Nobrega et al. (2016)
	Cinnamomum camphora	Li et al. (2018)		*p*-Cymene	Carrillohormaza et al. (2015)
	Coriandrum sativum	Ghalem et al. (2016)		Farnesol	Tolba et al. (2015)
	Commiphora myrrha	Ghalem et al. (2016)		Thymol	Kong et al. (2019)
	Hedychium spicatum	Ghalem et al. (2016)		Cinnamaldehyde	Ju et al. (2018)
	Origanum compactum	Chroho et al. (2022)		Salicylic acid	Kong et al. (2019)
	Origanum majorana	Lazrak et al. (2021)		Hexanal	Anusha et al. (2021), Zhang et al. (2019)
	Lupeol	Chavan et al. (2014)			
	Tetraterpenoid	Chavan et al. (2014)			
Inhibition of biofilm development	*Coriandrum sativum*	Trindade et al. (2015)	Inhibition of biofilm development	*Syzygium aromaticum*	Hiwandika et al. (2021)
	Croton cajucara	Sieniawska et al. (2022)		ρ-Cymene	Ahmad et al. (2013)
	Cymbopogon	Sahal et al. (2020)		γ-Terpinene	Souza et al. (2017)
	Citrus	Bang et al. (2000)		1–8-Cineole	Ahmad et al. (2013)
	Eucalyptus	Mieres-Castro et al. (2021)		Linalool	Guo et al. (2021)
	Laurus nobilis	Reis et al. (2020)		Terpinen-4-ol	Ahmad et al. (2013)
	Litsea	Haque et al. (2016)		Terpinolene	Fernandes, al. (2021)
	Melaleuca alternifolia	Battisti et al. (2021)		α-Terpineol	Laorenza and Harnkarnsujarit (2021)
	Mentha	Bound et al. (2016)		Eucarobustol E	Liu et al. (2017)
	Myrtus communis	Nowrouzi et al. (2022)		Eugenol	Ulanowska al. (2021)

(Continued)

TABLE 3.2
(Continued)

Principal effects	Essential oils or components	Reference	Principal effects	Essential oils or components	Reference
	Ocimum	Bound et al. (2016)		α-Terpinene	Trindade et al. (2015)
	Piper claussenianum	Mansur and Ferreira-Pereira (2021)			
	Rosmarinus officinalis	Garzoli et al. (2021)			
Anti-quorum sensing	Citrus	Cáceres et al. (2020)	Effect on micotoxins synthesis/ production	Cinnamomum	Najdabbasi et al. (2020)
	Juniperus communis	Haque et al. (2016)		*Origanum*	Tariq et al. (2019)
	Mentha piperita	Husain et al. (2015)		*Cymbopogon*	Tariq et al. (2019)
	Origanum	Haque et al. (2016)		Cider	Dao et al. (2021)
	Salvia sclarea	Haque et al. (2016)		Eugenol	Ulanowsk and Olas (2021)
	Limonene	Haque et al. (2016)		Thymus	Galovičová et al. (2021)
	Linalool	Haque et al. (2016)		*Mentha*	Tariq et al. (2019)
	α-Pinene	Haque et al. (2016)		*Ocimum sanctum*	Dao et al. (2021)
	Terpinene-4-ol	Haque et al. (2016)		*Rosmarinus officinalis*	Nazzaro et al. (2017)
				Satureja hortensis	Nazzaro et al. (2017)
				Zataria multiflora	Nazzaro et al. (2017)

Some EOs have more than one function. For example, citral damages the cell wall and cell membrane of *Fusarium moniliforme*, *F. solani*, *Alternaria alternata*, and *A. niger*, resulting in cytoplasmic leakage and inhibits the biosynthesis of DNA, RNA, proteins, peptidoglycans, and ergosterol of *C. albicans* (Jassal et al., 2021; Wang et al., 2019; Ju et al., 2020b). EOs from *Coriaria nepalensis* and *Coriandrum sativum* disrupt sterol biosynthesis. A reduction in ergosterol in fungal cell membrane can lead to changes in the metabolism, reproduction, and activity of fungi. Similarly, salicylic acid can inhibit the growth of *R. stoloniferus* and even kill fungi through the ergoziol synthesis pathway (Kong et al., 2019).

3.4.2 Effect on Nucleic Acids and Proteins

Citral, at a dose of 1.5 mg/l, has been shown to severely damage the DNA of *A. flavus*. The effect of citral on DNA has been studied using microscopic multichannel spectrophotometry, micro-laser scattering, super-resolution microscopic image analysis, and single-cell gel electrophoresis. The findings revealed that citral damages cells by

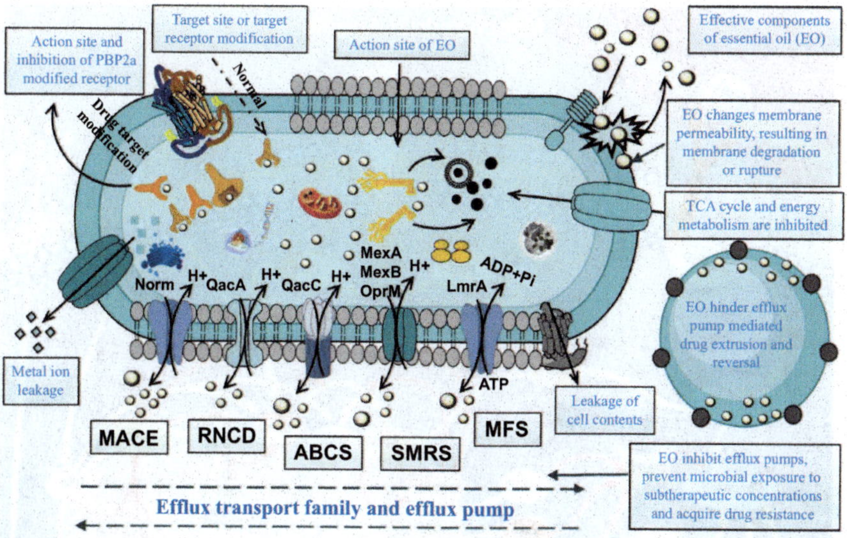

FIGURE 3.2 The possible action mode of EOs against fungi.

changing the permeability of the cell wall, plasma membrane, and mitochondrial membrane. When citral enters the cell, the biosynthesis of DNA, RNA, lipids, and proteins is inhibited, thereby impacting self-replication and contributing to cell death (Lopezmalo et al., 2002). The combination of citral and eugenol significantly ($P < 0.05$) impacts the conformation of nucleic acids and proteins in *A. niger* (Ju et al., 2020b).

3.4.3 Effect on Mitochondria

Mitochondria are the main sources of energy in cells. These organelles not only provide energy for cells, but also participate in cell differentiation, cell information transmission, and cellular apoptosis and have the ability to regulate cell growth and cell cycle (Kong et al., 2019). The mitochondrial cellular apoptosis mechanism is shown in Figure 3.3.

Some EOs affect mitochondrial function by inhibiting mitochondrial dehydrogenase activity, such as lactate dehydrogenase, malate dehydrogenase, and succinate dehydrogenase (Ma et al., 2022; Zhan et al., 2021). During hydrolysis by ATPase, a large number of hydrogen ions in mitochondria are pumped into the gap of the mitochondrial membrane, which disrupts the mitochondrial membrane potential (Cui et al., 2019). By altering mitochondrial function, EOs contribute to the accumulation of reactive oxygen species (ROS), which oxidize biological macromolecules and mediate cell death (Zheng et al., 2015; Dadi et al., 2019). Ju et al. (2020c) reported that the combination of eugenol and citral inhibits mitochondrial ATP synthesis in *P. roqueforti* by disturbing the tricarboxylic acid cycle. EOs from *Origanum compactum*, *Artemisia herba-alba*, and *Cinnamomum camphora* cause mitochondrial mutations in *Saccharomyces cerevisiae* (Kovacik et al., 2021; Agus, 2021; Alia et al., 2022). Terpenoids play a key role in reducing mitochondrial content, leading to changes in ROS and ATP levels (Haque et al., 2016).

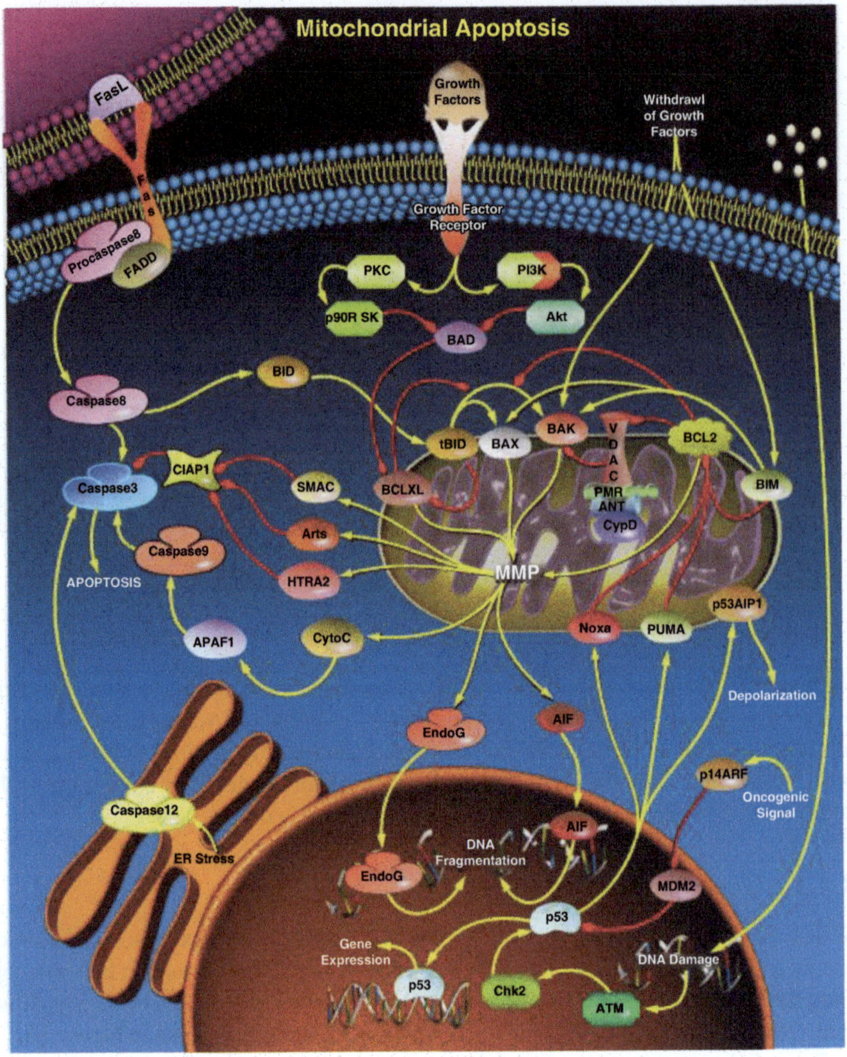

FIGURE 3.3 Schematic diagram of mitochondrial apoptosis mechanism.

3.4.4 Effect on Membrane Pumps

The fungal plasma membrane H^+-ATPase plays an important role in the physiology of fungal cells by supporting the transmembrane electrochemical proton gradient across the cell membrane, which is necessary for nutrient uptake (Orie et al., 2017). Additionally, H^+-ATPase regulates intracellular pH and cell growth and is involved in fungal pathogenicity through its effects on dimorphism, nutrient uptake, and medium acidification. Previous studies have shown that eugenol and thymol inhibit H+-ATPase activity. Thymol and carvacrol are the main chemical constituents of thyme EO. Thyme EO inhibits the expression of efflux pump genes *CDR1* and *MDR1* in *C. albicans*. Thyme EOs and fluconazole exhibit synergistic antifungal effects (Rivera-Yañez et al., 2022).

3.4.5 Effect on ROS Production

Bacterial nitric oxide synthase (BNOS) generate nitric oxide (NO) from arginine. NO generated from BNOS increases the resistance of bacteria to a broad spectrum of antibiotics, allowing the bacteria to survive and share environments with antibiotic-producing microorganisms. By mediating chemical modification of toxic compounds and reducing oxidative stress, NO-mediated resistance can be achieved. Therefore, inhibiting the activity of BNOS can increase the effectiveness of antimicrobial therapy (Shen et al., 2016). Few studies, however, have investigated NO synthesis in fungi.

Even though several antibiotics have different targets, the lethal effect is achieved through the production of ROS. EOs reduce NO levels and limit H_2O_2 production and NO synthase. Thymol has significant ($P < 0.05$) inhibitory effects on *A. flavus*, *Botrytis cinerea*, and *Rhizopus stolon* (Cotoras et al., 2013; Hou et al., 2020; Shen et al., 2016). ROS damage the mitochondria of *P. roqueforti* and *A. niger* (Ju et al., 2020b, 2020c). A schematic diagram of ROS formation in mitochondria is shown in Figure 3.4.

The active components of EOs may have synergistic or antagonistic effects. The combination of certain EOs increase their antifungal activity because pathogens are not likely to have resistance to two or more EOs (Stevic et al., 2014). The antifungal activity of terpenoids is associated with their functional groups, i.e., hydroxyl groups. Carvacrol and thymol damage cell membranes by interacting with sterols. When tested individually, carvacrol present in oregano and thyme EOs have strong

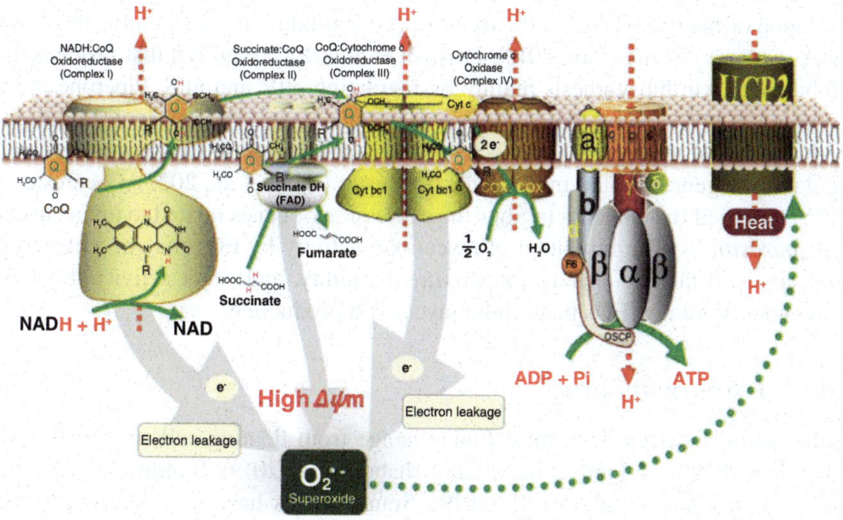

FIGURE 3.4 The generation of ROS in mitochondria.

antifungal activity (Sakkas et al., 2017). The structure of thymol is similar to that of caravol, but the position of the hydroxyl groups is different (Lima et al., 2013). However, these differences do not affect their activity. Thymol and eugenol damage the morphology of mycelia (Kong et al., 2019). Similar results have been observed with other monoterpenes, such as linalool, which damage the morphology of mycelia and inhibit the formation of biofilms (Dias et al., 2017; Nobrega et al., 2016). The antifungal activity of EOs and the ability to inhibit mycotoxin production are dependent on the chemical structure of EOs. Citronellal, a monoterpene and the main volatile component of citrus EOs, has antifungal activity (Ouyang et al., 2020; de Araújo-Filho José Vilemar et al., 2018).

3.4.6 Effect of Mycotoxins

Poisoning caused by mycotoxins is considered to be one of the ten highest risks to human health (Munkvold et al., 2021). Mycotoxins are classified as group 1 carcinogens by the International Agency for Research on Cancer (Baranyi et al., 2015). Chemically synthesized preservatives are often used to control fungal contamination. However, the excessive use of chemically synthesized preservatives results in increased fungal resistance and chemical residue concentrations (Shirazi et al., 2020; Muhialdin et al., 2020; Rmr et al., 2020). A large number of studies have confirmed that the effective components of plant essential oils or essential oils can effectively inhibit the synthesis of mycotoxins, for example, cinnamaldehyde, carvanol, thymol, menthol, perillyl alcohol, and citronellol (Jafarzadeh et al., 2022; Dixit et al., 2022; Soulaimani et al., 2021; Maurya et al., 2021).

The production of mycotoxins is closely related to the regulation of related genes. For example, the aflR gene is involved in aflatoxin synthesis. Methyl syringate present in *Betula platyphylla* EOs has a significant ($P < 0.05$) inhibitory effect on the toxin-producing genes of *A. parasitica* (Prakash et al., 2012). With increasing concentration of methyl syringate, the transcriptional inhibition rates of aflR, pksA, and omtA increase (Bruns et al., 2010). Prakash et al. (2012) reported that turmeric EO inhibits aflatoxin biosynthesis mainly by regulating aflR and aflS. Piperone inhibits the toxin-producing gene of *Fusarium graminearum*, and eugenol inhibits the production of aflatoxin B1. The mechanism of action of these EOs is via the down-regulation of genes located in the aflatoxin cluster (Xing et al., 2020). Caceres et al. (2016) observed that eugenol inhibits the expression of genes related to toxin clusters in *A. parasiticus*. The inhibition of mycotoxin activity by EOs may be affected by environmental factors such as temperature, humidity, and water activity. Table 3.3 shows some common EOs that inhibit mycotoxin production.

3.4.7 Effect on Biofilms

In the natural environment, most fungi change from floating cells to sessile cells, which form biofilms (Fang et al., 2020; Erhabor et al., 2019). Biofilms are dynamic and fluid structures (Fulaz et al., 2019). Some studies have revealed that biofilm formation is a highly regulated process. Cells in the sessile form have increased resistance to changes in the external environment, and the phenotype of different

TABLE 3.3
Common EOs with the Function of Inhibiting Mycotoxin

Essential oils name	Plant source	Position	Fungal species	Mycotoxin	Reference
Garlic essential oil	Garlic	Rhizome	*Aspergillus*	AFB1, FB1, ZEA	Bocate et al. (2021)
Clove essential oil	Clove	Whole plant	*Fusarium*	*Fusarium graminearum* and trichothecene mycotoxin	Wan et al. (2020)
Tea tree essential oil	Tea plant	Bark	*Penicillium griseofulvum* and *Penicillium verrucosum*	*P. griseofulvum*	Chidi et al. (2020)
Thyme oil	*Thyme vulgare* L.	Whole plant	*Aspergillus*	AFB1	Abdel-Fattah et al. (2010)
Zingiber officinale	*Zingiber officinale* Rosc	Rhizome	*Fusarium verticillioide*	FB1, FB2	Yamamotoribeiro et al. (2013)
Ginger	Ginger	Rhizome	*Aspergillus*	AFB1, STE, PAT	Vipin et al. (2017)
Rosemary essential oil	Rosemary	Whole plant	*Aspergillus*	AFB1	Ponzilacqua et al. (2019)
Ginger essential oil	Ginger	Rhizome	*Aflatoxin*	AFB1	Nerilo et al. (2020)
Thymus vulgaris	*Thymus Linn.*	Whole plant	*Aspergillus flavus*	AFB1, AFB2	Kohiyama et al. (2015)
Premna integrifolia	*Premna integrifolia* L.	Leaf	*Aspergillus*	AFB1	Singh et al. (2019)
Curcumin	*Curcuma longa* L.	Rhizome	*Aspergillus*	AFB1	Limaye et al. (2018)

planktonic cells is related to infection (Arnaouteli et al., 2021). The formation of fungal biofilms poses a threat to human health. For example, in the medical field, some pathogenic fungi adhere to the surface of catheters and prostheses, reaching the internal organs of patients. Biofilms confer fungal resistance against antifungal drugs due to the complex structure and specific metabolic system of biofilms (Sahal et al., 2020; Fernandes et al., 2022). Efflux pump genes, nutrients, quorum sensing molecules, and surface contact contribute to biofilm formation. In *Candida* spp., biofilms consist of dense yeast endosperm, hypha, pseudohypha, and extracellular matrix components. The type of components depends on the microorganism. For example, *Pneumocystis* spp. do not produce mycelium structures as part of their biofilm. Therefore, mycelium is not the main feature of fungal biofilms (Miranda et al., 2022; Min et al., 2020; Nazzaro et al., 2017). The general process of filamentous fungal biofilm formation is shown in Figure 3.5 (Harding et al., 2009).

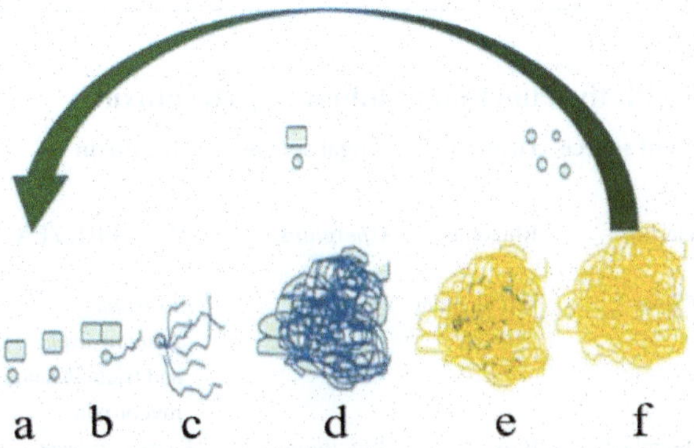

FIGURE 3.5 The general process of filamentous fungal biofilm formation.

The steps involved in the formation of filamentous fungal biofilms include (a) adsorption of propagules, involving the contact of spores, hyphae, or sporangia with the surface; (b) adhesion and secretion of adhesins and other reproductive structures by spores during germination; (c) elongation and colony formation of hyphal branches, which have a single layer produced by extracellular matrix; (d) formation of 3D dense mycelial networks and channels covered by extracellular matrix components; (e) final maturation, i.e., formation of fruiting bodies on fungi; and (f) dispersion (floating stage), in which conidia or mycelial segments are released to begin a new cycle (Figure 3.5). In the final stage, the biofilm can be released to a new infection site. In the process of biofilm formation, filamentous fungi secrete hydrophobic proteins, which participate in the adhesion of mycelium to hydrophobic surface.

Compared with conventional antifungal drugs, EOs have considerable potential as anti-biofilm agents. EOs are unlikely to contribute to fungal resistance, and they are relatively safe and easy to decompose compared with chemical synthetic drugs (Yu et al., 2019; Prakash et al., 2018; Fernandes et al., 2022). Tea tree EO inhibits the formation of *C. albicans* biofilms, and both tea tree and peppermint EOs damage the surface structure of *C. albicans* biofilms (Dąbrowska et al., 2021; Rajkowska et al., 2017). Rosemary EO strongly inhibits the biofilm development of *C. albicans* and *C. tropicalis*. Due to the importance of *Candida* spp. in the pathogenesis of human diseases, it is of great significance to develop film coatings containing EOs (Stringaro et al., 2014).

3.5 CONCLUDING REMARKS

With increasing knowledge of their antimicrobial mechanism, EOs are finding novel applications in several areas from medicine and agriculture to food technology and synthetic drugs. EOs are natural, safe, and pollution-free food additives. However, more research on EOs is needed. Specifically, there is a need to (1) evaluate the possible synergistic effects between EOs and other active components or between EOs and other antimicrobial agents; (2) determine the effective components in EOs; (3) assess the possible toxicity of EOs; (4) study the molecular target of EOs; and (5) establish a biological factory to obtain EOs with required chemical and biological characteristics. In the pharmaceutical and food industries, EOs play important roles as active substances and additives. The antifungal, antitoxin, and antibiofilm properties of EOs can be used to bridge traditional and modern applications.

ACKNOWLEDGEMENTS

This work was supported by the National Natural Science Foundation of China (32202192), Special fund for Taishan Scholars Project.

REFERENCES

Abdel-Fattah, S. M., Abosrea, Y. H., Shehata, F. E., Flourage, M. R., and Helal, A. D. 2010. The efficacy of thyme oil as antitoxicant of aflatoxin(s) toxicity in sheep. *J. South Am. Earth Sci.* 6: 948–960.

Agus, H. H. 2021. Terpene toxicity and oxidative stress. In *Toxicology*. New York: Academic Press. 33–42.

Ahmad, T., Belwal, T., Li, L., Ramola, S., Aadil, R. M., Abdullah, X. Y., and Zisheng, L. 2020. Utilization of wastewater from edible oil industry, turning waste into valuable products: A review. *Trends Food Sci. Technol.* 21–33.

Ahmady, S., Rezaei, M., and Khatony, A. 2019. Comparing effects of aromatherapy with lavender essential oil and orange essential oil on fatigue of hemodialysis patients: A randomized trial. *BMC Complementary Med. Ther.* 64–68.

Aisa, H. A., Xin, X. L., and Tang, D. 2020. Chemical constituents and their pharmacological activities of plants from *Cichorium* genus. *Chin. Herb. Med.* 12(3): 224–236.

Alia, D., Aicha, G., Youcef, H., Houria, B., and Mohammed-Réda, D. 2022. Essential oil (*Citrus aurantium* L.) induced changes in metabolic function of mitochondria isolated from yeast cells (*Saccharomyces cerevisiae*). *Ann. Romanian Soc. Cell Biol.* 26(01): 1925–1941.

Alsaraf, S., Hadi, Z., Allawati, W. M., Lawati, A. A., and Khan, S. A. 2020. Chemical composition, in vitro antibacterial and antioxidant potential of *Omani thyme* essential oil along with in silico studies of its major constituent. *Am. J. Sci.* 32(1): 1021–1028.

Andrade-Ochoa, S., Chacón-Vargas, K. F., Sánchez-Torres, L. E., Rivera-Chavira, B. E., Nogueda-Torres, B., and Nevárez-Moorillón, G. V. 2021. Differential antimicrobial effect of essential oils and their main components: Insights based on the cell membrane and external structure. *Membranes.* 11(6): 405.

Anusha, B., Thiribhuvanamala, G., and Prabakar, K. 2021. Induction of defense isoforms by hexanal and bacterial antagonists in mango. *J. Pharmacogn. Phytochem.* 10(2): 558–561.

Aprotosoaie, A. C., Pac, A., Hncianu, M., Miron, A., and Stnescu, U. 2010. The chemical profile of essential oils obtained from fennel fruits (*Foeniculum vulgare* mill.). *Farmacia.* 58(1): 46–53.

Arnaouteli, S., Bamford, N. C., Stanley-Wall, N. R., and Kovács, Á. T. 2021. *Bacillus subtilis* biofilm formation and social interactions. *Nat. Rev. Microbiol.* 19(9): 600–614.

Bang, K., Lee, D., Park, H., and Rhee, Y. H. 2000. Inhibition of fungal cell wall synthesizing enzymes by trans-cinnamaldehyde. *Biosci., Biotechnol., Biochem.* 64(5): 1061–1063.

Baranyi, N., Kocsubé, S., and Varga, J. 2015. Aflatoxins: Climate change and biodegradation. *Curr. Opin. Food Sci.* 5: 60–66.

Battisti, M. A., Caon, T., and de Campos, A. M. 2021. A short review on the antimicrobial micro-and nanoparticles loaded with *Melaleuca alternifolia* essential oil. *J. Drug Delivery Sci. Technol.* 63: 102283.

Bhavaniramya, S., Vishnupriya, S., Al-Aboody, M. S., Vijayakumar, R., and Baskaran, D. 2019. Role of essential oils in food safety: Antimicrobial and antioxidant applications. *Grain Oil Sci. Technol.* 2(2): 49–55.

Bocate, K. P., Evangelista, A. G., and Luciano, F. B. 2021. Garlic essential oil as an antifungal and anti-mycotoxin agent in stored corn. *LWT-Food Sci. Technol.* 147: 111600.

Borges, R. S., Ortiz, B. L., Pereira, A. C., Keita, H., and Carvalho, J. C. 2019. *Rosmarinus officinalis* essential oil: A review of its phytochemistry, anti-inflammatory activity, and mechanisms of action involved. *J. Ethnopharmacol.* 29–45.

Bound, D. J., Murthy, P. S., and Srinivas, P. 2016. 2,3-Dideoxyglucosides of selected terpene phenols and alcohols as potent antifungal compounds. *Food Chem.* 371–380.

Bousbia, N., Vian, M. A., Ferhat, M. A., Petitcolas, E., Meklati, B. Y., and Chemat, F. 2009. Comparison of two isolation methods for essential oil from rosemary leaves: Hydrodistillation and microwave hydrodiffusion and gravity. *Food Chem.* 114: 355–362.

Bruns, S., Seidler, M., Albrecht, D., Salvenmoser, S., Remme, N., Hertweck, C., and Muller, F. C. 2010. Functional genomic profiling of *Aspergillus fumigatus* biofilm reveals enhanced production of the mycotoxin gliotoxin. *Proteomics.* 10(17): 3097–3107.

Caceres, I., Khoury, R. E., Medina, A., Lippi, Y., Naylies, C., Atoui, A., and Puel, O. 2016. Deciphering the anti-aflatoxinogenic properties of eugenol using a large-scale q-PCR approach. *Toxins.* 8(5): 779–783.

Cáceres, M., Hidalgo, W., Stashenko, E., Torres, R., and Ortiz, C. 2020. Essential oils of aromatic plants with antibacterial, anti-biofilm and anti-quorum sensing activities against pathogenic bacteria. *Antibiotics.* 9(4): 147.

Cai, R., Hu, M., Zhang, Y., Niu, C., Yue, T., Yuan, Y., and Wang, Z. 2019. Antifungal activity and mechanism of citral, limonene and eugenol against *Zygosaccharomyces rouxii*. *LWT-Food Sci. Technol.* 106: 50–56.

Carrillohormaza, L., Mora, C., Alvarez, R., Alzate, F., and Osorio, E. 2015. Chemical composition and antibacterial activity against *Enterobacter cloacae* of essential oils from *Asteraceae* species growing in the Páramos of Colombia. *Ind. Crops Prod.* 108–115.

Charfi, S., Boujida, N., Abrini, J., and Senhaji, N. S. 2019. Study of chemical composition and antibacterial activity of *Moroccan thymbra* capitata essential oil and its possible use in orange juice conservation. *Mater. Today: Proc.* 706–712.

Chaudhari, A. K., Singh, V. K., Das, S., Prasad, J., Dwivedy, A. K., and Dubey, N. K. 2020. Improvement of in vitro and in situ antifungal, AFB1 inhibitory and antioxidant activity of *Origanum majorana* L. essential oil through nanoemulsion and recommending as novel food preservative. *Food Chem. Toxicol.* 143: 111536.

Chagonda, L. S., Makanda, C. D., and Chalchat, J. C. 2000. The essential oils of *Ocimum canum* Sims (basilic camphor) and *Ocimum urticifolia* Roth from Zimbabwe. *Flavour Fragrance J.* 15(1): 23–26.

Chen, J., Tang, C., Zhang, R., Ye, S., Zhao, Z., Huang, Y., and Yang, D. 2020. Metabolomics analysis to evaluate the antibacterial activity of the essential oil from the leaves of *Cinnamomum camphora* (Linn.) Presl. *J. Ethnopharmacol.* 253: 112652.

Cheraif, K., Bakchiche, B., Gherib, A., Bardaweel, S. K., and Ghareeb, M. A. 2020. Chemical composition, antioxidant, anti-tyrosinase, anti-cholinesterase and cytotoxic activities of essential oils of six Algerian plants. *Molecules.* 25(7): 1710.

Chidi, F., Bouhoudan, A., and Khaddor, M. 2020. Antifungal effect of the tea tree essential oil (*Melaleuca alternifolia*) against *Penicillium griseofulvum* and *Penicillium verrucosum*. *J. King Saud Univ. Sci.* 32(3): 2041–2045.

Chroho, M., Bouymajane, A., Aazza, M., Oulad El Majdoub, Y., Cacciola, F., Mondello, L., and Bouissane, L. 2022. Determination of the phenolic profile, and evaluation of biological activities of hydroethanolic extract from aerial parts of *Origanum compactum* from morocco. *Molecules.* 27(16): 5189.

Costa, W. K., de Oliveira, A. M., da Silva Santos, I. B., Silva, V. B. G., da Silva, E. K. C., de Oliveira Alves, J. V., and da Silva, M. V. 2022. Antibacterial mechanism of *Eugenia stipitata* McVaugh essential oil and synergistic effect against *Staphylococcus aureus*. *S. Afr. J. Bot.* 147: 724–730.

Cotoras, M., Castro, P., Vivanco, H., Melo, R., and Mendoza, L. 2013. Farnesol induces apoptosis-like phenotype in the phytopathogenic fungus *Botrytis cinerea*. *Mycologia*. 105(1): 28–33.

Cui, H., Zhang, C., Li, C., and Lin, L. 2019. Antibacterial mechanism of oregano essential oil. *Ind. Crops Prod.* 139: 111498.

Cui, H., Zhang, C., Li, C., and Lin, L. 2020. Inhibition of *Escherichia coli* O157:H7 biofilm on vegetable surface by solid liposomes of clove oil. *LWT-Food Sci. Technol.* 117: 108656.

da Silva Bomfim, N., Kohiyama, C. Y., Nakasugi, L. P., Nerilo, S. B., Mossini, S. A. G., Romoli, J. C. Z., and Machinski Jr, M. 2020. Antifungal and antiaflatoxigenic activity of rosemary essential oil (*Rosmarinus officinalis* L.) against *Aspergillus flavus*. *Food Addit. Contam: Part A*. 37(1): 153–161.

Dąbrowska, M., Zielińska-Bliźniewska, H., Kwiatkowski, P., Łopusiewicz, Ł., Pruss, A., Kostek, M., and Sienkiewicz, M. 2021. Inhibitory Effect of Eugenol and trans-Anethole Alone and in Combination with Antifungal Medicines on *Candida albicans* Clinical Isolates. *Chem. Biodiversity.* 18(5): e2000843.

Dadi, R., Rabah, A., Traore, M., Mielcarek, C., and Kanaev, A. 2019. Antibacterial activity of ZnO and CuO nanoparticles against gram positive and gram-negative strains. *Mater. Sci. Eng., C.* 66: 340–351.

Dao, H. T., Vu, D. H., Le Dang, Q., and Dai Lam, T. 2021. Application of botanical pesticides in organic agriculture production: Potential and challenges. *Vietnam Journal of Science and Technology*. 59(6): 676–710.

Das, S., Kumar Singh, V., Kumar Dwivedy, A., Kumar Chaudhari, A., Upadhyay, N., Singh, A., and Dubey, N. K. 2020. Assessment of chemically characterised *Myristica fragrans* essential oil against fungi contaminating stored scented rice and its mode of action as novel aflatoxin inhibitor. *Nat. Prod. Chem. Res.* 34(11): 1611–1615.

de Araújo-Filho José Vilemar, Ribeiro, W. L. C., André Weibson P. P., Cavalcante Géssica S., Guerra, M. d. C. M., and Muniz, C. R. 2018. Effects of eucalyptus citriodora essential oil and its major component, citronellal, on *Haemonchus contortus* isolates susceptible and resistant to synthetic anthelmintics. *Ind. Crops Prod.* 124: 294–299.

Debonne, E., Baert, H., Eeckhout, M., Devlieghere, F., and Van Bockstaele, F. 2019. Optimization of composite dough for the enrichment of bread crust with antifungal active compounds. *LWT-Food Sci. Technol.* 417–422.

Dias, I. J., Trajano, E. R. I. S., Castro, R. D., Ferreira, G. L. S., Medeiros, H. C. M., and Gomes, D. Q. C. 2017. Antifungal activity of linalool in cases of *Candida* spp. isolated from individuals with oral candidiasis. *Braz. J. Microbiol.* 78: 368–374.

Ding, X., Zhao, L., Khan, I. M., Yue, L., Zhang, Y., and Wang, Z. 2022. Emerging chitosan grafted essential oil components: A review on synthesis, characterization, and potential application. *Carbohydr. Polym.* 120011.

Dixit, N. M., Kalagatur, N. K., Poda, S., Kadirvelu, K., Behara, M., and Mangamuri, U. K. 2022. Application of *Syzygium aromaticum*, *Ocimum sanctum*, and *Cananga odorata* essential oils for management of Ochratoxin A content by *Aspergillus ochraceus* and *Penicillium verrucosum*: An in vitro assessment in maize grains. *NIScPR-CSIR*. 172–182.

Du, Y., Zhou, H., Yang, L., Jiang, L., Chen, D., Qiu, D., and Yang, Y. 2022. Advances in biosynthesis and pharmacological effects of *Cinnamomum camphora* (L.) Presl essential oil. *Forests*. 13(7): 1020.

Dutta, S., Kundu, A., Saha, S., Prabhakaran, P., and Mandal, A. 2020. Characterization, antifungal properties and in silico modelling perspectives of *Trachyspermum ammi* essential oil. *LWT-Food Sci. Technol.* 131:109786.

Erhabor, C. R., Erhabor, J. O., and Mcgaw, L. J. 2019. The potential of south African medicinal plants against microbial biofilm and quorum sensing of foodborne pathogens: A review. *S. Afr. J. Bot.* 214–231.

Falleh, H., Jemaa, M. B., Saada, M., and Ksouri, R. 2020. Essential oils: A promising eco-friendly food preservative. *Food Chem.* 330: 127268.

Fang, K., Park, O. J., and Hong, S. H. 2020. Controlling biofilms using synthetic biology approaches. *Biotechnol. Adv.* 107518.

Feng, G., Li, S., Xiao, L., and Song, W. 2021. Energy absorption performance of honeycombs with curved cell walls under quasi-static compression. *Int. J. Mech. Sci.* 210: 106746.

Fernandes, L., Costa, R., Henriques, M., and Rodrigues, M. E. 2022. Simulated vaginal fluid: *Candida* resistant strains' biofilm characterization and vapor phase of essential oil effect. *J. Med. Vet. Mycol.* 101329.

Fernandes, L., Gonçalves, B., Costa, R., Fernandes, Â., Gomes, A., Nogueira-Silva, C., and Henriques, M. 2022. Vapor-phase of essential oils as a promising solution to prevent *Candida* vaginal biofilms caused by antifungal resistant strains. *Healthcare.* 10(9): 1649.

Fernandes, L. M., Lacerda, M. C., Cavalcanti, Y. W., and de Almeida, L. D. F. D. 2021. Cinnamaldehyde and α-terpineol inhibit the growth of planktonic cultures of *Candida albicans* and non albicans. *Research, Society and Development.* 10(10): e554101019027–e554101019027.

Flamini, G., Smaili, T., Zellagui, A., Gherraf, N., and Cioni, P. L. 2013. Effect of growth stage on essential-oil yield and composition of *Daucus sahariensis. Chem. Biodiversity.* 10: 2014–2020.

Foddai, M., Marchetti, M., Ruggero, A., Juliano, C., and Usai, M. 2019. Evaluation of chemical composition and anti-inflammatory, antioxidant, antibacterial activity of essential oil of *Sardinian santolina* Corsica Jord. & Fourr. *Saudi J. Biol. Sci.* 26(5): 930–937.

Fulaz, S., Vitale, S., Quinn, L., and Casey, E. 2019. Nanoparticle—biofilm interactions: The role of the EPS matrix. *Trends Microbiol.* 27(11): 915–926.

Garnier, L., Penland, M., Thierry, A., Maillard, M. B., and Jérme, M. 2020. Antifungal activity of fermented dairy ingredients: Identification of antifungal compounds. *Int. J. Food Microbiol.* 322: 108574–108588.

Galovičová, L., Borotová, P., Valková, V., Vukovic, N. L., Vukic, M., Štefániková, J., and Kačániová, M. 2021. *Thymus vulgaris* essential oil and its biological activity. *Plants.* 10(9): 1959.

Garzoli, S., Laghezza Masci, V., Franceschi, S., Tiezzi, A., Giacomello, P., and Ovidi, E. 2021. Headspace/GC–MS analysis and investigation of antibacterial, antioxidant and cytotoxic activity of essential oils and hydrolates from *Rosmarinus officinalis* L. and *Lavandula angustifolia* Miller. *Foods.* 10(8): 1768.

Guo, F., Chen, Q., Liang, Q., Zhang, M., Chen, W., Chen, H., and Chen, W. 2021. Antimicrobial activity and proposed action mechanism of linalool against *Pseudomonas fluorescens. Frontiers in Microbiology.* 12: 562094.

Ghasemi, G., Alirezalu, A., Ishkeh, S. R., and Ghosta, Y. 2021. Phytochemical properties of essential oil from *Artemisia sieberi* Besser (Iranian accession) and its antioxidant and antifungal activities. *Nat. Prod. Chem. Res.* 35(21): 4154–4158.

Haloui, T., Farah, A., Lebrazi, S., Fadil, M., and Alaoui, A. B. 2015. Effect of harvesting period and drying time on the essential oil yield of *Pistacia lentiscus* L. leaves. *Der Pharma Chemica.* 7: 320–324.

Hammer, K. A., Carson, C. F., and Riley, T. V. 2004. Antifungal effects of *Melaleuca alternifolia* (tea tree) oil and its components on *Candida albicans*, *Candida glabrata* and *Saccharomyces cerevisiae. J. Antimicrob. Chemother.* 53(6): 1081–1085.

Haque, E., Irfan, S., Kamil, M., Sheikh, S., Hasan, A., Ahmad, A., and Mir, S. S. 2016. Terpenoids with antifungal activity trigger mitochondrial dysfunction in *Saccharomyces cerevisiae. Microbiology.* 85(4): 436–443.

Harding, M. W., Marques, L. L., Howard, R. J., and Olson, M. E. 2009. Can filamentous fungi form biofilms? *Trends Microbiol.* 17(11): 475–480.

He, Q., Zhang, L., Song, L., Zhang, X., Liu, D., Hu, Y., and Guo, M. 2021. Inactivation of *Staphylococcus aureus* using ultrasound in combination with thyme essential oil nanoemulsions and its synergistic mechanism. *LWT-Food Sci. Technol.* 147: 111574.

Heydari, S., and Pirzad, A. 2021. Efficiency of Funneliformis mosseae and *Thiobacillus* sp. on the secondary metabolites (essential oil, seed oil and mucilage) of *Lallemantia iberica* under salinity stress. *J. Hortic. Sci. Biotechnol.* 96(2): 249–259.

Hiwandika, N., Sudrajat, S. E., and Rahayu, I. 2021. Antibacterial and antifungal activity of clove extract (*Syzygium aromaticum*): Review. *Eureka Herba Indonesia.* 2(2): 86–94.

Hou, H., Zhang, X., Zhao, T., and Zhou, L. 2020. Effects of *Origanum vulgare* essential oil and its two main components, carvacrol and thymol, on the plant pathogen *Botrytis cinerea. PeerJ.* 8: e9626.

Huang, F., Kong, J., Ju, J., Zhang, Y., Guo, Y., and Cheng, Y. 2019. Membrane damage mechanism contributes to inhibition of trans-cinnamaldehyde on *Penicillium italicum* using surface-enhanced Raman spectroscopy. *Sci. Rep.* 9(1): 1–10.

Huang, J., Qian, C., Xu, H., and Huang, Y. 2018. Antibacterial activity of artemisia asiatica essential oil against some common respiratory infection causing bacterial strains and its mechanism of action in haemophilus influenzae. *Microb. Pathog.* 114: 470–475.

Husain, F. M., Ahmad, I., Khan, M. S., Ahmad, E., Tahseen, Q., Khan, M. S., and Alshabib, N. A. 2015. Sub-MICs of *Mentha piperita* essential oil and menthol inhibits AHL mediated quorum sensing and biofilm of Gram-negative bacteria. *Front. Terr. Microbiol.* 420–420.

Ivanov, M., Kannan, A., Stojković, D. S., Glamočlija, J., Calhelha, R. C., Ferreira, I. C., and Soković, M. 2021. Camphor and eucalyptol—anticandidal spectrum, antivirulence effect, efflux pumps interference and cytotoxicity. *Int. J. Mol. Sci.* 22(2): 483.

Jafarzadeh, S., Hadidi, M., Forough, M., Nafchi, A. M., and Mousavi Khaneghah, A. 2022. The control of fungi and mycotoxins by food active packaging: A review. *Crit. Rev. Food Sci. Nutr.* 1–19.

Jafri, H., and Ahmad, I. 2020. *Thymus vulgaris* essential oil and thymol inhibit biofilms and interact synergistically with antifungal drugs against drug resistant strains of *Candida albicans* and *Candida tropicalis. Journal de Mycologie Medicale.* 30(1): 100911.

Jassal, K., Kaushal, S., Rashmi, and Rani, R. 2021. Antifungal potential of guava (*Psidium guajava*) leaves essential oil, major compounds: Beta-caryophyllene and caryophyllene oxide. *Arch. Phytopathol. Plant Prot.* 54(19–20): 2034–2050.

Ju, J., Xie, Y., Yu, H., Guo, Y., Cheng, Y., Chen, Y., and Yao, W. 2020a. Synergistic properties of citral and eugenol for the inactivation of foodborne molds in vitro and on bread. *LWT-Food Sci. Technol.* 80(1): 154–169.

Ju, J., Xie, Y., Yu, H., Guo, Y., Cheng, Y., Zhang, R., and Yao, W. 2020b. Synergistic inhibition effect of citral and eugenol against *Aspergillus niger* and their application in bread preservation. *Food Chem.* 310: 125974.

Ju, J., Xie, Y., Yu, H., Guo, Y., and Yao, W. 2020c. Major components in lilac and *Litsea cubeba* essential oils kill *Penicillium roqueforti* through mitochondrial apoptosis pathway. *Ind. Crops Prod.* 149: 112349.

Khosravi, A. R., Shokri, H., and Saffarian, Z. 2020. Anti-fungal activity of some native essential oils against emerging multi-drug resistant human nondermatophytic moulds. *J. Herb. Med. Toxicol.* 100370.

Kohiyama, C. Y., Yamamoto Ribeiro, M. M., Galerani Mossini, S. A., Bando, E., Bomfim, N. D. S., and Nerilo, S. B. 2015. Antifungal properties and inhibitory effects upon aflatoxin production of *Thymus vulgaris* l. by *Aspergillus flavus* link. *Food Chem.* 173(173): 1006–1010.

Kong, J., Xie, Y., Yu, H., Guo, Y., Cheng, Y., Qian, H., and Yao, W. 2021. Synergistic antifungal mechanism of thymol and salicylic acid on *Fusarium solani. LWT-Food Sci. Technol.* 140: 110787.

Kong, J., Zhang, Y., Ju, J., Xie, Y., Guo, Y., and Cheng, Y. 2019. Antifungal effects of thymol and salicylic acid on cell membrane and mitochondria of *Rhizopus stolonifer* and their application in postharvest preservation of tomatoes. *Food Chem.* 285: 380–388.

Kovacik, A., Hlebova, M., Hleba, L., Jambor, T., and Kovacikova, E. 2021. Potential effect of thyme and oregano and oregano essential oils to antimicrobial activity and yeast enzymatic antioxidative system. *J. Microbiol., Biotechnol. Food Sci.* 11(3): e5583–e5583.

Krauze, M., Cendrowska-Pinkosz, M., Matusevičius, P., Stępniowska, A., Jurczak, P., and Ognik, K. 2021. The effect of administration of a phytobiotic containing cinnamon oil and citric acid on the metabolism, immunity, and growth performance of broiler chickens. *Animals.* 11(2): 399.

Krishnamoorthy, R., Gassem, M. A., Athinarayanan, J., Periyasamy, V. S., and Alshatwi, A. A. 2021. Antifungal activity of nanoemulsion from Cleome viscosa essential oil against food-borne pathogenic *Candida albicans. Saudi J. Biol. Sci.* 28(1): 286–293.

Kujur, A., Kumar, A., Yadav, A., and Prakash, B. 2020. Antifungal and aflatoxin B1 inhibitory efficacy of nanoencapsulated pelargonium graveolens l. essential oil and its mode of action. *LWT-Food Sci. Technol.* 109619.

Laorenza, Y., and Harnkarnsujarit, N. 2021. Carvacrol, citral and α-terpineol essential oil incorporated biodegradable films for functional active packaging of Pacific white shrimp. *Food Chem.* 363: 130252.

Lazrak, J., El Assiri, E. H., Arrousse, N., El-Hajjaji, F., Taleb, M., Rais, Z., and Hammouti, B. 2021. *Origanum compactum* essential oil as a green inhibitor for mild steel in 1 M hydrochloric acid solution: Experimental and Monte Carlo simulation studies. *Mat. Today: Proc.* 45: 7486–7493.

Li, B., Zheng, K., Lu, J., Zeng, D., Xiang, Q., and Ma, Y. 2022b. Antibacterial characteristics of oregano essential oil and its mechanisms against *Escherichia coli* O157: H7. *J. Food Meas. Charact.* 1–10.

Li, L., Shi, C., Yin, Z., Jia, R., Peng, L., and Kang, S. 2014. Antibacterial activity of α-terpineol may induce morphostructural alterations in *Escherichia coli. Braz. J. Microbiol.* 45(4): 1409–1413.

Li, Q., Xu, L., Wu, H., Liu, J., Lin, J., and Guan, X. 2018. Differential proteome analysis of the extracts from the xylem of *Cinnamomum camphora* inhibiting *Coriolus versicolor. Holzforschung.* 72(6): 459–466.

Liao, S., Yang, G., Huang, S., Li, B., Li, A., and Kan, J. 2022. Chemical composition of *Zanthoxylum schinifolium* Siebold & Zucc. essential oil and evaluation of its antifungal activity and potential modes of action on *Malassezia restricta. Ind. Crops Prod.* 180: 114698.

Lima, I. O., Pereira, F. D., De Oliveira, W. A., Lima, E. D., Menezes, E. A., Cunha, F. A., and Diniz, M. D. 2013. Antifungal activity and mode of action of carvacrol against *Candida albicans* strains. *J. Essent. Oil Res.* 25(2): 138–142.

Lima, R. C., Carvalho, A. P. A. D., Vieira, C. P., Moreira, R. V., and Conte-Junior, C. A. 2021. Green and healthier alternatives to chemical additives as cheese preservative: Natural antimicrobials in active nanopackaging/coatings. *Polymers.* 13(16): 2675.

Limaye, A., Yu, R. C., Chou, C. C., and Liu, J. R. 2018. Protective and detoxifying effects conferred by dietary selenium and curcumin against AFB 1 -mediated toxicity in livestock: A review. *Toxins.* 10: 25.

Liu, R. H., Shang, Z. C., Li, T. X., Yang, M. H., and Kong, L. Y. 2017. In vitro antibiofilm activity of eucarobustol E against *Candida albicans. Antimicrob. Agents Chemother.* 61(8): e02707–e02716.

Lopezmalo, A., Alzamora, S. M., and Palou, E. 2002. *Aspergillus flavus* dose—response curves to selected natural and synthetic antimicrobials. *Int. J. Food Microbiol.* 73(2): 213–218.

Lopezreyes, J. G., Spadaro, D., Prelle, A., Garibaldi, A., and Gullino, M. L. 2013. Efficacy of plant essential oils on postharvest control of rots caused by fungi on different stone fruits in vivo. *J. Food Prot.* 76(4): 631–639.

Ma, W., Zhao, L., Johnson, E. T., Xie, Y., and Zhang, M. 2022. Natural food flavour (E)-2-hexenal, a potential antifungal agent, induces mitochondria-mediated apoptosis in *Aspergillus flavus* conidia via a ROS-dependent pathway. *Int. J. Food Microbiol.* 370: 109633.

Ma, W., Zhao, L., Zhao, W., and Xie, Y. 2019. (E)-2-Hexenal, as a potential natural antifungal compound, inhibits *Aspergillus flavus* spore germination by disrupting mitochondrial energy metabolism. *J. Agric. Food Chem.* 67(4): 1138–1145.

Macwan, S. R., Dabhi, B. K., Aparnathi, K. D., and Prajapati, J. B. 2016. Essential oils of herbs and spices: Their antimicrobial activity and application in preservation of food. *Int. J. Curr. Microbiol. Appl. Sci.* 5(5): 885–901.

Mansur, E., and Ferreira-Pereira, A. 2021. Levy Tenorio Sousa Domingos1, 6, Flaviane Gomes Pereira2, Daniel Clemente de Moraes1, 7, Ronaldo Marquete3, Marco Eduardo do Nascimento Rocha4, 8, Davyson de Lima Moreira4, 9. *Rodriguésia.*72: e00432020.

Maurya, A., Prasad, J., Das, S., and Dwivedy, A. K. 2021. Essential oils and their application in food safety. *Frontiers in Sustainable Food Systems.* 5: 133.

Mieres-Castro, D., Ahmar, S., Shabbir, R., and Mora-Poblete, F. 2021. Antiviral activities of eucalyptus essential oils: Their effectiveness as therapeutic targets against human viruses. *Pharmaceuticals.* 14(12): 1210.

Min, K., Neiman, A. M., and Konopka, J. B. 2020. Fungal pathogens: shape-shifting invaders. *Trends Microbiol.* 28(11): 922–933.

Miranda, A. C., Leães, G. F., and Copetti, M. V. 2022. Fungal biofilms: Insights for the food industry. *Curr. Opin. Food Sci.* 46: 100846. Moazeni, M., Davari, A., Shabanzadeh, S., Akhtari, J., Saeedi, M., Mortyeza-Semnani, K., and Nokhodchi, A. 2021. In vitro antifungal activity of *Thymus vulgaris* essential oil nanoemulsion. *J. Herb. Med.* 28: 100452.

Mobasseri, G., Juteh, C. S., Ooi, P. T., and Thong, K. L. 2019. The emergence of colistin-resistant *Klebsiella pneumoniae* strains from swine in Malaysia. *J. Global Antimicrob. Resist.* 17: 227–232.

Mohammed, M. J., Anand, U., Altemimi, A. B., Tripathi, V., Guo, Y., and Pratap-Singh, A. 2021. Phenolic composition, antioxidant capacity and antibacterial activity of white wormwood (Artemisia herba-alba). *Plants.* 10(1): 164.

Mohsen, A., Abdel-Gawwad, H. A., and Ramadan, M. 2020. Performance, radiation shielding, and anti-fungal activity of alkali-activated slag individually modified with zinc oxide and zinc ferrite nano-particles. *Constr. Build. Mater.* 257: 119584.

Mortazavi, N., and Aliakbarlu, J. 2019. Antibacterial effects of ultrasound, cinnamon essential oil, and their combination against *Listeria monocytogenes* and *Salmonella typhimurium* in milk. *J. Food Sci.* 84(12): 3700–3706.

Muhialdin, B. J., Algboory, H. L., Kadum, H., Mohammed, N. K., Saari, N., Hassan, Z., and Hussin, A. S. 2020. Antifungal activity determination for the peptides generated by Lactobacillus plantarum TE10 against *Aspergillus flavus* in maize seeds. *Food Control.* 109: 106898.

Muhoza, B., Qi, B., Harindintwali, J. D., Koko, M. Y. F., Zhang, S., and Li, Y. 2021. Encapsulation of cinnamaldehyde: an insight on delivery systems and food applications. *Crit. Rev. Food Sci. Nutr.* 1–23.

Munkvold, G. P., Proctor, R. H., and Moretti, A. 2021. Mycotoxin production in *Fusarium* according to contemporary species concepts. *Annu. Rev. Phytopathol.* 59: 373–402.

Muslim, S. N., and Hussin, Z. S. 2020. Chemical compounds and synergistic antifungal properties of *Thymus kotschanus* essential oil plus ketoconazole against *Candida* spp. *Gene Reports.* 21: 100916.

Najdabbasi, N., Mirmajlessi, S. M., Dewitte, K., Landschoot, S., Mänd, M., Audenaert, K., and Haesaert, G. 2020. Biocidal activity of plant-derived compounds against *Phytophthora infestans*: An alternative approach to late blight management. *Crop Protection.* 138: 105315.

Nazzaro, F., Fratianni, F., Coppola, R., and De Feo, V. 2017. Essential oils and antifungal activity. *Pharmaceuticals*. 10(4): 1–20.

Nerilo, S. B., Romoli, J. C. Z., Nakasugi, L. P., Zampieri, N. S., Mossini, S. A. G., Rocha, G. H. O., Micotti da, E. G., Abreu Filho, B. A., and Machinski Jr., M. 2020. Antifungal activity and inhibition of aflatoxins production by *Zingiber officinale* Roscoe essential oil against *Aspergillus flavus* in stored maize grains. *Ciência Rural*. 50(6): e20190779.

Nobrega, R. D., Teixeira, A. P., De Oliveira, W. A., Lima, E. D., and Lima, I. O. 2016. Investigation of the antifungal activity of carvacrol against strains of *Cryptococcus neoformans*. *Pharm. Biol.* 54(11): 2591–2596.

Nowrouzi, I., Mohammadi, A. H., and Manshad, A. K. 2022. Preliminary evaluation of a natural surfactant extracted from *Myrtus communis* plant for enhancing oil recovery from carbonate oil reservoirs. *J. Pet. Explor. Prod. Technol.* 12(3): 783–792.

Orie, N. N., Warren, A. R., Basaric, J., Lau-Cam, C, PiTka-Ottlik, M., and MOchowski, J. 2017. In vitro assessment of the growth and plasma membrane h+-atpase inhibitory activity of Ebselen and structurally related selenium and sulfur-containing compounds in *Candida albicans*. *J Biochem Mol Toxicol*. e21892.

Oueslati, M. H., Abutaha, N., Alghamdi, F. A., Nehdi, I. A., Nasr, F. A., Mansour, L., and Harrath, A. H. 2020. Analysis of the chemical composition and in vitro cytotoxic activities of the essential oil of the aerial parts of *Lavandula atriplicifolia* Benth. *Am. J. Anim. Vet. Sci.* 32(2): 1476–1481.

Ouyang, Q., Okwong, R. O., Chen, Y., and Tao, N. 2020. Synergistic activity of cinnamaldehyde and citronellal against green mold in citrus fruit. *Postharvest Biol. Technol.* 162: 111095.

Pizzolitto, R. P., Jacquat, A. G., Usseglio, V. L., Achimon, F., Cuello, A. E., Zygadlo, J. A., and Dambolena, J. S. 2020. Quantitative-structure-activity relationship study to predict the antifungal activity of essential oils against *Fusarium verticillioides*. *Food Control*. 250–261.

Ponzilacqua, B., Rottinghaus, G. E., Landers, B. R., and Oliveira, C. F. 2019. Effects of medicinal herb and Brazilian traditional plant extracts on in vitro mycotoxin decontamination. *Food Control*. 100: 24–27.

Prakash, A., Baskaran, R., Paramasivam, N., and Vadivel, V. 2018. Essential oil based nanoemulsions to improve the microbial quality of minimally processed fruits and vegetables: A review. *Food Res. Int.* 509–523.

Rajkowska, K., Otlewska, A., Kunicka-Styczyńska, A., and Krajewska, A. 2017. *Candida albicans* impairments induced by peppermint and clove oils at sub-inhibitory concentrations. *Int. J. Mol. Sci.* 18(6): 1307.

Ranjbar, A., Ramezanian, A., Shekarforoush, S., Niakousari, M., and Eshghi, S. 2022. Antifungal activity of thymol against the main fungi causing pomegranate fruit rot by suppressing the activity of cell wall degrading enzymes. *LWT-Food Sci. Technol.* 161: 113303.

Rao, J., Chen, B., and Mcclements, D. J. 2019. Improving the efficacy of essential oils as antimicrobials in foods: Mechanisms of action. *Annu. Rev. Food Sci. Technol.* 10(1): 365–387.

Reis, P. M. L., Mezzomo, N., Aguiar, G. P. S., Hotza, D., Ribeiro, D. H. B., Ferreira, S. R. S., and Hense, H. 2020. Formation, stability and antimicrobial activity of laurel leaves essential oil (*Laurus nobilis* L.) particles in suspension obtained by SFEE. *J. Supercrit. Fluids*. 166: 105032.

Rivera-Yañez, C. R., Ruiz-Hurtado, P. A., Reyes-Reali, J., Mendoza-Ramos, M. I., Vargas-Díaz, M. E., Hernández-Sánchez, K. M., and Rivera-Yañez, N. 2022. Antifungal activity of mexican propolis on clinical isolates of *Candida* species. *Molecules*. *27*(17): 5651.

Rmr, K., Shinde, R. M., Wasule, D. L., Gaharwar, A. M., Parlavar, N. D., and Patle, K. P. 2020. Extraction and characterization of phytochemicals from *Cochlospermum* spp. and its antifungal activity. *Int. J. Chem. Stud.* 8(1): 410–413.

Rodenak-Kladniew, B., Castro, A., Stärkel, P., Galle, M., and Crespo, R. 2020. 1, 8-Cineole promotes G0/G1 cell cycle arrest and oxidative stress-induced senescence in HepG2 cells and sensitizes cells to anti-senescence drugs. *Life Sci.* 243: 117271.

Sahal, G., Woerdenbag, H. J., Hinrichs, W. L., Visser, A., Tepper, P. G., Quax, W. J., and Bilkay, I. S. 2020. Antifungal and biofilm inhibitory effect of *Cymbopogon citratus* (lemongrass) essential oil on biofilm forming by *Candida tropicalis* isolates; an in vitro study. *J. Ethnopharmacol.* 246: 112188.

Saharkhiz, M. J., Omidbaigi, R., and Sefidkon, F. 2005. The effects of different harvest stages on the essential oil content and composition of ajowan (*Trachyspermum ammi* Sprague) cultivated in Iran. *J. Essent. Oil-Bear. Plants.* 8: 300–303.

Sakkas, H., and Papadopoulou, C. 2017. Antimicrobial activity of basil, oregano, and thyme essential oils. *J. Microbiol. Biotechnol.* 27(3): 429–438.

Salehi, B., Sharifirad, J., Quispe, C., Llaique, H., Villalobos, M., Smeriglio, A., and Martins, N. 2019. Insights into eucalyptus genus chemical constituents, biological activities and health-promoting effects. *Trends Food Sci. Technol.* 609–624.

Şanli, A., and Karadoğan, T. 2017. Geographical impact on essential oil composition of endemic *Kundmannia anatolica* Hub-Mor. (Apiaceae). *Afr. J. Tradit., Complementary Altern. Med.* 14: 131–137.

Sharma, A. D., and Kaur, I. 2022. Targeting UDP-glycosyltransferase, glucosamine-6-phosphate synthase and chitin synthase by using bioactive 1, 8 cineole for "aspergillosis" fungal disease mutilating COVID-19 patients: Insights from molecular docking, pharmacokinetics and in-vitro studies. *Chem. Afr.* 1–12.

Shen, Q., Zhou, W., Li, H., Hu, L., and Mo, H. 2016. Ros involves the fungicidal actions of thymol against spores of *Aspergillus flavus* via the induction of nitric oxide. *PLoS One.* 11(5): e0155647.

Shirazi, M., Abid, M., Hussain, F., Abbas, A., and Sitara, U. 2020. Antifungal activity of some medicinal plant extracts against soil-borne phytopathogens. *Pak. J. Bot.* 52(2): 970–977.

Sieniawska, E., Trifan, A., and Greige-Gerges, H. 2022. Special issue: Isolation and utilization of essential oils: As antimicrobials and boosters of antimicrobial drug activity. *Processes Isolation and Utilization of Essential Oils.* 10: 309.

Silvestre, W. P., Livinalli, N. F., Baldasso, C., and Tessaro, I. C. 2019. Pervaporation in the separation of essential oil components: A review. *Trends Food Sci. Technol.* 42–52.

Singh, C., Prakash, C., Mishra, P., Tiwari, K. N., Mishra, S. K., More, R. S., Kumar, V., and Singh, J. 2019. Hepatoprotective efficacy of *Premna intengrifolia* L. leaves against aflatoxin B 1 -induced toxicity in mice. *Toxicology.* 166: 88–100.

Singh, B., Singh, J. P., Kaur, A., and Yadav, M. P. 2021. Insights into the chemical composition and bioactivities of citrus peel essential oils. *Food Res. Int.* 143: 110231.

Sirelkhatim, N., Parveen, A., Lajeunesse, D., Yu, D., and Zhang, L. 2019. Polyacrylonitrile nanofibrous mat from electrospinning: Born with potential anti-fungal functionality. *Eur. Polym. J.* 176–180.

Sirgamalla, R., Kommakula, A., Konduru, S., Ponakanti, R., Devaram, J., and Boda, S. 2020. Cupper-catalyzed an efficient synthesis, characterization of 2-Substituted benzoxazoles, 2-Substituted benzothiazoles derivatives and their anti-fungal activity. *Chem. Data Collect.* 198–222.

Smigielski, K., Prusinowska, R., Raj, A., Sikora, M., Woliñska, K., and Gruska, R. 2011. Effect of drying on the composition of essential oil from *Lavandula angustifolia*. *J. Essent. Oil-Bear. Plants.* 14: 532–542.

Soulaimani, B., Varoni, E., Iriti, M., Mezrioui, N. E., Hassani, L., and Abbad, A. 2021. Synergistic anticandidal effects of six essential oils in Combination with Fluconazole or Amphotericin B against four clinically isolated *Candida* Strains. *Antibiotics.* 10(9): 1049.

Souza, M. E., Lopes, L. Q. S., Bonez, P. C., Gündel, A., Martinez, D. S. T., Sagrillo, M. R., and Santos, R. C. V. 2017. *Melaleuca alternifolia* nanoparticles against *Candida* species biofilms. *Microb. Pathog.* 104: 125–132.

Stevic, T., Beric, T., Savikin, K., Sokovic, M., Gođevac, D., Dimkic, I., and Stankovic, S. 2014. Antifungal activity of selected essential oils against fungi isolated from medicinal plant. *Ind. Crops Prod.* 116–122.

Stringaro, A., Vavala, E., Colone, M., Pepi, F., Mignogna, G., Garzoli, S., and Angiolella, L. 2014. Effects of Mentha suaveolens essential oil alone or in combination with other drugs in *Candida albicans*. *J. Evidence-Based Complementary Altern. Med.* 1–9.

Tariq, S., Wani, S., Rasool, W., Shafi, K., Bhat, M. A., Prabhakar, A., and Rather, M. A. 2019. A comprehensive review of the antibacterial, antifungal and antiviral potential of essential oils and their chemical constituents against drug-resistant microbial pathogens. *Microb. Pathog.* 134: 103580.

Tian, F., Woo, S. Y., Lee, S. Y., and Chun, H. S. 2018. P-cymene and its derivatives exhibit antiaflatoxigenic activities against *Aspergillus flavus* through multiple modes of action. *Appl. Biol. Chem.* 61(5): 489–497.

Tohidi, B., Rahimmalek, M., and Trindade, H. 2019. Review on essential oil, extracts composition, molecular and phytochemical properties of Thymus species in Iran. *Ind. Crops Prod.* 89–99.

Tolba, H., Moghrani, H., Benelmouffok, A., Kellou, D., and Maachi, R. 2015. Essential oil of *Algerian eucalyptus* citriodora: Chemical composition, antifungal activity. *Journal De Mycologie Medicale.* 25(4): e128–e133.

Tomaś, N., Myszka, K., Wolko, Ł., Nuc, K., Szwengiel, A., Grygier, A., and Majcher, M. 2021. Effect of black pepper essential oil on quorum sensing and efflux pump systems in the fish-borne spoiler *Pseudomonas psychrophila* KM02 identified by RNA-seq, RT-qPCR and molecular docking analyses. *Food Control.* 130: 108284.

Trindade, L. A., Oliveira, J. D., De Castro, R. D., and Lima, E. D. 2015. Inhibition of adherence of *C. albicans* to dental implants and cover screws by *Cymbopogon nardus* essential oil and citronellal. *Clin. Oral Implants Res.* 19(9): 2223–2231.

Ulanowska, M., and Olas, B. 2021. Biological properties and prospects for the application of eugenol: A review. *Int. J. Mol. Sci.* 22(7): 3671.

Vipin, A. V., Raksha, R. K., Nawneet, K. K., Anu, A. K., and Venkateswaran, G. 2017. Protective effects of phenolics rich extract of ginger against aflatoxin B 1 -induced. Biomed. *Pharmacother.* 91: 415–424.

Wan, J., Jin, Z., Zhong, S., Schwarz, P., Chen, B., and Rao, J. 2020. Clove oil-in-water nanoemulsion: Mitigates growth of *Fusarium gramine* arum and trichothecene mycotoxin production during the malting of *Fusarium* infected barley. *Food Chem.* 312: 126120.

Wang, L., Jiang, N., Wang, D., and Wang, M. 2019. Effects of essential oil citral on the growth, mycotoxin biosynthesis and transcriptomic profile of *Alternaria alternata. Toxins.* 11(10): 553.

Wang, Y., Lin, W., Yan, H., Neng, J., Zheng, Y., Yang, K., and Sun, P. 2021. iTRAQ proteome analysis of the antifungal mechanism of citral on mycelial growth and OTA production in *Aspergillus ochraceus*. *J. Sci. Food Agric.* 101(12): 4969–4979.

Weisany, W., Amini, J., Samadi, S., Hossaini, S., Yousefi, S., and Struik, P. C. 2019. Nano silver-encapsulation of *Thymus daenensis* and *Anethum graveolens* essential oils enhances antifungal potential against strawberry anthracnose. *Ind. Crops Prod.* 141: 111808.

Wu, H., Zhao, F., Li, Q., Huang, J., and Ju, J. 2022. Antifungal mechanism of essential oil against foodborne fungi and its application in the preservation of baked food. *Crit. Rev. Food Sci. Nutr.* 1–13.

Wulandari, Y. W., Anwar, C., and Supriyadi, S. 2019. Effects of drying time on essential oil production of kaffir lime (Citrus hystrix DC) leaves at ambient temperature. *Mater. Sci. Eng., Proc. Conf.* 633: 012011.

Xiang, F., Zhao, Q., Zhao, K., Pei, H., and Tao, F. 2020. The efficacy of composite essential oils against aflatoxigenic fungus *Aspergillus flavus* in maize. *Toxins.* 12(9): 562.

Xiao, Y., Liu, Z., Gu, H., Yang, F., Zhang, L., and Yang, L. 2021. Improved method to obtain essential oil, asarinin and sesamin from *Asarum heterotropoides* var. mandshuricum using microwave-assisted steam distillation followed by solvent extraction and antifungal activity of essential oil against *Fusarium* spp. *Ind. Crops Prod.* 162: 113295.

Xu, Y., Chu, Y., Feng, X., Gao, C., and Tang, X. 2020. Effects of zein stabilized clove essential oil pickering emulsion on the structure and properties of chitosan-based edible films. *Int. J. Biol. Macromol.* 156: 111–119.

Xu, Y., Wei, J., Wei, Y., Han, P., Dai, K., Zou, X., and Shao, X. 2021. Tea tree oil controls brown rot in peaches by damaging the cell membrane of *Monilinia fructicola*. *Postharvest Biol. Technol.* 175: 111474.

Yamamotoribeiro, M. M., Grespan, R., Kohiyama, C. Y., Ferreira, F. D., Mossini, S. A., Silva, E. L., and Junior, M. M. 2013. Effect of *Zingiber officinale* essential oil on *Fusarium verticillioides* and fumonisin production. *Food Chem.* 141(3): 3147–3152.

Yan, J., Wu, H., Shi, F., Wang, H., Chen, K., Feng, J., and Jia, W. 2021. Antifungal activity screening for mint and thyme essential oils against *Rhizopus stolonifer* and their application in postharvest preservation of strawberry and peach fruits. *J. Appl. Microbiol.* 130(6): 1993–2007.

Yang, Z., He, Q., Ismail, B. B., Hu, Y., and Guo, M. 2022. Ultrasonication induced nanoemulsification of thyme essential oil: Optimization and antibacterial mechanism against *Escherichia coli*. *Food Control.* 133: 108609.

Yu, B., Li, C., Gu, L., Zhang, L., Wang, Q., Zhang, Y., and Zhao, G. 2022. Eugenol protects against *Aspergillus fumigatus* keratitis by inhibiting inflammatory response and reducing fungal load. *Eur. J. Pharmacol.* 924: 174955.

Yu, Z., Tang, J., Khare, T., and Kumar, V. 2019. The alarming antimicrobial resistance in eskapee pathogens: Can essential oils come to the rescue. *Fitoterapia.* 140: 104433.

Zhan, J., He, F., Cai, H., Wu, M., Xiao, Y., Xiang, F., and Li, S. 2021. Composition and antifungal mechanism of essential oil from *Chrysanthemum morifolium* cv. Fubaiju. *J. Funct. Foods.* 87: 104746.

Zhan, X., Tan, Y., Lv, Y., Fang, J., Zhou, Y., Gao, X., and Shi, C. 2022. The antimicrobial and antibiofilm activity of oregano essential oil against *Enterococcus faecalis* and its application in chicken breast. *Foods.* 11(15): 2296.

Zhang, K. C., Wei, L. P., Shen, H., and Jiang, L. K. 2011. Comparative study of inhibitive effect of cinnamaldehyde and citral upon *Aspergillus niger* growth. *Chin J Microecol*, 23(2): 141–143.

Zhang, L. L., Zhang, L. F., Hu, Q. P., Hao, D. L., and Xu, J. G. 2017. Chemical composition, antibacterial activity of cyperus rotundus rhizomes essential oil against *Staphylococcus aureus* via membrane disruption and apoptosis pathway. *Food Control.* S0956713517302591.

Zhang, X., Zhou, D., Cao, Y., Zhang, Y., Xiao, X., Liu, F., and Yu, Y. 2022. Synergistic inactivation of *Escherichia coli* O157: H7 and *Staphylococcus aureus* by gallic acid and thymol and its potential application on fresh-cut tomatoes. *Food Microbiol.* 102: 103925.

Zhang, Y. B., Xiao, Y. L., Wang, Y., Ping, J., Siew, Y. Q. 2016. Antibacterial activity and mechanism of cinnamon essential oil against *Escherichia coli* and *Staphylococcus aureus*. *Food Control.* 59: 282–289.

Zhao, C., Yang, X., Tian, H., and Yang, L. 2018. An improved method to obtain essential oil, flavonols and proanthocyanidins from fresh *Cinnamomum japonicum* Sieb. Leaves using solvent-free microwave-assisted distillation followed by homogenate extraction. *Arabian J. Chem.* 13: 2041–2052.

Zhao, Y., Yang, Y. H., Ye, M., Wang, K. B., Fan, L. M., and Su, F. W. 2021. Chemical composition and antifungal activity of essential oil from *Origanum vulgare* against *Botrytis cinerea*. *Food Chem.* 365: 130506.

Zheng, S., Jing, G., Wang, X., Ouyang, Q., Jia, L., and Tao, N. 2015. Citral exerts its antifungal activity against *Penicillium digitatum* by affecting the mitochondrial morphology and function. *Food Chem*. 76–81.

Zhu, Y., Li, C., Cui, H., and Lin, L. 2020. Plasma enhanced-nutmeg essential oil solid liposome treatment on the gelling and storage properties of pork meat batters. *J. Food Eng*. 266: 109696.

4 Essential Oils in Vapor Phase as Antimicrobial Agents

4.1 INTRODUCTION

With the increasing demand for green, safe, and healthy products, there is a need for more natural foods without chemical additives (Ju et al., 2020; Tânia et al., 2020). Researchers have evaluated alternatives to traditional food chemical preservatives. Secondary metabolites produced by plants represent a promising alternative, because several compounds produced by plants possess antimicrobial properties (Sadeer et al., 2022; Caleja et al., 2020).

Essential oils and their components have extensive antibacterial, antifungal, antioxidant, and insecticidal properties. Due to the fact that they are plant extracts that are environment-friendly, EOs are considered to be relatively safe (Jun-Hyung et al., 2020; Saxena et al., 2020; Ju et al., 2018). As a result, there is increasing interest in EOs. Several in vitro studies have reported that EOs inhibit foodborne microorganisms (Farahmandfar et al., 2020; Benyoucef et al., 2020; María et al., 2020; Mulat et al., 2020; Najjaa et al., 2020). However, higher EO concentrations are required in foods than in vitro to achieve similar antimicrobial effects. It is noteworthy that high EO concentrations in foods may generate off-flavors and negatively impact food quality (Ju et al., 2019; Ju, Xie, Yu, Guo, Cheng, Zhang et al., 2020). Therefore, if EOs are used as natural preservatives, their effects on food quality should be considered.

There are potential methods to reduce the concentration of EOs required to inhibit or inactivate microorganisms. For example, EOs in vapor phase are used in food packaging to inhibit the growth of microorganisms (Sumalan et al., 2020; Bouche et al., 2001). EOs in vapor phase have generated encouraging results in food preservation. Considering that EOs represent a mixture of several active compounds, it is not always possible to identify the compounds that lead to microbial inactivation (Gavahian et al., 2020; Arzola-Alvarez et al., 2020; Hayatgheib et al., 2020). In this chapter, we reported the main active components in EOs and the corresponding mechanism of action. In addition, we described different techniques for evaluating the effectiveness of EOs in vapor phase. The antimicrobial effects of EOs in vapor phase and the specific applications in food preservation are summarized.

4.2 CHEMICAL COMPOSITION OF ESSENTIAL OILS

EOs, which are complex volatile compounds, can be extracted from flowers, leaves, roots, bark, fruits, and seeds by distillation, fractionation, steam, or pressing (Ju et al., 2022; Araujo et al., 2020). Individual EOs are identified by the name of the plant from

DOI: 10.1201/9781003329268-4

which they are extracted. The unique fragrance of plants may be attributed to their EOs. The molecular weight of EOs is <300 (Smaoui et al., 2022). Currently, ~1,340 plants with antimicrobial properties and 30,000 effective ingredients have been identified (Ghabraie et al., 2016). In plants, EOs are used as defense agents against microbial infection. EOs contain 20–60 components in different proportions and concentrations. Compared with other trace components, there are usually two or three components at relatively high concentrations (20–90%) in EOs (Hou et al., 2022).

EOs with antimicrobial properties and the content of active components are shown in Table 4.1. Considering that phenols and terpenes are the main antimicrobial substances in EOs, they were selected as the focus of this section.

TABLE 4.1
The Source of EOs with Antibacterial Properties and the Content of Main Active Components

Plant species	Main components	Content (%)	Inhibited microorganism	Reference
Thyme EO	Thymol	10–64	*Listeria monocytogenes*	Cho et al. (2020)
	γ-Terpinene	2–31	*Staphylococcus aureus*	Kot et al. (2019)
Origanum EO	Carvol	30	*Pseudomonas aeruginosa*	Guliani et al. (2021)
Rosemary EO	1,8-Cineole	46.6	*Staphylococcus aureus, Enterococcus faecalis, Streptococcus pyogenes*	Lima et al. (2021)
	α-Pinene	11.8	*Saccharomyces cerevisiae*	Kumar et al. (2021)
	Limonene	9.3	*Candida albicans*	Ahmedi et al. (2022)
Maryulan EO	Terpenol	20.8	*Escherichia coli*	Yue et al. (2021)
	α-Terpinene	9.2	*Staphylococcus aureus*	de Morais Oliveira et al. (2018)
Lemon grass EO	Citronellal	45.7	*Staphylococcus aureus*	Zhang et al. (2022a)
	Laurene	3.9	*Staphylococcus aureus*	Alarif et al. (2012)
Fennel EO	*trans*-Anethole	56.4	*Salmonella*	Roby et al. (2013)
	Fenchone	8.3	*Escherichia coli*	Roby et al. (2013)
	Piperonol methyl	5.2	–	Roby et al. (2013)
Eucalyptus EO	1,8-Cineole	4.5–70.4	*Staphylococcus aureus*	Elaissi et al. (2011)
Clove EO	Eugenol	17.5	*Staphylococcus aureus*	Bezerra et al. (2022)
	trans-Cinnamaldehyde	68.4	*Agrobacterium tumefaciens*	Lee et al. (2020)
Cinnamon EO	Limonene	13.2	*Candida tropicalis*	Yu et al. (2022)
	Limonene	59.2	*Staphylococcus aureus*	de Araújo et al. (2020)
Fingered citron EO	Linalyl acetate	16.8	*Klebsiella pneumoniae*	Yang et al. (2021)
	Linalool	9.5	*Listeria monocytogenes*	Wang et al. (2020a)
Black pepper EO	Piperine	33.5	*Staphylococcus aureus, Escherichia coli*	Wang et al. (2020b)

4.2.1 Phenols

The antimicrobial properties of phenols in EOs are determined by their chemical structure (Tian et al., 2020). Common phenolic compounds in EOs include thymol, eugenol, and carvacrol. Antibacterial properties are mainly attributed to carvacrol and thymol in thyme EOs and to eugenol in clove EOs. For example, related studies have shown that nanoemulsions containing thyme essential oil can significantly ($P < 0.05$) damage the cell membranes of *Escherichia coli O157:H7* and *Staphylococcus aureus*. Thymol is the key to play an antibacterial role in thyme essential oil (He et al., 2022). Similarly, carvanol can also significantly ($P < 0.05$) destroy the cell structure and membrane permeability of *E. coli O157:H7* and *S. aureus*, resulting in bacterial cell content leakage (Zhang et al., 2022b). In addition, eugenol has also been proved to have significant ($P < 0.05$) inhibitory effects on *E. coli O157:H7*, *Salmonella*, and *Vibrio parahaemolyticus* (Ashrafudoulla et al., 2020; Bezerra et al., 2022; Devi et al., 2010).

The antimicrobial activity of phenols is associated with their hydroxyl groups, with benzene rings enhancing their activity (Álvarez-Martínez et al., 2021). However, the presence of free hydroxyl groups and electron delocalization in carvacrol may affect its antimicrobial activity. 4-Isopropyltoluene has a benzene ring without any hydroxyl groups; therefore, it is not an effective antibacterial agent. Carvacrol acetate and carvacrol methyl ether have lower antimicrobial activity than carvacrol due to the absence of free hydroxyl groups, which limit their ability to alter cell membrane integrity. In addition, the relative position of hydroxyl groups on the phenol ring affects antimicrobial efficacy (Lopresti et al., 2019). Compared to thymol and eugenol, carvacrol has higher antibacterial activity against *E. coli*, *S. aureus*, and *Bacillus cereus*.

4.2.2 Terpenoids

Terpenoids are natural compounds in EOs, which are derived from isoprene units. Terpenoids can be classified into monoterpenoids, sesquiterpenoids, and diterpenoids based on the number of isoprene units (Masyita et al., 2022). Monoterpenoids with antimicrobial activity include myrcene, citronellol, geraniol, nerol, citral, perillaldehyde, menthol, and carvone. Sesquiterpenoids with antimicrobial activity include farnesol, germacrone, and guaiazulene. Diterpenoids with antimicrobial activity include ginkgolides, paclitaxel, and stevioside (Tariq et al., 2019; Ahmad et al., 2021).

The antimicrobial activity of most terpenoids is related to the presence of functional groups. In general, terpenes containing alcohols and aldehydes have higher antimicrobial effects than terpenes containing carbonyl groups. For example, studies have evaluated the antimicrobial activity of 60 terpenoids with different functional groups against *Pseudomonas aeruginosa*, *E. coli*, *S. aureus*, and *Candida albicans*. The results revealed that the antimicrobial activity of these terpenoids depended not only on their functional groups but also on hydrogen bonding and solubility (Mastelic et al., 2005; Jagatap et al., 2021).

The chemical composition of EOs depends on the environment (geography and climate), harvest time, plant age and part, and extraction method (da Silva et al., 2021; Jayasundara et al., 2021; Ilardi et al., 2020).

4.3 DETECTION METHODS OF VOLATILE COMPOUNDS IN ESSENTIAL OILS

EOs can be extracted from plants by various methods. The most commonly used extraction methods include distillation, freeze pressing extraction, oil separation, solvent extraction, immersion, cold pressing, supercritical fluid extraction, water distillation, microwave-assisted extraction, and ultrasonic assisted extraction (Kant and Kumar, 2022; Katekar et al., 2022; Feyzy et al., 2017). The combination of chromatography and other methods allows the identification of the main chemical components of EOs. Gas chromatography (GC) is one of the most effective analytical methods, because it can separate very small amounts of substances and is usually used in quantitative analysis (Feriotto et al., 2018). Amiri et al. (2021) used GC to determine the content of allicin in garlic EO. The qualitative analysis of EOs can also be accomplished by using the mass spectra obtained by gas chromatography and mass spectrometry (GC-MS) and the retention time of the corresponding spectra. Several researchers have used GC-MS to qualitatively and quantitatively analyze the composition of EOs. For example, Belhachemi et al. (2022) analyzed the main active components in eucalyptus essential oil by GC-MS. The results showed that the main active components in eucalyptus essential oil were andresol, *p*-cymene, and α-pinene. Cebi et al. (2021) analyzed and quantified the active components in *Turkish rose* essential oil by GC-MS. The authors confirm that this technique is a reliable, stable, rapid, accurate, and low-cost analysis technique.

EOs are volatile and easily affected by environmental changes. Therefore, headspace sampling (HS) is an appropriate choice when EOs are selectively introduced into gas chromatograph. According to the literature, a solvent-free extraction method, which requires a small number of samples, was introduced in the 1990s. This method is called solid-phase microextraction (SPME) (Wang et al., 2015). The silica fiber coated with the stationary phase is directly exposed to the top space of the sample bottle, and the target analyte is adsorbed from the sample matrix to the fiber coating. Following extraction, the fiber layer is transferred to the heated sample inlet of GC-MS for analysis. In recent years, this method has been widely used to determine the volatile compounds in EOs (Table 4.2).

4.4 INTERACTION BETWEEN ESSENTIAL OILS AND THEIR COMPONENTS

The use of different EOs or components can produce four possible effects: synergistic, additive, irrelevant, or antagonistic (Ju et al., 2022). To improve the antimicrobial effects of EOs and to minimize the emergence of bacterial resistance, researchers have evaluated the efficacy of different mixtures or combinations of EOs.

The effect of different combinations of EOs depends on the types of EOs and the microorganisms. For example, the same combination of EOs may have different effects on different microbial strains. Similarly, different combinations of EOs have different effects on the same microbial strain (Ju et al., 2020a). Cinnamon and clove EOs have antagonistic effects on *E. coli*, but synergistic effects on monocytic *Bacillus lester* and *B. cereus* (Gutierrez et al., 2009). Eugenol and citral EOs have

TABLE 4.2
Detection of Volatile Compounds in EOs by Selected Techniques

Essential oil	Volatile compounds	Technique name	Reference
Neroli	Linalool, limonene, α-pinene	GC-C/P-IRMS	Cuchet et al. (2021)
Ruellia paniculata L.	Cariophylla-4(12)-8-(13)-dien-5β-ol, (β)-caryophyllene, caryophyllene oxide	GC-MS	Vasconcelos et al. (2021)
Blanket-leaf	Sesquiterpenes (β)-caryophyllene, caryophyllene oxide, caryophyll-4 (12),8(13)-dien-5ß-ol	GC-MS	Kafil et al. (2018)
Achillea eriophora	Eugenol, camphene, α-terpineol	GC-MS	Piryaei et al. (2021)
Origanum compactum	Carvacrol, *p*-cymene, γ-terpinene	GC-MS	Jeldi et al. (2022)
Thymus	Thymol, arvacrol	GC-MS	Labiad et al. (2022)
Cymbopogon nardus	β-Citronellal, (E)-geranial, geranyl acetate	GC-MS	Giménez-Martínez et al. (2022)
Clove buds	Eugenol, β-caryophellene, eugenyl acetate	GC-MS	Chen et al. (2022)
Nepeta leucophylla	Caryophyllene, germacrene D, α-pinene, α-selinene	SPME GC-MS	Sharma et al. (2021)
Cinnamomum reticulatum Hay	L-α-Terpineol, (−)-bornyl acetate	GC-MS/MS	Li et al. (2022)
Eucalyptus camaldulensis	Eucalyptol (1,8-cineol), *p*-cymene, α-pinene, γ-terpinene	GC-MS and GC-FID	Belhachemi et al. (2022)
Turkish rose	Linalool, citronellol	GC-MS	Cebi et al. (2021)
Artemisiae argyi Folium	Eucalyptol, α-thujone, camphor, borneol, bornyl acetate, eugenol, β-caryophyllene, caryophyllene oxide	GC-MS/MS	Hou et al. (2021)
Crithmum maritimum L.	Hydroxycinnamic acids	GC-MS and HPLC-DAD-MS/MS	Alves-Silva et al. (2020)
Piper mikanianum (Kunth) Steud	Phenylpropanoids, terpenes	GC-MS	Carneiro et al. (2020)
Mikania cordifolia	Limonene	GC-MS–FID	de Araújo et al. (2020)
Betel leaves	Eugenol, estragole, linalool, α-copaene, anethole, chavicol	GC-MS	Madhumita et al. (2019)
Nepeta leucophylla	Caryophyllene, germacrene D, α-pinene, α-selinene, viridiflorol	GC-MS and HS-SPME GC-MS	Sharma et al. (2021)

significant ($P < 0.05$) synergistic inhibition on *Penicillium roqueforti*, while the combination of cinnamaldehyde and citral has additive effects on the strain (Ju, Xie, Yu, Guo, Cheng, Chen et al., 2020). Therefore, it is difficult to predict the antimicrobial effect of different combinations of EOs. However, the synergistic effect of EOs may potentially reduce high EO concentrations in practical applications.

The combination of different EOs may have different effects as aforementioned. However, when mixed in proportion, EO mixtures have stronger antimicrobial activity than the single EO (Costa et al., 2022; Huang et al., 2021; Ayaz et al., 2019). It has been reported that trace components in EOs may be very important in conferring antimicrobial activity. Phenols have significant ($P < 0.05$) antimicrobial activity; therefore, most of the current research has focused on phenols. For example, the combination of thymol and salicylic acid and of eugenol and citral have significant ($P < 0.05$) synergistic effects on *Rhizopus solani* (Kong et al., 2019) and *P. roqueforti* (Ju et al., 2020a), respectively. Similarly, the combination of thymol and 4-isopropyltoluene has a synergistic effect on *B. cereus*. This finding may be attributed to the presence of a hydroxyl group on the benzene ring of thymol, which increases the antimicrobial activity of 4-isopropyltoluene (Delgado et al., 2004).

4.5 ANTIMICROBIAL ACTIVITY OF ESSENTIAL OILS IN VAPOR PHASE

It has been reported that EOs have greater antimicrobial activity in vapor phase than in liquid phase (Zimmermann et al., 2022; Kulkarni et al., 2022; Ács et al., 2018). Ju et al. (2020b) concluded that the inhibitory effect of cinnamaldehyde on *P. roqueforti* and *Aspergillus niger* was higher in vapor phase than in the liquid state. In the liquid phase, EOs associate to form micelles, thereby inhibiting the combination of EOs and microorganisms (Boukhatem et al., 2014). EOs represent a mixture of several compounds, the proportion of which is relatively stable in the liquid state. However, each compound has different volatility. Therefore, when EOs are introduced into the unsaturated environment, the volatile compounds diffuse at different rates according to their molecular weight until they reach equilibrium in the closed environment (Reyes-Jurado et al., 2020). EOs in vapor phase affect different stages of microbial growth, e.g., mold germination, hypha growth, and spore formation. EOs in vapor phase inactivate conidia transmitted in the air, which is a key part of inhibiting fungal transmission, because conidia in the air are stable in the presence of heat, light, and chemicals.

4.6 EVALUATION OF THE ANTIMICROBIAL ACTIVITY OF ESSENTIAL OILS IN VAPOR PHASE

The antimicrobial activity of EOs in vapor phase is effective at high concentrations and during a short period of time. Gas-phase technology is mainly based on the steam generated by EOs. Even though several methods are available to assess the antimicrobial activity of EOs in vapor phase, there is currently no standard method.

4.6.1 Inverted Petri Dish Method

In the inverted Petri dish method, a sterile filter paper containing a known volume of EOs is transferred to the inner surface of the top cover of the culture dish, and the petri dish containing the microorganisms is inverted and incubated. Following incubation, the survival rate of microorganisms is measured. Different concentrations of EOs can be used at different incubation temperatures. The volume between the agar surface and the lid can be determined by corresponding calculations so as to obtain the concentration of EOs. This method is based on the volatile characteristics of EOs, and the inhibition effect is achieved by the rapid contact of volatile compounds with microorganisms. This technique is mainly used for the determination of MIC. If EOs are added to a small filter paper tray, the inhibition area generated depends not only on the activity of the EOs, but also on their evaporation rate. If it is necessary to test the activity of multiple EOs on a filter paper tray, this method may be of limited use.

4.6.2 Boxes of Different Materials to Create the Volatile Space of Essential Oils

This technique is based on sealed boxes of different materials. The volume of the box is determined based on the experimental requirements. A known concentration of EOs and microorganisms growing on agar are placed in the box. Based on the microorganisms, the box can be incubated at different temperatures. EOs can inhibit microorganisms by volatilization. Even though this method can be used in actual food and model systems, it is particularly suitable for mold, because they grow slowly. The method can test different microorganisms; however, the consumption of EOs is quite high and the standardization of the "box" remains a challenge.

4.6.3 Agar Plug Method

In 2017, Amat et al. (2017) developed a novel technique called the agar plug method. The technique involves embedding microtubules containing a specific concentration of EOs into an agar plate. The agar plug (13 mm in diameter) containing the bacterial suspension is subsequently placed on the microtubule cover and incubated at 37°C for 24 h. After incubation, the growth of bacteria in the agar plug is visually inspected. Alternatively, the agar plug is removed, and the bacteria are placed in 5 ml broth for ~10 min and vortexed for 30 s for counting.

4.6.4 Other Methods

Seo et al. (2015) developed another method to determine MIC of EOs in vapor phase. The device consists of an upper chamber with seven holes, each of which contains a solid medium. The target microorganism is evenly distributed on the medium. Similarly, the device has seven lower chambers that allow the evaporation of EOs. The equipment is made of polycarbonate after high-pressure sterilization. Its center and four corners are tightly sealed by nuts and bolts. The main advantage of this technique is that it improves efficiency. Additionally, the method allows the simultaneous evaluation of different concentrations of EOs in vapor phase.

From the previous technology, an autoclaved polycarbonate vial was developed, which includes upper and lower holes (Lee et al., 2018). There is solid medium in the upper hole, and EOs in vapor phase in the lower hole. To prevent the leakage of EOs, an O-ring is placed at the joint of the upper and lower holes.

In these gas-phase techniques, the EOs are not in direct contact with microorganisms or foods. Therefore, the effectiveness of these methods depends on the volatility, type, and concentration of compounds in EOs. It can be seen that from extraction and characterization to use as an antimicrobial, proper management of EOs is very important (Kuorwel et al., 2013).

4.7 POTENTIAL APPLICATIONS OF ESSENTIAL OILS IN VAPOR PHASE

The potential applications of EOs in vapor phase are still under study. Table 4.3 summarizes the effects of EOs in vapor phase against microorganisms. *Listeria monocytogenes* and *S. aureus* are the two most studied microorganisms, and cinnamon EOs and clove EOs are the two most studied EOs.

As previously reported, there is a direct relationship between the chemical composition of EOs and their biological activity, which means that the main components of EOs contribute to their antimicrobial efficacy. Reyes-Jurado et al. (2020) studied the inhibitory effects of thyme, lemon grass, pepper, coriander, cinnamon, eucalyptus, lemon, and lavender EOs in vapor phase. The results revealed that cinnamon, lemon grass, and thyme EOs had the highest antibacterial activity. Furthermore, the authors reported that phenols had strong antifungal activity in vapor phase and that antifungal activity of EOs were in the following order: alcohol > ketone > ether > oxide.

EOs in vapor phase have a lower inhibitory effect on microorganisms in food systems than in vitro. Therefore, the antimicrobial effect of EOs in vapor phase on microorganisms is systematically being evaluated (Lee et al., 2018). EOs in vapor phase have a positive effect when combined with active packaging (Laorenza and Harnkarnsujarit, 2021; Tao et al., 2021; Serrano et al., 2005). For example, plastic bags containing carbon dioxide, oxygen, and eucalyptus EOs not only maintain the appearance of mushrooms, but also significantly ($P < 0.05$) reduce the rate of mushroom decay (Liu and Xia, 2012). Similarly, packaging with tea tree EOs significantly ($P < 0.05$) extend the shelf life of raspberries (El-Wakil et al., 2022) and packaging containing eucalyptus and cinnamon EOs significantly ($P < 0.05$) inhibit microbial growth and extend the shelf life of strawberries (Wang et al., 2017). Active packaging with eugenol, menthol, and thymol can significantly ($P < 0.05$) reduce the growth of yeast and mold during cherry storage and preserve the color of the fruit (Serrano et al., 2005). Recently, Aguilar González et al. (2017) demonstrated the in vitro efficacy of mustard EOs in vapor phase in vitro against *A. niger* contamination in tomatoes. The authors suggested that mustard EOs may be a good substitute for traditional synthetic antimicrobial agents. In addition, Ju et al. (2020b) concluded that active packaging containing eugenol and citral can significantly ($P < 0.05$) prolong the shelf

TABLE 4.3
The Case Studies of Different EO Vapors against Microorganisms

Microorganism tested	Essential oil	MIC	Reference
Aspergillus niger	*Melaleuca rhaphiophylla*	LC_{50} (90.55 and 72.88 of substance l^{-1} of air)	Zimmermann et al. (2022)
	Clove	0.83 µl/ml	Ju, Xie, Yu, Guo, Cheng, Zhang et al. (2020)
Colletotrichum musae	*Monarda fistulosa*	4 µl	Kulkarni et al. (2022)
Penicillium expansum	Mustard	1.6 µg/ml	Clemente et al. (2019)
	Thyme	160–200 µl of thyme EO/l of air	Reyes-Jurado et al. (2022)
Penicillium roqueforti	Cinnamon	0.21 µl/ml	Ju, Xie, Yu, Guo, Cheng, Chen et al. (2020)
Mucor spp.	Cinnamon	0.42 µl/ml	Ju et al. (2018)
	Satureja montana	0.033 µl/ml	Nedorostova et al. (2009)
Staphylococcus aureus	*Origanum majorana*	0.53 µl/ml	
	Cinnamon	36.0 µl/ml	Inouye et al. (2006)
	Clove	27.0 µl/ml	
	Eucalyptus globulus	2.25 mg/ml	Tyagi et al. (2011a)
	Clove	18.0 µg/l	Goni et al. (2009)
Escherichia coli	*Origanum majorana*	0.26 µl/ml	Nedorostova et al. (2009)
	Cinnamon	18.0 µg/l	Goni et al. (2009)
Listeria monocytogenes	Oregano	78.1 µl/l	Lee et al. (2018)
	Thyme	78.1 µl/l	
	Cinnamon bark	78.1 µl/l	
	Cinnamon leaf	156.3 µl/l	
	Clove bud	156.3 µl/l	
	Carrot seed	625 µl/l	
	Basil	625 µl/l	
	Cinnamon	54.0 µg/l	Goni et al. (2009)
	Clove	18.0 µg/l	
	Thymus vulgaris	0.26 µl/ml	Nedorostova et al. (2009)
	Satureja montana	0.26 µl/ml	
Trichophyton mentagrophytes	Oregano	2.0 µl/ml	Ahsan et al. (2012)
	Lavender	16.0 µl/ml	
	Clove	4.0 µl/ml	
Saccharomyces cerevisiae	*Mentha piperita*	1.13 mg/ml	Tyagi et al. (2011b)
Candida albicans	*Mentha piperita*	1.13 mg/ml	
Penicillium digitatum	*Mentha piperita*	2.25 mg/ml	
Aspergillus flavus	Orange peel	8,000 mg/l	Velazquez-Nu nez et al. (2013)
	Eucalyptus globulus	4.5 mg/ml	Tyagi et al. (2011a)
Aspergillus niger	*Eucalyptus globulus*	9.0 mg/ml	
Pseudomonas fluorescens	*Mentha piperita*	1.125 mg/ml	Tyagi and Malik (2010)
	Eucalyptus globulus	2.25 mg/ml	
	Mentha arvensis	0.567 mg/ml	

(*Continued*)

TABLE 4.3
(Continued)

Microorganism tested	Essential oil	MIC	Reference
Salmonella enteritidis	*Thymus vulgaris*	0.033 μl/ml	Nedorostova et al. (2009)
Yersinia enterocolitica	Cinnamon	18.0 μg/ml	Goni et al. (2009)
	Clove	9.0 μg/ml	
Bacillus subtilis	*Eucalyptus globulus*	2.25 mg/ml	Tyagi and Malik (2011)
Bacillus cereus	Cinnamon	18.0 μg/l	Nandi et al. (2013)

life of bread. The combination of eugenol and citral had synergistic inhibitory effects against *P. roqueforti* and *A. niger.*

4.8 CONCLUDING REMARKS

Several EOs and their compounds have antimicrobial activity. The volatile compounds are mainly responsible for the antimicrobial activity of EOs in vapor phase. It should be emphasized that the combination of EOs in vapor phase and active packaging has shown synergistic antimicrobial effects and control microbial growth in food. Future studies should assess the sensory properties of foods exposed to EOs in vapor phase.

ACKNOWLEDGEMENTS

This work was supported by the National Natural Science Foundation of China (32202192), Special fund for Taishan Scholars Project.

REFERENCES

Ács, K., Balázs, V. L., Kocsis, B., Bencsik, T., Böszörményi, A., and Horváth, G. 2018. Antibacterial activity evaluation of selected essential oils in liquid and vapor phase on respiratory tract pathogens. *BMC Complementary Altern. Med.* 18(1): 1–9.

Aguilar-González, A. E., Enrique, P., and López-Malo, A. 2017. Response of *Aspergillus niger* inoculated on tomatoes exposed to vapor phase mustard essential oil for short or long periods and sensory evaluation of treated tomatoes. *J. Food Qual.* 1–7.

Ahmad, A., Elisha, I. L., van Vuuren, S., and Viljoen, A. 2021. Volatile phenolics: A comprehensive review of the anti-infective properties of an important class of essential oil constituents. *Phytochemistry*. 190: 112864.

Ahmedi, S., Pant, P., Raj, N., and Manzoor, N. 2022. Limonene inhibits virulence associated traits in *Candida albicans*: In-vitro and in-silico studies. *Phytomedicine Plus*. 2(3): 100285.

Amiri, N., Afsharmanesh, M., Salarmoini, M., Meimandipour, A., Hosseini, S. A., and Ebrahimnejad, H. 2021. Nanoencapsulation (in vitro and in vivo) as an efficient technology to boost the potential of garlic essential oil as alternatives for antibiotics in broiler nutrition. *Animal*. 15(1): 100022.

Ahsan, K., Rohit, M., Rita, G., Awadhesh, K., Ashok, B., and Anupam, D. 2012. Therapeutic effects of essential oil from waste leaves of *Psidium guajava* L. against cosmetic embarrassment using phylogenetic approach. *Am. J. Plant Sci.* 3(6): 8–16.

Alarif, W. M., Al-Lihaibi, S. S., Ayyad, S. E. N., Abdel-Rhman, M. H., and Badria, F. A. 2012. Laurene-type sesquiterpenes from the Red Sea red alga *Laurencia obtusa* as potential antitumor—antimicrobial agents. *European Journal of Medicinal Chemistry*. 55: 462–466.

Álvarez-Martínez, F. J., Barrajón-Catalán, E., Herranz-López, M., and Micol, V. 2021. Antibacterial plant compounds, extracts and essential oils: An updated review on their effects and putative mechanisms of action. *Phytomedicine*. 90: 153626.

Alves-Silva, J. M., Guerra, I., Gonçalves, M. J., Cavaleiro, C., Cruz, M. T., Figueirinha, A., and Salgueiro, L. 2020. Chemical composition of *Crithmum maritimum* L. essential oil and hydrodistillation residual water by GC-MS and HPLC-DAD-MS/MS, and their biological activities. *Ind. Crops Prod.* 149: 112329.

Amat, S., Baines, D., and Alexander, T. W. 2017. A vapour phase assay for evaluating the antimicrobial activities of essential oils against bovine respiratory bacterial pathogens. *Lett. Appl. Microbiol.* 65(6): 489–495.

Araujo, C. A., Nascimento, A. L. S., Camara, C. A. G. D., and Moraes, M. M. D. 2020. Composition of the essential oil of *Podranea ricasoliana*. *Chem. Nat. Compd.* 56(4): 551–3556.

Arzola-Alvarez, C., Hume, M. E., Anderson, R. C., Latham, E. A, Ruiz-Barrera, O., and Castillo-Castillo, Y. 2020. Influence of sodium chlorate, ferulic acid, and essential oils on *Escherichia coli* and porcine fecal microbiota. *Asian J. Anim. Sci.* 98(3): 192–201.

Ashrafudoulla, M., Mizan, M., Ha, J. W., Si, H. P., and Ha, S. D. 2020. Antibacterial and antibiofilm mechanism of eugenol against antibiotic resistance vibrio parahaemolyticus. *Food Microbiol.* 91: 103500.

Ayaz, M., Ullah, F., Sadiq, A., Ullah, F., Ovais, M., Ahmed, J., and Devkota, H. P. 2019. Synergistic interactions of phytochemicals with antimicrobial agents: Potential strategy to counteract drug resistance. *Chem. Biol. Interact.* 308: 294–303.

Belhachemi, A., Maatoug, M. H., and Canela-Garayoa, R. 2022. GC-MS and GC-FID analyses of the essential oil of *Eucalyptus camaldulensis* grown under greenhouses differentiated by the LDPE cover-films. *Ind. Crops Prod.* 178: 114606.

Benyoucef, F. Dib, M., Tabti, B., Zoheir, A., Costa, J., and Muselli, A. 2020. Synergistic effects of essential oils of *Ammoides verticillata* and *Satureja candidissima* against many pathogenic microorganisms. *Anti-Infective Agents*. 18(1): 72–78.

Bezerra, S. R., Bezerra, A. H., de Sousa Silveira, Z., Macedo, N. S., dos Santos Barbosa, C. R., Muniz, D. F., and da Cunha, F. A. B. 2022. Antibacterial activity of eugenol on the IS-58 strain of *Staphylococcus aureus* resistant to tetracycline and toxicity in *Drosophila melanogaster*. *Microbial Pathogenesis*. 164: 105456.

Bouche, M., Van Bocxlaer, J., Rolly, G., Versichelen, L., Struys, M., Mortier, E., and De Leenheer, A. P. 2001. Quantitative determination of vapor-phase compound A in sevoflurane anesthesia using gas chromatography-mass spectrometry. *Sep. Tech. Clin. Chem.* 47(2): 281–291.

Boukhatem, M. N., Ferhat, M. A., Kameli, A., Saidi, F., and Mekarnia, M. 2014. Liquid and vapour phase antibacterial activity of *Eucalyptus globulus* essential oil susceptibility of selected respiratory tract pathogens. *Am. J. Infect. Dis.* 10(3): 105–117.

Caleja, C., Barros, L., Barreira, J. C. M., Sokovi, M., Calhelha, R. C., and Bento, A. 2020. *Castanea sativa* male flower extracts as an alternative additive in the Portuguese pastry delicacy "pastel de nata". *Food Funct.* 11: 90–112.

Carneiro, J. N. P., da Cruz, R. P., Campina, F. F., do Socorro Costa, M., Dos Santos, A. T. L., Sales, D. L., and Morais-Braga, M. F. B. 2020. GC/MS analysis and antimicrobial activity of the *Piper mikanianum* (Kunth) Steud. essential oil. *Food Chem. Toxicol.* 135: 110987.

Cebi, N., Arici, M., and Sagdic, O. 2021. The famous *Turkish rose* essential oil: Characterization and authenticity monitoring by FTIR, Raman and GC—MS techniques combined with chemometrics. *Food Chem.* 354: 129495.

Chen, Y., Xu, F., Pang, M., Jin, X., Lv, H., Li, Z., and Lee, M. 2022. Microwave-assisted hydrodistillation extraction based on microwave-assisted preparation of deep eutectic solvents coupled with GC-MS for analysis of essential oils from clove buds. *Sustainable Chem. Pharm.* 27: 100695.

Cho, Y., Kim, H., Beuchat, L. R., and Ryu, J. H. 2020. Synergistic activities of gaseous oregano and thyme thymol essential oils against *Listeria monocytogenes* on surfaces of a laboratory medium and radish sprouts. *Food Microbiology.* 86: 103357.

Clemente, I., Aznar, M., and Nerín, C. 2019. Synergistic properties of mustard and cinnamon essential oils for the inactivation of foodborne moulds in vitro and on Spanish bread. *Int. J. Food Microbiol.* 298: 44–50.

Costa, W. K., de Oliveira, A. M., da Silva Santos, I. B., Silva, V. B. G., da Silva, E. K. C., de Oliveira Alves, J. V., and da Silva, M. V. 2022. Antibacterial mechanism of *Eugenia stipitata* McVaugh essential oil and synergistic effect against *Staphylococcus aureus. South African Journal of Botany*. 147: 724–730.

Cuchet, A., Anchisi, A., Schiets, F., Clément, Y., Lantéri, P., Bonnefoy, C., and Casabianca, H. 2021. Determination of enantiomeric and stable isotope ratio fingerprints of active secondary metabolites in neroli (*Citrus aurantium* L.) essential oils for authentication by multidimensional gas chromatography and GC-C/P-IRMS. *J. Chromatogr. B.* 1185: 123003.

da Silva, B. D., Bernardes, P. C., Pinheiro, P. F., Fantuzzi, E., and Roberto, C. D. 2021. Chemical composition, extraction sources and action mechanisms of essential oils: Natural preservative and limitations of use in meat products. *Meat Science.* 176: 108463.

de Araújo, A. C. J., Freitas, P. R., dos Santos Barbosa, C. R., Muniz, D. F., Rocha, J. E., da Silva, A. C. A., and Coutinho, H. D. M. 2020. GC-MS-FID characterization and antibacterial activity of the *Mikania cordifolia* essential oil and limonene against MDR strains. *Food Chem. Toxicol.* 136: 111023.

de Morais Oliveira-Tintino, C. D., Tintino, S. R., Limaverde, P. W., Figueredo, F. G., Campina, F. F., da Cunha, F. A., and da Silva, T. G. 2018. Inhibition of the essential oil from *Chenopodium ambrosioides* L. and α-terpinene on the NorA efflux-pump of *Staphylococcus aureus. Food Chem.* 262: 72–77.

Delgado, B., Pablo S. Fernández, Palop, A., and Periago, P. M. 2004. Effect of thymol and cymene on *Bacillus cereus* vegetative cells evaluated through the use of frequency distributions. *Food Microbiol.* 21(3): 327–334.

Devi, K. P., Nisha, S. A., Sakthivel, R., and Pandian, S. K. 2010. Eugenol (an essential oil of clove) acts as an antibacterial agent against *Salmonella typhi* by disrupting the cellular membrane. *J. Ethnopharmacol.* 130(1): 107–115.

Elaissi, A., Salah, K. H., Mabrouk, S., Larbi, K. M., Chemli, R., and Harzallah-Skhiri, F. 2011. Antibacterial activity and chemical composition of 20 eucalyptus species' essential oils. *Food Chem.* 129(4): 1427–1434.

El-Wakil, A. E. A. A., Moustafa, H., and Youssef, A. M. 2022. Antimicrobial low-density polyethylene/low-density polyethylene-grafted acrylic acid biocomposites based on rice bran with tea tree oil for food packaging applications. *J. Thermoplast. Compos. Mater.* 35(7): 938–956.

Farahmandfar, R., Tirgarian, B., Dehghan, B., and Nemati, A. 2020. Changes in chemical composition and biological activity of essential oil from Thomson navel orange (*Citrus sinensis* L. Osbeck) peel under freezing, convective, vacuum, and microwave drying methods. *Food Sci. Nutr.* 8(1): 124–138.

Feriotto, G., Marchetti, N., Costa, V., Beninati, S., Tagliati, F., and Mischiati, C. 2018. Chemical composition of essential oils from *Thymus vulgaris*, *Cymbopogon citratus*, and *Rosmarinus officinalis*, and their effects on the HIV-1 tat protein function. *Chem. Biodiversity.* 15(2): 1–10.

Feyzy, E., Eikani, M., Golmohammad, F., and Tafaghodina. B. 2017. Extraction of essential oil from *Bunium persicum* (Boiss.) by instant controlled pressure drop. *J. Chromatogr. A*. 1530: 59–67.

Gavahian, M., Sastry, S. K., Farhoosh, R., and Farahnaky, A. 2020. Ohmic heating as a promising technique for extraction of herbal essential oils: Understanding mechanisms, recent findings, and associated challenges. *Adv. Food Nutr. Res*. 24(7): 532–541.

Ghabraie, M., Dang, K., Vu, L., Tata, S., Salmieri, Y., and Lacroix. M. 2016. Antimicrobial effect of essential oils in combinations against five bacteria and their effect on sensorial quality of ground meat. *LWT-Food Sci. Technol*. 66: 332–341.

Giménez-Martínez, P., Ramirez, C., Mitton, G., Arcerito, F. M., Ramos, F., Cooley, H., and Maggi, M. 2022. Lethal concentrations of *Cymbopogon nardus* essential oils and their main component citronellal on *Varroa destructor* and *Apis mellifera*. *Exp. Parasitol*. 108279.

Guliani, A., Verma, M., Kumari, A., and Acharya, A. 2021. Retaining the 'essence' of essential oil: Nanoemulsions of citral and carvone reduced oil loss and enhanced antibacterial efficacy via bacterial membrane perturbation. *J. Drug. Deliv. Sci. Technol*. 61: 102243.

Gutierrez, J., Barryryan, C., and Bourke, P. 2009. Antimicrobial activity of plant essential oils using food model media: Efficacy, synergistic potential and interactions with food components. *Food Microbiol*. 26(2): 142–150.

Hayatgheib, N., Fournel, C., Calvez, S., Pouliquen, H., and Moreau, E. 2020. In vitro antimicrobial effect of various commercial essential oils and their chemical constituents on *Aeromonas salmonicida* subsp. salmonicida. *J Appl. Environ. Microbiol*. 129: 137–145.

He, Q., Zhang, L., Yang, Z., Ding, T., Ye, X., Liu, D., and Guo, M. 2022. Antibacterial mechanisms of thyme essential oil nanoemulsions against *Escherichia coli* O157: H7 and *Staphylococcus aureus*: Alterations in membrane compositions and characteristics. *Innovative Food Sci. Emerging Technol*. 75: 102902.

Hou, M. Z., Chen, L. L., Chang, C., Zan, J. F., and Du, S. M. 2021. Pharmacokinetic and tissue distribution study of eight volatile constituents in rats orally administrated with the essential oil of *Artemisiae argyi* Folium by GC—MS/MS. *J. Chromatogr. B*. 1181: 122904.

Hou, T., Sana, S. S., Li, H., Xing, Y., Nanda, A., Netala, V. R., and Zhang, Z. 2022. Essential oils and its antibacterial, antifungal and anti-oxidant activity applications: A review. *Food Biosci*. 101716.

Huang, Z., Pang, D., Liao, S., Zou, Y., Zhou, P., Li, E., and Wang, W. 2021. Synergistic effects of cinnamaldehyde and cinnamic acid in cinnamon essential oil against *S. pullorum*. *Ind. Crops Prod*. 162: 113296.

Ilardi, V., Badalamenti, N., and Bruno, M. 2020. Chemical composition of the essential oil from different vegetative parts of *Foeniculum vulgare* subsp. piperitum (Ucria) Coutinho (Umbelliferae) growing wild in Sicily. *Nat. Prod. Res*. 1–11.

Jagatap, V. R., Ahmad, I., and Patel, H. M. 2021. Recent updates in natural terpenoids as potential anti-mycobacterial agents. *Indian J. Tuberc*. 69(3): 282–304.

Jayasundara, N. D. B., and Arampath, P. 2021. Effect of variety, location and maturity stage at harvesting, on essential oil chemical composition, and weight yield of *Zingiber officinale* roscoe grown in Sri Lanka. *Heliyon*. 7(3): e06560.

Jeldi, L., Taarabt, K. O., Mazri, M. A., Ouahmane, L., and Alfeddy, M. N. 2022. Chemical composition, antifungal and antioxidant activities of wild and cultivated *Origanum compactum* essential oils from the municipality of Chaoun, Morocco. *S. Afr. J Bot*. 147: 852–858.

Ju, J., Chen, X., Xie, Y., Yu, H., Cheng, Y., Qian, H., and Yao, W. 2019. Simple microencapsulation of plant essential oil in porous starch granules: Adsorption kinetics and antibacterial activity evaluation. *J. Food Process. Preserv*. 43(10): e14156.

Ju, J., Guo, Y., Cheng, Y., and Yaoc, W. 2022. Analysis of the synergistic antifungal mechanism of small molecular combinations of essential oils at the molecular level. *Ind. Crops Prod.* 188: 115612.

Ju, J., Xie, Y., Yu, H., Guo, Y., Cheng, Y., Chen, Y., and Yao, W. 2020. Synergistic properties of citral and eugenol for the inactivation of foodborne molds in vitro and on bread. *LWT-Food Sci. Technol.* 122: 109063.

Ju, J., Xie, Y., Yu, H., Guo, Y., Cheng, Y., Qian, H., and Yao, W. 2020a. Analysis of the synergistic antifungal mechanism of eugenol and citral. *LWT-Food Sci. Technol.* 123: 109128.

Ju, J., Xie, Y., Yu, H., Guo, Y., Cheng, Y., Qian, H., and Yao, W. 2020b. A novel method to prolong bread shelf life: Sachets containing essential oils components. *LWT-Food Sci. Technol.* 125: 109744.

Ju, J., Xie, Y., Yu, H., Guo, Y., Cheng, Y., Qian, H., and Yao, W. 2022. Synergistic interactions of plant essential oils with antimicrobial agents: A new antimicrobial therapy. *Crit. Rev. Food Sci. Nutr.* 62(7): 1740–1751.

Ju, J., Xie, Y., Yu, H., Guo, Y., Cheng, Y., Zhang, R., and Yao, W. 2020. Synergistic inhibition effect of citral and eugenol against *Aspergillus niger* and their application in bread preservation. *Food Chem.* 310: 125974.

Ju, J., Xu, X., Xie, Y., Guo, Y., Cheng, Y., Qian, H., and Yao, W. 2018. Inhibitory effects of cinnamon and clove essential oils on mold growth on baked foods. *Food Chem.* 85: 850–855.

Jun-Hyung, T., Renaud, C. Q. R., Maia, T., Bernier, U. R., Kenneth, L., and Bloomquist, J. R. 2020. Screening for enhancement of permethrin toxicity by plant essential oils against adult females of the yellow fever mosquito (Diptera: Culicidae). *J. Med. Entomol.* 14(1): 665–674.

Kafil, Z., Babashpour-Asl, M., and Piryaei, M. 2018. Determination of essential oils composition of blanket-leaf (Stachys byzantina C. Koch.) by microwave assisted extraction coupled to headspace single-drop microextraction. *Nat. Prod. Res.* 32(21): 2621–2624.

Kant, R., and Kumar, A. 2022. Review on essential oil extraction from aromatic and medicinal plants: Techniques, performance and economic analysis. *Sustainable Chem. Pharm.* 30: 100829.

Katekar, V. P., Rao, A. B., and Sardeshpande, V. R. 2022. Review of the rose essential oil extraction by hydrodistillation: An investigation for the optimum operating condition for maximum yield. *Sustainable Chem. Pharm.* 29: 100783.

Kong, J., Zhang, Y., Ju, J., Xie, Y., Guo, Y., and Cheng, Y. 2019. Antifungal effects of thymol and salicylic acid on cell membrane and mitochondria of *Rhizopus stolonifer* and their application in postharvest preservation of tomatoes. *Food Chem.* 285: 380–388.

Kot, B., Wierzchowska, K., Piechota, M., Czerniewicz, P., and Chrzanowski, G. 2019. Antimicrobial activity of five essential oils from lamiaceae against multidrug-resistant *Staphylococcus aureus*. *Nat. Prod. Res.* 33(24): 3587–3591.

Kulkarni, S. A., Sellamuthu, P. S., Nagarajan, S. K., Madhavan, T., and Sadiku, E. R. 2022. Antifungal activity of wild bergamot (Monarda fistulosa) essential oil against postharvest fungal pathogens of banana fruits. *S. Afr. J. Bot.* 144: 166–174.

Kumar, A., Dev, K., and Sourirajan, A. 2021. Essential Oils of *Rosmarinus officinalis* L., *Cymbopogon citratus* (DC.) Stapf., and the phyto-compounds, delta-carene and alpha-pinene mediate cell cycle arrest at G2/M transition in budding yeast *Saccharomyces cerevisiae*. *S. Afr. J. Bot.* 141: 296–305.

Kuorwel, K. K., Cran, M. J., Sonneveld, K., Miltz, J., and Bigger, S. W. 2013. Migration of antimicrobial agents from starch-based films into a food simulant. *LWT-Food Sci. Technol.* 50(2): 432–438.

Labiad, M. H., Belmaghraoui, W., Ghanimi, A., El-Guezzane, C., Chahboun, N., Harhar, H., and Tabyaoui, M. 2022. Biological properties and chemical profiling of essential oils of Thymus (vulgaris, algeriensis and broussonettii) grown in Morocco. *Chem. Data Collect.* 37: 100797.

Laorenza, Y., and Harnkarnsujarit, N. 2021. Carvacrol, citral and α-terpineol essential oil incorporated biodegradable films for functional active packaging of Pacific white shrimp. *Food Chem.* 363: 130252.

Lee, G., Kim, Y., Kim, H., Beuchat, L. R., and Ryu, J. 2018. Antimicrobial activities of gaseous essential oils against *Listeria monocytogenes* on a laboratory medium and radish sprouts. *Int. J. Food Microbiol.* 265(1): 49–54.

Lee, J. E., Jung, M., Lee, S. C., Huh, M. J., Seo, S. M., and Park, I. K. 2020. Antibacterial mode of action of trans-cinnamaldehyde derived from cinnamon bark (Cinnamomum verum) essential oil against Agrobacterium tumefaciens. *Pestic. Biochem. Physiol.* 165: 104546.

Li, Y., Cao, X., Sun, J., Zhang, W., Zhang, J., Ding, Y., and Liu, Y. 2022. Characterization of chemical compositions by a GC—MS/MS approach and evaluation of antioxidant activities of essential oils from *Cinnamomum reticulatum* Hay, *Leptospermum petersonii* Bailey, and *Juniperus formosana* Hayata. *Arabian J. Chem.* 15(2): 103609.

Lima, T. S., Silva, M. F. S., Nunes, X. P., Colombo, A. V., Oliveira, H. P., Goto, P. L., and Siqueira-Moura, M. P. 2021. Cineole-containing nanoemulsion: Development, stability, and antibacterial activity. *Chem. Phys. Lipids.* 239: 105113.

Lopresti, F., Botta, L., Scaffaro, R., Bilello, V., Settanni, L., and Gaglio, R. 2019. Antibacterial biopolymeric foams: Structure—property relationship and carvacrol release kinetics. *Eur. Polym. J.* 121: 109298.

Lu, S., and Xia, Q. 2012. Effects of combined treatments with modified-atmosphere packaging on shelf-life improvement of food products. *Prog. Food Preserv.* 67–109.

Madhumita, M., Guha, P., and Nag, A. 2019. Extraction of betel leaves (Piper betle L.) essential oil and its bio-actives identification: Process optimization, GC-MS analysis and anti-microbial activity. *Ind. Crops Prod.* 138: 111578.

María Ruiz-Rico, Moreno, Y., and José M. Barat. 2020. In vitro antimicrobial activity of immobilised essential oil components against helicobacter pylori. *World J. Microbiol. Biotechnol.* 36(1): 1–9.

Mastelic, J., Politeo, O., Jerkovic, I., and Radosevic, N. 2005. Composition and antimicrobial activity of *Helichrysum italicum* essential oil and its terpene and terpenoid fractions. *Chem. Nat. Compd.* 41(1): 35–40.

Masyita, A., Sari, R. M., Astuti, A. D., Yasir, B., Rumata, N. R., Emran, T. B., and Simal-Gandara, J. 2022. Terpenes and terpenoids as main bioactive compounds of essential oils, their roles in human health and potential application as natural food preservatives. *Food Chem: X.* 100217.

Mulat, M., Khan, F., Muluneh, G., and Pandita, A. 2020. Phytochemical profile and antimicrobial effects of different medicinal plant: Current knowledge and future perspectives. *Curr. Tradit. Med.* 6(1): 24–42.

Najjaa, H., Chekki, R., Elfalleh, W., Tlili, H., Jaballah, S., and Bouzouita, N. 2020. Freeze-dried, oven-dried, and microencapsulation of essential oil from *Allium sativum* as potential preservative agents of minced meat. *Food Sci. Nutr.* 8: 1995–2003.

Nandi, S., Saleh-e-In, M. M., Rahim, M. M., Bhuiyan, M. N. H., Sultana, N., Ahsan, M. A., and Roy, S. K. 2013. Quality composition and biological significance of the Bangladeshi and China ginger (*Zingiber officinale* Rosc.). *J. Microbiol., Biotechnol. Food Sci.* 2(5): 2283–2290.

Nedorostova, L., Kloucek, P., Kokoska, L., Stolcova, M., and Pulkrabek. J. 2009. Antimicrobial properties of selected essential oils in vapour phase against foodborne bacteria. *Food Control.* 20(2): 157–166.

Piryaei, M., and Babashpour-Asl, M. 2021. Carbon nanotube/layered double hydroxide nanocomposite as a fibre coating for determination the essential oils of *Achillea eriophora* DC with the headspace solid-phase microextraction. *Nat. Prod. Res.* 35(7): 1217–1220.

Reyes-Jurado, F., Bárcena-Massberg, Z., Ramírez-Corona, N., López-Malo, A., and Palou, E. 2022. Fungal inactivation on Mexican corn tortillas by means of thyme essential oil in vapor-phase. *Curr. Res. Food Sci.* 5: 629–633.

Reyes-Jurado, F., Navarro-Cruz, A. R., Ochoa-Velasco, C. E., Palou, E., López-Malo, A., and Ávila-Sosa, R. 2020. Essential oils in vapor phase as alternative antimicrobials: A review. *Crit. Rev. Food Sci. Nutr.* 60(10): 1641–1650.

Roby, M. H., Sarhan, M. A., Selim, K. A., and Khalel, K. I. 2013. Antioxidant and antimicrobial activities of essential oil and extracts of fennel (*Foeniculum vulgare* L.) and chamomile (*Matricaria chamomilla* L.). *Ind. Crops Prod.* 437–445.

Sadeer, N. B., Zengin, G., and Mahomoodally, M. F. 2022. Biotechnological applications of mangrove plants and their isolated compounds in medicine-a mechanistic overview. *Crit. Rev. Biotechnol.* 1–22.

Saxena, A., Sharma, L., and Maity, T. 2020. Enrichment of edible coatings and films with plant extracts or essential oils for the preservation of fruits and vegetables. *Biopolym-Based Formulations.* 34: 859–880.

Seo, H. S., Beuchat, L. R., Kim, H., and Ryu. J. H. 2015. Development of an experimental apparatus and protocol for determining antimicrobial activities of gaseous plant essential oils. *Int J Food Microbiol.* 215: 95–100.

Serrano, M., Martinezromero, D., Castillo, S., Guillen, F., and Valero, D. 2005. The use of natural antifungal compounds improves the beneficial effect of MAP in sweet cherry storage. *Innovative Food Sci. Emerging Technol.* 6(1): 115–123.

Sharma, A., Bhardwaj, G., and Cannoo, D. S. 2021. Antioxidant potential, GC/MS and headspace GC/MS analysis of essential oils isolated from the roots, stems and aerial parts of *Nepeta leucophylla. Biocatal. Agric. Biotechnol.* 32: 101950.

Smaoui, S., Hlima, H. B., Tavares, L., Ennouri, K., Braiek, O. B., Mellouli, L., and Khaneghah, A. M. 2022. Application of essential oils in meat packaging: A systemic review of recent literature. *Food Control.* 132: 108566.

Sumalan, R. M., Kuganov, R., Obistioiu, D., Popescu, I., and Cocan, I. 2020. Assessment of mint, basil, and lavender essential oil vapor-phase in antifungal protection and lemon fruit quality. *Molecules.* 25(8): 1831–1844.

Tânia C. S. P. Pires, Maria Inês Dias, Márcio Carocho, João C. M. Barreira, Santos-Buelga, C., and Barros, L. 2020. Extracts from vaccinium myrtillus l. fruits as a source of natural colorants: Chemical characterization and incorporation in yogurts. *Food Funct.* 11: 189–197.

Tao, R., Sedman, J., and Ismail, A. 2021. Antimicrobial activity of various essential oils and their application in active packaging of frozen vegetable products. *Food Chem.* 360: 129956.

Tariq, S., Wani, S., Rasool, W., Shafi, K., Bhat, M. A., Prabhakar, A., and Rather, M. A. 2019. A comprehensive review of the antibacterial, antifungal and antiviral potential of essential oils and their chemical constituents against drug-resistant microbial pathogens. *Microb. Pathog.* 134: 103580.

Tyagi, A. K., and Malik, A. 2010. Antimicrobial action of essential oil vapours and negative air ions against *Pseudomonas fluorescens. Int. J. Food Microbiol.* 143(3): 205–210.

Tyagi, A. K., and Malik, A. 2011. Antimicrobial potential and chemical composition of *Eucalyptus globulus* oil in liquid and vapour phase against food spoilage microorganisms. *Food Chem.* 126(1): 228–235.

Teissedre, P. L., and Waterhouse, A. L. 2000. Inhibition of oxidation of human low-density lipoproteins by phenolic substances in different essential oils varieties. *J. Agric. Food Chem.* 48(9): 3801–3515.

Tian, B., Harrison, R., Morton, J., and Jaspers, M. 2020. Influence of skin contact and different extractants on extraction of proteins and phenolic substances in sauvignon Blanc grape skin. *Aust. J. Grape Wine Res.* 26(2): 180–186.

Vasconcelos, A. A., Veras, I. N. D. S., Vasconcelos, M. A. D., Andrade, A. L., Dos Santos, H. S., Bandeira, P. N., and Teixeira, E. H. 2021. Chemical composition determination and evaluation of the antibacterial activity of essential oils from *Ruellia asperula* (Mart. Ex Ness) Lindau and *Ruellia paniculata* L. against oral streptococci. *Nat. Prod. Res.* 1–5.

Velazquez-Nu~nez, M., Avila-Sosa, R., Palou, E., and Lopez-Malo, A. 2013. Antifungal activity of orange (*Citrus sinensis* var. Valencia) peel essential oil applied by direct addition or vapor contact. *Food Control.* 31(1): 1–4.

Wang, C. Y. 2003. Maintaining postharvest quality of raspberries with natural volatile compounds. *Int. J. Food Sci. Technol.* 38(8): 869–875.

Wang, F., You, H., Guo, Y., Wei, Y., Xia, P., Yang, Z., and Yang, D. 2020a. Essential oils from three kinds of fingered citrons and their antibacterial activities. *Ind. Crops Prod.* 147: 112172.

Wang, H., Fisher, T., Wieprecht, E., and Moller, D. 2015. A predictive method for volatile organic compounds emission from soil: Evaporation and diffusion behavior investigation of a representative component of crude oil. *Sci. Total Environ.* 530: 38–44.

Wang, X., Shen, Y., Thakur, K., Han, J., Zhang, J. G., Hu, F., and Wei, Z. J. 2020b. Antibacterial activity and mechanism of ginger essential oil against *Escherichia coli* and *Staphylococcus aureus*. *Molecules*. 25(17): 3955.

Wang, Y., Xia, Y., Zhang, P., Ye, L., Wu, L., and He, S. 2017. Physical characterization and pork packaging application of chitosan films incorporated with combined essential oils of cinnamon and ginger. *Food Bioprocess Technol.* 10(3): 503–511.

Yang, S. K., Yusoff, K., Ajat, M., Yap, W. S., Lim, S. H. E., and Lai, K. S. 2021. Antimicrobial activity and mode of action of terpene linalyl anthranilate against carbapenemase-producing *Klebsiella pneumoniae*. *J. Pharm. Anal.* 11(2): 210–219.

Yu, H., Lin, Z. X., Xiang, W. L., Huang, M., Tang, J., Lu, Y., and Liu, L. 2022. Antifungal activity and mechanism of d-limonene against foodborne opportunistic pathogen *Candida tropicalis*. *LWT-Food Sci. Technol.* 159: 113144.

Yue, L., Zheng, M., Wang, M., Khan, I. M., Wang, B., Ma, X., and Xia, W. 2021. A general strategy to synthesis chitosan oligosaccharide-O-Terpenol derivatives with antibacterial properties. *Carbohydr. Res.* 503: 108315.

Zhang, M., Li, H., Agyekumwaa, A. K., Yu, Y., and Xiao, X. 2022a. Effects of citronellal on growth and enterotoxins production in *Staphylococcus aureus* ATCC 29213. *Toxicon.* 213: 92–98.

Zhang, X., Zhou, D., Cao, Y., Zhang, Y., Xiao, X., Liu, F., and Yu, Y. 2022b. Synergistic inactivation of *Escherichia coli* O157: H7 and *Staphylococcus aureus* by gallic acid and thymol and its potential application on fresh-cut tomatoes. *Food Microbiol.*102: 103925.

Zimmermann, R. C., Poitevin, C. G., Bischoff, A. M., Beger, M., da Luz, T. S., Mazarotto, E. J., and Zawadneak, M. A. 2022. Insecticidal and antifungal activities of *Melaleuca rhaphiophylla* essential oil against insects and seed-borne pathogens in stored products. *Ind. Crops Prod.* 182: 114871.

5 Synergistic Antibacterial Effect of Essential Oils and Antibacterial Agent

5.1 INTRODUCTION

In recent years, food safety has become a hot issue of global public health. Inappropriate or overuse of antimicrobial agents leads to the emergence and spread of drug-resistant foodborne pathogens, and also the emergence of multidrug-resistant pathogens is a global threat (Bhattacharjee et al., 2022; Shah et al., 2022; Lemons et al., 2020).

Foodborne pathogens usually refer to the pathogenic bacteria introduced in the process of food processing and circulation. These pathogens can secrete toxic substances, which directly or indirectly lead to disease or poisoning. The common pathogens causing food poisoning are *Escherichia coli*, *Salmonella*, *Shigella*, *Listeria monocytogenes*, *Vibrio parahaemolyticus*, *Streptococcus haemolyticus*, and *Staphylococcus aureus* (Mihalache et al., 2022). In recent years, more and more microorganisms that can cause human disease or poisoning have been found. These microorganisms are not only harmful to human health, but also an important cause of foodborne disease outbreaks (Levy et al., 2022).

One of the most effective strategies against multidrug-resistant pathogens is the combined use of plant active compounds and antibiotics. Plant essential oils are natural antibacterial agents extracted from plants. They are usually a complex mixture of many different kinds of compounds. Each EO contains 20–80 different molecular types (Ju et al., 2019; Rao et al., 2019; Zhang et al., 2019). A large number of studies have confirmed that EOs have significant ($P < 0.05$) antibacterial activity (Hou et al., 2022; Ju et al., 2018a). The antibacterial activity of specific EOs mainly depends on their concentration and chemical composition, among which terpene, terpenoids, phenols, and aldehydes are the main components with antibacterial activity (Ju et al., 2019; Álvarez-Martínez et al., 2021; Reyesjurado et al., 2019).

The purpose of combined antimicrobial agents applications is to reduce the drug resistance and toxicity of microorganisms and achieve synergistic antibacterial effects (Ju et al., 2018b; Liu et al., 2019). Combining natural products and antimicrobial agents reduces minimum inhibitory concentrations (MICs) and increases the sensitivity of multidrug-resistant bacteria to the antibiotics (Newman and Cragg, 2016; Shah et al., 2019). The mechanisms of several compounds have been tested for their ability to improve microbial resistance. Among them, essential oils or plant extracts have been found to be effective against almost all targets. For example, thymol and carvol increase the permeability of the bacterial outer membrane; eugenol

DOI: 10.1201/9781003329268-5

and citral inhibit β-lactamase; and sage acid inhibits bacterial efflux pumps (Abreu et al., 2012; Hemaiswarya et al., 2008; Ju et al., 2020).

This chapter has three objectives: first, to analyze the main factors affecting the antimicrobial activity of EOs; second, to introduce the evaluation methods of drug synergy in detail; and third, to summarize and discuss the efficacy of EOs and their active components in combination with antimicrobial agents against pathogenic microorganisms.

5.2 FACTORS AFFECTING THE ANTIMICROBIAL ACTIVITY OF ESSENTIAL OILS

Many factors affect the antibacterial activity of EOs, including the type and quantity of microorganisms, culture conditions, and the presence of dispersants and emulsifiers (Coimbra et al., 2022). Therefore, when evaluating the antimicrobial efficacy of EOs in different systems, it is necessary to provide accurate details of these factors.

5.2.1 Microorganisms

5.2.1.1 Species of Microorganism

In general, the inhibitory effects of EOs or plant extracts are greater on Gram-positive bacteria than on Gram-negative bacteria (Wu et al., 2022; Tariq et al., 2019; Ahmad et al., 2021). This may be due to the fact that the outer membrane of Gram-negative bacteria contains more lipopolysaccharide, which is almost impermeable to lipophilic compounds, thereby allowing the bacteria to resist the infiltration of EO active components (Al-Reza et al., 2009; Kotzekidou et al., 2008). In addition, hydrolases in the periplasmic space of Gram-negative bacteria help to degrade the active components of EOs. In contrast, Gram-positive bacteria do not have this natural barrier, allowing the lipophilic active molecules in EOs to come into direct contact with the phospholipid bilayer of the cell membrane, increasing the permeability of the cell membrane and causing the contents of the cell to leak (Gao et al., 2010). Although Gram-negative bacteria have this natural protective shell, EOs can still inhibit or kill them. This is due to the presence of porin in the outer membrane of Gram-negative bacteria, which provides a wide enough channel to allow small molecular weight compounds to pass through (Kotzekidou et al., 2008; Seow et al., 2014).

5.2.1.2 Number of Microorganisms

The antimicrobial activity of EOs depends, to some extent, on the number of microorganisms present, although no correlation has been found between the number of microorganisms and the bioactivity of EOs. Generally speaking, the greater the amount of the microorganism inoculated, the higher the MIC of the EO, and some active components tend to have better biological activity at low concentrations of microorganisms (Kalemba and Kunicka, 2003; Lambert et al., 2001). In addition, the tolerance of microorganisms to EOs also differs at different growth stages. In general, microbes in the exponential growth phase are more tolerant to EOs. Therefore, it is usually necessary to select strains in the exponential growth phase when exploring the antimicrobial activity of EOs.

5.2.1.3 Stress Response of Microorganisms

The stress response of microorganisms mainly refers to changes that occur in microorganisms when they are affected by physical, chemical, or biological factors. There have been few studies on the effects of EO-related stress on the survival of microorganisms, and those studies that do exist have reached different conclusions. Tea tree EO has been reported to have a significant ($P < 0.05$) inhibitory effect on *E. coli*, *S. aureus*, and *Salmonella* (Mcmahon et al., 2006). However, exposure for 3 days to sublethal concentrations increased the resistance of these microorganisms to gentamicin. In contrast, it has been reported that tea tree EO did not alter the resistance of *S. aureus*, *E. coli*, or *Staphylococcus epidermidis* to antibiotics (Katherine et al., 2012). In addition, it has also been reported that treatment of *L. monocytogenes* with oregano *S. epidermidis* stress did not lead to resistance (Luz, Gomes Neto, Tavares, Nunes et al., 2012).

Although the authors of these studies have come to different conclusions, this does not mean that microorganisms do not initiate their own protective mechanisms under stress. The stress response often leads to cross-resistance of microorganisms (Luz, Gomes Neto, Tavares, Magnani et al., 2012). Therefore, EOs are still an option to replace some antibiotics to not only target microorganisms but also reduce resistance in some drug-resistant bacteria.

5.2.2 Dispersants and Emulsifiers

Most EOs are insoluble in water, and only more water-soluble components can be uniformly diffused into different systems (Bölek et al., 2022; Gorjian et al., 2022). Therefore, in order to improve the solubility of EOs, many dispersants and surfactants, such as anhydrous ethyl alcohol, Tween, dimethyl sulfoxide, agar, and lecithin, have been used to ensure maximum contact with microorganisms. Emulsifiers can slow the separation of EOs and water, thereby increasing the contact area between EOs and bacteria (Reis et al., 2021; Kang, 2022; Duarte et al., 2020). For example, it has been reported that the dispersion of citral, geraniol, and linalool in Tween-20 can improve their antimicrobial activity (Kim et al., 1995). Ethanol and dimethyl sulfoxide both are polar solvents, which can enhance contact with other polar materials and improve the antibacterial activity of EOs (Chalova et al., 2010).

5.2.3 Cultivation Conditions

5.2.3.1 Time

Longer exposure times usually result in stronger antimicrobial activity. However, there may be a maximum, that is, a point at which antibacterial activity no longer increases over time. EOs can be divided into two types depending on the length of time that they produce antibacterial activity: slow-acting and fast-acting. Some antimicrobial agents, such as cinnamaldehyde, carvanol, and geraniol, are considered fast acting because they can inactivate *E. coli O157:H7* and *Salmonella* in as little as 5 min (Alves et al., 2020; Chuang et al., 2021), whereas slow-acting compounds usually take 30–60 min to show antibacterial activity.

5.2.3.2 Temperature

The effect of temperature on the antibacterial activity of EOs cannot be ignored. This is because different microbial species have different optimal growth temperatures, above or below which the activity of the microorganism is reduced. For example, the same concentration of lemon EO completely inhibits the growth of bacteria at 30°C for 30 days, but is only active for 2–3 days at 20°C (Moleyar and Narasimham, 1992). In contrast, it has been reported that lowering the culture temperature seems to improve inhibitory effects on *E. coli* (Rivas et al., 2010).

5.2.3.3 Oxygen Content

Oxygen also has an effect on the antimicrobial activity of EOs. On the one hand, oxygen causes a series of chemical reactions with the active components in EOs. For example, under aerobic conditions, thymol exhibits stronger antibacterial activity. However, tea tree EO is converted into stronger antibacterial compounds under anaerobic conditions (Kalemba and Kunicka, 2003). On the other hand, different oxygen concentrations may affect the growth and metabolism of microorganisms. EOs combined with modified atmosphere packaging at high concentrations of carbon dioxide (40%) have been shown to significantly inhibit the growth of bacteria (Matan et al., 2006).

5.2.3.4 Acidity and Alkalinity

In general, the sensitivity of bacteria to EOs increases with decreasing pH. However, fungi do not seem to be as sensitive as bacteria. The combination of low pH and EOs may have a synergistic inhibitory effect on bacteria. This may be due to the fact that, at lower pH values, EOs do not decompose and have stronger hydrophobic effects, making it easier for them to combine with bacterial cell membranes (Rivas et al., 2010). However, acidic conditions are not always effective for all microbes, with the exception of *E. coli O157:H7* (Friedman et al., 2004).

5.3 METHODS FOR EVALUATING THE SYNERGISTIC EFFECT OF DRUGS

Because it is very difficult to develop new antibiotics, the synergistic effects of different combinations of antimicrobials provides a promising strategy to solve the problem of microbial drug resistance. Combined EOs and antibiotics may have one of three different effects: additive, synergistic, or antagonistic (Delaquis et al., 2002). Additive effects provide the basis for quantitative analysis of the effects of combined drugs. Synergistic effects of drug combinations exceed those of the individual drugs. Antagonism refers to weakened effects as a result of the combination of drugs (Zhu et al., 2021). However, it is worth noting that different evaluation criteria or evaluation methods may produce different results. Table 5.1 summarizes some common

TABLE 5.1
The Evaluation Method of Drug Combination and Its Advantages and Disadvantages

Evaluation method	Advantages	Disadvantages	References
Algebraic sum	The method is very simple, and it is easy to judge the nature of the combination of drugs	It can only be used to judge the combination of drugs with simple linear relationship	Levine (1978)
Q value method	This method can be directly used to compare the original dose effect level and the operation is simple	The amount of information is small, only qualitative	Zolfpour-Arokhlo et al. (2014), Abt and Paul (2013)
Burgi formula method	This method is commonly used to measure the final effect of combined drugs	Because it does not need to consider the dose–response relationship and the mode of action of the combination of drugs, its application is limited	Burgi (2016), Guo et al. (2015)
Chou-Talalay	It can be used in combination of multiple drugs or in combination of unsteady dose ratio, and can be described qualitatively and quantitatively	This method cannot give the drug map with nonconstant ratio, only the equivalent line map after standardization	Chou and Rideout (1987)
Finney Harmonic average method	When the combined action of the two drugs is similar, the equivalent line can be derived by this method	It may be too simple to determine the experimental results by probability operation, unless a large number of experiments are carried out, the results are not very reliable	Finney (1952)
Webb Fractional product method	It is the simplest method and is widely used at present	It is only suitable for the additive calculation of nonrepellent drugs	Ribo and Rogers (2010)
Reaction surface method	It can display two-dimensional and three-dimensional atlas with reliable results. According to the spectrum, the best joint mode can be obtained	The mathematical model is complex and the workload is heavy. It needs relevant statistical knowledge and S-curve relationship of drug effect	Minto et al. (2000)
Relative effect method	The combined effect is determined by the value of the two drugs acting alone and the actual value of the single drug	The confidence interval cannot be calculated. In most cases, it lacks reliability and can only be analyzed qualitatively	Pradhan and Kim (2014)

(*Continued*)

TABLE 5.1
(Continued)

Evaluation method	Advantages	Disadvantages	References
Mapping analysis	This method does not need to consider the type of action between the two drugs. It is a relatively general analysis method	A large amount of data is needed to draw dose–response curve, and a fixed ratio of drug combination is needed, so it will be limited in practical application	Zheng and Sun (2000)
Parametric method	Data can be analyzed systematically. Abundant data and reliable results	It needs high statistical knowledge and a lot of work. It is suitable for data that can be fitted by Hill equation	Wang et al. (2013)
Logistic regression model analysis	This method needs to design appropriate experimental factors, and the method is not mature, and the relevant reports of using this method are rare	With the help of computer, it is easy to judge the nature of combined drug use	Gennings et al. (2002)
Weight matching method	This method can be used to judge the interaction between six drugs and six concentrations, and can show the intensity of the interaction between drugs	It is necessary to carry out pre-experiment. When the effective dose range of the drug is small, the experiment design should be changed	Qing and Rui (1999)
Orthogonal *t*-value method	The principle is clear, easy to understand, simple to calculate, and easy to analyze the synergistic or antagonistic effect between the two drugs	The combination of drugs can be made according to or without drugs, and then the nature of the combination can be judged. It is limited in practical application	Jin and Gong (2011)

methods and their advantages and disadvantages. Researchers can choose one or more methods according to the needs of the experiment.

5.4 THE INTERACTION BETWEEN ESSENTIAL OILS AND ITS ACTIVE COMPONENTS

The synergism or antagonism of EO combinations depends on the types of EOs or the microorganisms. Different EOs or microorganisms may have different effects. Thymol and carvol have synergistic inhibitory effects on *E. coli O157:H7*, *S. aureus*, *L. monocytogenes*, *Saccharomyces cerevisiae*, and *Aspergillus niger* (Guarda et al., 2011). In another study, the combination of cinnamon and clove EOs antagonized the growth of *E. coli* and showed a synergistic effect on *L. monocytogenes*, *Bacillus cereus*, and *Yersinia enterocolitica* (Moleyar and Narasimham, 1992). Therefore, it is difficult to directly predict the antimicrobial efficacy of an EOs mixture. However, some studies have concluded that when mixed in proportion, the whole EOs shows stronger antibacterial activity than the main components. For example, *Ocimum basilicum* EOs have a stronger inhibitory effect on *Lactobacillus campylobacter* and *S. cerevisiae* than its main components linalool or methyl piperol (Lachowicz et al., 1998). The antibacterial activity of conifer EOs against *L. monocytogenes* was significantly ($P < 0.05$) higher than that of its main components (Mourey and Canillac, 2002). This finding shows that the trace components in the EOs may be very important for microbial activity to produce a synergistic effect.

Recently, it has been found that cinnamic acid and cinnamaldehyde in cinnamon essential oil have synergistic inhibitory effect on *E. coli*. This combination can produce synergistic damage to the cell membrane of *E. coli* (Huang et al., 2021). In addition, Moroccan *Artemisia herba-alba*, *Lavandula angustifolia*, and *Rosmarinus officinalis* essential oils have synergistic bacteriostatic effects on *S. aureus*, *E. coli*, and *Pseudomonas aeruginosa* at low concentrations (Moussii et al., 2020). Similarly, any three groups of the following four essential oils (cinnamon essential oil, tea tree essential oil, lemon grass essential oil, and oregano essential oil) have synergistic inhibitory effect on *S. aureus* (Oh et al., 2022).

5.5 COMBINED APPLICATION OF ESSENTIAL OILS AND ANTIMICROBIAL AGENTS

5.5.1 The Main Causes of Bacterial Drug Resistance

There are two main causes of drug resistance in bacteria. First, when bacteria come into contact with drugs, they produce enzymes or receptors to protect themselves (Ayaz et al., 2019), rendering them resistant to the antimicrobial agents. A common example is the alternative penicillin binding protein (PBP2a) produced by methicillin-resistant *S. aureus* (MRSA). This altered protein is encoded by the mecA gene and reduces affinity for β-lactam, penicillins, and cephalosporins (Leonard and Markey, 2008). A second cause of bacterial drug resistance is β-lactamase, which catalyzes the hydrolysis of penicillins, cephalosporins, and other β-lactam antibiotics (Ubukata et al., 1989). More than 200 kinds of β-lactamases have been described in the literature and are widely found in a variety of Gram-positive and Gram-negative bacteria. This enzyme can hydrolyze both penicillins and cephalosporins (Shimizu et al., 2001).

5.5.2 The Potential of Combined Essential Oils and Food Additives

In general, the use of a small number of multiple preservatives is more effective than the use of a large number of single preservatives because this can increase the action targets of preservatives on microorganisms, and at the same time avoid the emergence of drug-resistant microorganisms. The combination of different EOs and food additives may be an effective method for food preservation (Ju et al., 2017).

It has been reported that the combination of peppermint EOs and methylparaben has a synergistic inhibitory effect on *P. aeruginosa* (Parsaeimehr et al., 2010). In addition, oregano EOs combined with propyl hydroxybenzoate showed synergistic inhibitory effect on *S. aureus* (Patrone et al., 2010). In recent years, a large number of studies on the combination of essential oils and nisin have been carried out and it has been proved that the combination has significant synergistic antibacterial effect. For example, oregano essential oil combined with nisin has a significant ($P < 0.05$) synergistic inhibitory effect on microorganisms in fish fillets, especially on lactic acid bacteria (Zhang et al., 2021). Perilla essential oil combined with nisin has significant ($P < 0.05$) inhibitory effect on *S. aureus*, *E. coli*, *Salmonella enteritidis*, and *Pseudomonas tolaasii* (Wang et al., 2021). Similarly, garlic essential oil combined with nisin has a synergistic inhibitory effect on *L. monocytogenes* (Somrani et al., 2020; Rohani et al., 2021).

Related to the application of food preservation industry, the combination of thyme EOs and chitosan can significantly ($P < 0.05$) inhibit the growth of *P. aeruginosa*, yeast, and mold and prolong the shelf life of ready-to-eat chicken (Giatrakou et al., 2010). Similarly, the combination of thyme EOs and nisin has a synergistic inhibitory effect on *L. monocytogenes* and can effectively prolong the shelf life of beef (Solomakos et al., 2008b). In addition, the combination of tea polyphenols and nisin can effectively prolong the shelf life of chilled sea bass (Antonio et al., 2009). The use of electrolytic NaCl solution and thymol (0.5%) on carp fillets showed stronger antibacterial and antioxidant effects (Mahmoud et al., 2006). Star anise combined with polylysine and nisin can significantly ($P < 0.05$) prolong the shelf life (Liu et al., 2020) of strawberries. The combination of garlic essential oil, propyl isothiocyanate, and nisin can be used as a biological preservative to extend the shelf life of fresh sausages (Araújo et al., 2018). All of this means that there are unlimited possibilities for combining essential oils with food additives.

5.5.3 The Potential of Combined Essential Oils and Antibiotics

The combined use of EOs and antibiotics is a potential strategy to solve the problem of microbial drug resistance. Phytochemicals can inhibit or kill bacteria at low concentrations, thereby minimizing any potentially toxic effects. The combination of *E. stipitate* essential oil and antibiotics has synergistic bacteriostatic effect on *S. aureus* (Costa et al., 2022). The combination of *Litsea cubeba* essential oil and tetracycline or oxytetracycline hydrochloride has a synergistic inhibitory effect on *V. parahaemolyticus* (Li et al., 2022). *Hymenaea rubriflora* essential oil combined with gentamicin and fluconazole has synergistic inhibitory effect on *S. aureus* and *Bacillus subtilis* (da Silva et al., 2021). Active plant components extracted from *Artemisia argyi* were used against *S. aureus* in combination with ciprofloxacin and norfloxacin. The results showed that the MICs of ciprofloxacin and norfloxacin decreased by fivefold and tenfold, respectively. Synergistic effects of *geranium* EO and prickly ash EO with fluoroquinolone antibiotics have also been demonstrated. In addition, salicylic acid, a phenolic compound found in many plants, has been shown to significantly ($P < 0.05$) inhibit the growth of *S. aureus* when combined with ciprofloxacin. Volatile oil from myrtle leaves and amphotericin B also have synergistic inhibitory effects on *C. albicans* and *Aspergillus*. EOs have also been shown to be effective against antibiotic-resistant bacteria. When oxacillin was used in combination with *A. argyi* EO against oxacillin-resistant *S. aureus* CCARM3511, the MIC of *A. argyi* EO decreased significantly by fourfold to eightfold.

Tetracycline antibiotics, including chlortetracycline, oxytetracycline, tetracycline, doxycycline, and minocycline, are relatively cheap and effective broad-spectrum antibiotics produced by actinomycetes. However, with the emergence of drug resistance in bacteria, the role of these antibiotics is gradually decreasing. *S. aureus* employs two mechanisms to resist these antibiotics: (1) active efflux of tetracycline from the cell via the transporting proteins Tet (K) or Tet (L) belonging to the major facilitator superfamily (MFS) and (2) the ribosomal protection proteins Tet (M) and Tet (O) (Trzcinski et al., 2010). In contrast, three mechanisms are involved in erythromycin resistance: (1) use of energy-dependent efflux, (2) production of inactivating enzymes, and (3) alteration of 23S rRNA methylases (Wang et al., 2008). The first mechanism involves the macrolide efflux pumps Msr(A) and/or Msr(B), which belong to the ATP-binding cassette (ABC) transporter family and export 14-membered macrolide antibiotics from bacterial cells (Leclercq, 2002).

In a study in which an ethanol mixture extracted from mango peel was combined with erythromycin and tetracycline to explore the effect of these combinations on drug-resistant *S. aureus*, it was observed that the MICs of erythromycin and tetracycline decreased fourfold (de Oliveira et al., 2011). Similarly, baicalein isolated from thyme leaves significantly ($P < 0.05$) decreased the MIC of tetracycline against MRSA from 4 to 0.06 μg/ml after 16 h (Fujita et al., 2005). The antibacterial activity of tetracycline was also significantly ($P < 0.05$) increased eightfold by rosemary EO. These results are comparable to those reported for reserpine (Oluwatuyi et al., 2004).

Aminoglycosides, such as kanamycin, streptomycin, gentamicin, and amikacin, are broad-spectrum antibiotics used to treat *S. aureus* infections (Ramirez and Tolmasky, 2010). Mechanisms of resistance to these antibiotics involve the production of aminoglycoside-modifying enzymes. It has been shown that there is a significant enhancement effect between *Zanthoxylum bungeanum* EOs and aminoglycosides (Ramirez and Tolmasky, 2010). More recently, studies have focused on inhibiting the synergistic effects of drug-resistant pathogenic microorganisms (Dhara and Tripathi, 2020; Gao et al., 2020). Therefore, synergistic combinations of drugs may be beneficial to the public, especially young and elderly patients with low immune functioning. In addition to the aforementioned EOs, we also list some commonly used phytochemicals that have synergistic antibacterial activity with antibiotics in Figure 5.1.

5.6 CONCLUDING REMARKS

In this chapter, the main factors affecting the antibacterial activity of EOs and the possible antibacterial mechanism of EOs were introduced in detail. Then the potential applications of combined EOs and antimicrobial agents are analyzed to address the problem of microbial drug resistance. However, future research needs to establish

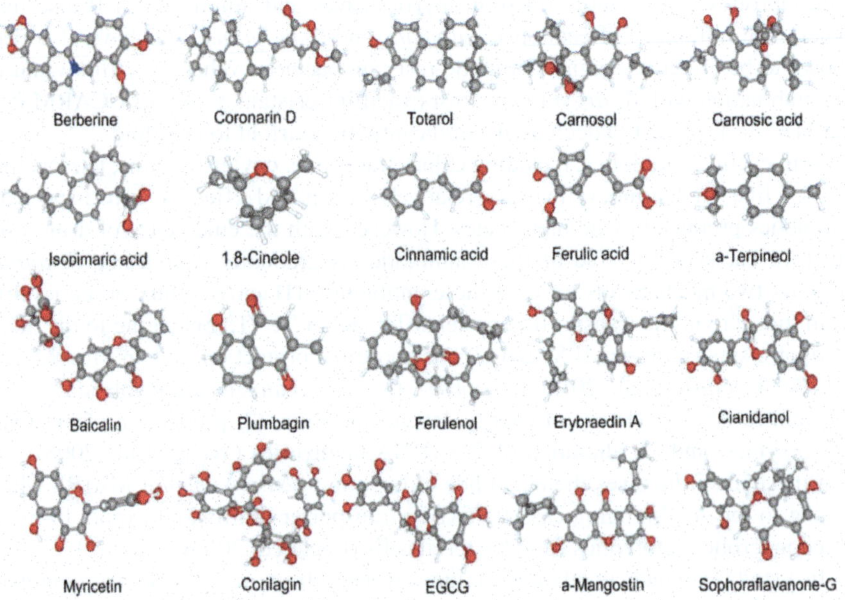

FIGURE 5.1 List of plant chemical components with synergistic antibacterial effect combined with antibiotics.

the validity of each combination for different microbial species, improve the efficacy and specificity of each combination, and develop high-throughput screening methods. In order to develop more effective combinations, it will be necessary to optimize the proportion and dose of each combination. In addition, more sophisticated instruments combined with computer technology are needed to identify and evaluate the effects of synergistic drug combinations and their corresponding contribution rates. Finally, safer, cheaper, and more effective natural drugs against foodborne pathogens may be developed using genetic engineering, genomics, proteomics, and metabolomics.

ACKNOWLEDGEMENTS

This work was supported by the National Natural Science Foundation of China (32202192), Special fund for Taishan Scholars Project.

REFERENCES

Abreu, A. C., Mcbain, A. J., and Simoes, M. 2012. Plants as sources of new antimicrobials and resistance-modifying agents. *Nat. Prod. Rep.* 29(9): 1007–1021.

Ahmad, A., Elisha, I. L., van Vuuren, S., and Viljoen, A. 2021. Volatile phenolics: A comprehensive review of the anti-infective properties of an important class of essential oil constituents. *Phytochemistry*. 190: 112864.

Al-Reza, S. M., Rahman, A., and Kang, S. C. 2009. Chemical composition and inhibitory effect of essential oil and organic extracts of *Cestrum nocturnum* L. on food-borne pathogens. *Int. J. Food Sci. Tech.* 44(6): 1176–1182.

Álvarez-Martínez, F. J., Barrajón-Catalán, E., Herranz-López, M., and Micol, V. 2021. Antibacterial plant compounds, extracts and essential oils: An updated review on their effects and putative mechanisms of action. *Phytomedicine*. 90: 153626.

Alves, D., Cerqueira, M. A., Pastrana, L. M., and Sillankorva, S. 2020. Entrapment of a phage cocktail and cinnamaldehyde on sodium alginate emulsion-based films to fight food contamination by *Escherichia coli* and *Salmonella enteritidis*. *Food Res. Int.* 128: 108791.

Antonio, C. M., Abriouel, H., Lopez, R. L., Omar, N. B., Valdivia, E., and Galvez, A. 2009. Enhanced bactericidal activity of enterocin AS-48 in combination with essential oils, natural bioactive compounds and chemical preservatives against *Listeria monocytogenes* in ready-to-eat salad. *Food Chem. Toxicol.* 47(9): 2216–2223.

Araújo, M. K., Gumiela, A. M., Bordin, K., Luciano, F. B., and de Macedo, R. E. F. 2018. Combination of garlic essential oil, allyl isothiocyanate, and nisin Z as bio-preservatives in fresh sausage. *Meat Sci.* 143: 177–3183.

Ayaz, M., Ullah, F., Sadiq, A., Ullah, F., and Devkota, H. P. 2019. Synergistic interactions of phytochemicals with antimicrobial agents: Potential strategy to counteract drug resistance. *Chem.-Biol. Interact.* 308: 294–303.

Bhattacharjee, R., Nandi, A., Mitra, P., Saha, K., Patel, P., Jha, E., and Suar, M. 2022. Theragnostic application of nanoparticle and CRISPR against food-borne multi-drug resistant pathogens. *Mater. Today Bio.* 100291.

Bölek, S., Tosya, F., and Akçura, S. 2022. Effects of Santolina chamaecyparissus essential oil on rheological, thermal and antioxidative properties of dark chocolate. *Int J Gastron Food Sci.* 27: 100481.

Burgi, K. W. 2016. *Reflection matrix method for controlling light after reflection from a diffuse scattering surface*. Theses and Dissertations.

Chalova, V. I., Crandall, P. G., and Ricke, S. C. 2010. Microbial inhibitory and radical scavenging activities of cold-pressed Terpeneless Valencia orange (*Citrus sinensis*) oil in different dispersing agents. *J. Sci. Food Agr.* 90(5): 870–876.

Chou, T. C., and Rideout, D. 1987. Synergism and antagonism in chemotherapy. *New York Academic Press*. 37–64.

Chuang, S., Sheen, S., Sommers, C. H., and Sheen, L. Y. 2021. Modeling the reduction of *Salmonella* and *Listeria monocytogenes* in ground chicken meat by high pressure processing and trans-cinnamaldehyde. *LWT-Food Sci. Technol.* 139: 110601.

Coimbra, A., Ferreira, S., and Duarte, A. P. 2022. Biological properties of *Thymus zygis* essential oil with emphasis on antimicrobial activity and food application. *Food Chem.* 133370.

Costa, W. K., de Oliveira, A. M., da Silva Santos, I. B., Silva, V. B. G., da Silva, E. K. C., de Oliveira Alves, J. V., and da Silva, M. V. 2022. Antibacterial mechanism of *Eugenia stipitata* McVaugh essential oil and synergistic effect against *Staphylococcus aureus*. *S. Afr. J. Bot.* 147: 724–730.

da Silva, G. C., de Veras, B. O., de Assis, C. R. D., Navarro, D. M. D. A. F., Diniz, D. L. V., Brayner dos Santos, F. A., and dos Santos Correia, M. T. 2021. Chemical composition, antimicrobial activity and synergistic effects with conventional antibiotics under clinical isolates by essential oil of *Hymenaea rubriflora* Ducke (FABACEAE). *Nat. Prod. Res.* 35(22): 4828–4832.

De Oliveira, S. M., Falcaosilva, V. S., Siqueirajunior, J. P., Costa, M. J., and Diniz, M. D. 2011. Modulation of drug resistance in *Staphylococcus aureus* by extract of mango (*Mangifera indica* L., Anacardiaceae) peel. *Res. Rev. J. Pharmacogn. Phytochem.* 21(1): 190–193.

Delaquis, P. J., Stanich, K., Girard, B., and Mazza, G. 2002. Antimicrobial activity of individual and mixed fractions of dill, cilantro, coriander and eucalyptus essential oils. *Int. J. Food Microbiol.* 74(1–2): 101–109.

Dhara, L., and Tripathi, A. 2020. Cinnamaldehyde: A compound with antimicrobial and synergistic activity against ESBL-producing quinolone-resistant pathogenic enterobacteriaceae. *Eur. J. Clin. Microbiol. Infect. Dis*. 39(1): 65–73.

Duarte, J. L., Bezerra, D. C., da Conceicao, E. C., Mourão, R. H., and Fernandes, C. P. 2020. Self-nano-emulsification of chamomile essential oil: A novel approach for a high value phytochemical. *Colloid Interface Sci. Commun*. 34: 100225.

Finney, D. J. 1952. *Probit analysis*, 2nd ed. Cambridge: Cambridge University Press. 146–153.

Friedman, M., Henika, P. R., Levin, C. E., and Mandrell, R. E. 2004. Antibacterial activities of plant essential oils and their components against *Escherichia coli* O157:H7 and *Salmonella* enterica in apple juice. *J. Agri. Food Chemistry*. 52(19): 6042–6048.

Fujita, M., Shiota, S., and Kuroda, T. 2005. Remarkable synergies between baicalein and tetracycline, and baicalein and betalactams against methicillin-resistant *Staphylococcus aureus*. *Microbiol Immunol*. 49: 391–396.

Gao, C., Tian, C., Lu, Y., Xu, J., Luo, J., and Guo, X. 2010. Essential oil composition and antimicrobial activity of *Sphallerocarpus gracilis* seeds against selected food-related bacteria. *Food Control*. 22(3–4): 517–522.

Gao, L., Sun, Y., Yuan, M., Li, M., and Zeng, T. 2020. In vitro and in vivo study on the synergistic effect of minocycline and azoles against pathogenic fungi. *Antimicrob. Agents Chemother*. 64: 150–162.

Gennings, C., Carter, W. H., Campain, J. A., Bae, D. S., and Yang, R. S. H. 2002. Statistical analysis of interactive cytotoxicity in human epidermal keratinocytes following exposure to a mixture of four metals. *Am. J. Agric. Biol. Sci*. 7(1): 58–73.

Giatrakou, V., Ntzimani, A., and Savvaidis, I. N. 2010. Combined chitosanthyme treatments with modified atmosphere packaging on a ready-to-cook poultry product. *J. Food Protect*. 73(4): 663–669.

Gorjian, H., Mihankhah, P., and Khaligh, N. G. 2022. Influence of Tween nature and type on physicochemical properties and stability of spearmint essential oil (*Mentha spicata* L.) stabilized with basil seed mucilage nanoemulsion. *J. Mol. Liq*. 119379.

Guarda, A., Rubilar, J. F., Miltz, J., and Galotto, M. J. 2011. The antimicrobial activity of microencapsulated thymol and carvacrol. *Int. J. Food Microbiol*. 146(2): 144–150.

Hemaiswarya, S., Kruthiventi, A. K., and Doble, M. 2008. Synergism between natural products and antibiotics against infectious diseases. *Phytomedicine*. 15(8): 639–652.

Hou, T., Sana, S. S., Li, H., Xing, Y., Nanda, A., Netala, V. R., and Zhang, Z. 2022. Essential oils and its antibacterial, antifungal and anti-oxidant activity applications: A review. *Food Biosci*. 101716.

Huang, Z., Pang, D., Liao, S., Zou, Y., Zhou, P., Li, E., and Wang, W. 2021. Synergistic effects of cinnamaldehyde and cinnamic acid in cinnamon essential oil against *S. pullorum*. *Ind. Crops Prod*. 162: 113296.

Jin, W., and Gong, W. 2011. Disintegration of orthogonal t value method on prescription of anti-fatigue health wine. *Ra Harmaal Jornal*. 120–131.

Ju, J., Chen, X., Xie, Y., Yu, H., Guo, Y., Cheng, Y., and Yao, W. 2019. Application of essential oil as a sustained release preparation in food packaging. *Trends Food Sci. Technol*. 92: 22–32.

Ju, J., Xie, Y., Guo, Y., Cheng, Y., Qian, H., and Yao, W. 2018a. Application of edible coating with essential oil in food preservation. *Crit. Rev. Food Sci. Nutr*. 01–62.

Ju, J., Xie, Y., Guo, Y., Cheng, Y., Qian, H., and Yao, W. 2018b. Application of starch microcapsules containing essential oil in food preservation. *Crit. Rev. Food Sci. Nutr*. 1–45.

Ju, J., Xie, Y., Yu, H., Guo, Y., Cheng, Y., Zhang, R., and Yao, W. 2020. Major components in Lilac and *Litsea cubeba* essential oils kill *Penicillium roqueforti* through mitochondrial apoptosis pathway. *Ind. Crop. Prod*. 149: 112349.

Ju, J., Xu, X., Xie, Y., Guo, Y., Cheng, Y., and Qian, H. 2017. Inhibitory effects of cinnamon and clove essential oils on mold growth on baked foods. *Food Chem*. 240: 850–855.

Kalemba, D., and Kunicka, A. 2003. Antibacterial and antifungal properties of essential oils. *Curr. Med. Chem*. 10(10): 813–829.

Kang, J. H. 2022. Understanding inactivation of *Listeria monocytogenes* and *Escherichia coli* O157: H7 inoculated on romaine lettuce by emulsified thyme essential oil. *Food Microbiology*. 105: 104013.

Katherine, A., Christine, F., and Carson, V. R. 2012. Effects of *Melaleuca alternifolia* (tea tree) essential oil and the major monoterpene component terpinen-4-ol on the development of single-and multistep antibiotic resistance and antimicrobial susceptibility. *Antimicrob. Agents Chemother.* 56(2): 909–915.

Kim, J. M., Marshall, M. R., Cornell, J. A., Preston III, J. F., and Wei, C. I. 1995. Antibacterial activity of carvacrol, citral, and geraniol against *Salmonella typhimurium* in culture medium and on fish cubes. *J. Food Sci.* 60(6): 1364–1368.

Kotzekidou, P., Giannakidis, P., and Boulamatsis, A. 2008. Antimicrobial activity of some plant extracts and essential oils against foodborne pathogens in vitro and on the fate of inoculated pathogens in chocolate. *LWT-Food Sci. Tech.* 41(1): 119–127.

Lachowicz, K. J., Jones, G. P., Briggs, D. R., Bienvenu, F. E., and Coventry, M. J. 1998. The synergistic preservative effects of the essential oils of sweet basil (*Ocimum basilicum* L.) against acidtolerant food microflora. *Lett. Appl. Microbiol.* 26(3): 209–214.

Lambert, R. J., Skandamis, P. N., Coote, P. J., and Nychas, G. E. 2001. A study of the minimum inhibitory concentration and mode of action of oregano essential oil, thymol and carvacrol. *J. Appl. Microbiol.* 91(3): 453–462.

Leclercq, R. 2002. Mechanisms of resistance to macrolides and lincosamides: Nature of the resistance elements and their clinical implications. *Clin Infect Dis.* 34: 482–492.

Lemons, A. R., McClelland, T. L., Martin, S. B., Lindsley, W. G., and Green, B. J. 2020. Inactivation of the multi-drug-resistant pathogen *Candida auris* using ultraviolet germicidal irradiation. *J. Hosp. Infect.* 105(3): 495–501.

Leonard, F. C., and Markey, B. K. 2008. Meticillin-resistant *Staphylococcus aureus* in animals: A review. *Veterinary Journal.* 175(1): 27–36.

Levine, R. R. 1978. *Pharmacology: Drug actions & reactions*, 2nd ed. Boston: Little Brown Co. 284–285.

Levy, N., Hashiguchi, T. C. O., and Cecchini, M. 2022. Food safety policies and their effectiveness to prevent foodborne diseases in catering establishments: A systematic review and meta-analysis. *Food Res. Int.* 111076.

Li, A., Shi, C., Qian, S., Wang, Z., Zhao, S., Liu, Y., and Xue, Z. 2022. Evaluation of antibiotic combination of *Litsea cubeba* essential oil on Vibrio parahaemolyticus inhibition mechanism and anti-biofilm ability. *Microbial Pathogenesis.* 105574.

Liu, Q., Zhang, M., Bhandari, B., Xu, J., and Yang, C. 2020. Effects of nanoemulsion-based active coatings with composite mixture of star anise essential oil, polylysine, and nisin on the quality and shelf life of ready-to-eat Yao meat products. *Food Control.* 107: 106771.

Liu, R., Gao, H., Chen, H., Fang, X., and Wu, W. 2019. Synergistic effect of 1-methylcyclopropene and carvacrol on preservation of red pitaya (*Hylocereus polyrhizus*). *Food Chem.* 588–595.

Luz, I. D. S., Gomes Neto, N. J. G., Tavares, A. G., Magnani, M., and Souza, E. L. D. 2012. Exposure of *Listeria monocytogenes* to sublethal amounts of *Origanum vulgare* L. essential oil or carvacrol in a food-based medium does not induce direct or cross protection. *Food Res. Int.* 48(2): 667–672.

Luz, I. D. S., Gomes Neto, N. J., Tavares, A. G., Nunes, P. C., Magnani, M., and De Souza, E. L. 2012. Evidence for lack of acquisition of tolerance in *Salmonella* enterica serovar typhimurium ATCC 14028 after exposure to subinhibitory amounts of *Origanum vulgarel* essential oil and carvacrol. *Appl. Environ. Microbiol.* 78(14): 5021–5024.

Mahmoud, B. S. M., Yamazaki, K., Miyashita, K., Kawai, Y., Shin, I. S., and Suzuki, T. 2006. Preservative effect of combined treatment with electrolyzed NaCl solutions and essential oil compounds on carp fillets during convectional air-drying. *Int. J. Food Microbiol.* 106(3): 331–337.

Matan, N., Rimkeeree, H., Mawson, A. J., Chompreeda, P., Haruthaithanasan, V., and Parker, M. 2006. Antimicrobial activity of cinnamon and clove oils under modified atmosphere conditions. *Int. J. Food Microbiol.* 107(2): 180–185.

Mcmahon, M. A., Blair, I. S., Moore, J. E., and Mcdowell, D. A. 2006. Habituation to sub-lethal concentrations of tea tree oil (*Melaleuca alternifolia*) is associated with reduced susceptibility to antibiotics in human pathogens. *J. Antimicrob. Chemother*. 59(1): 125–127.

Mihalache, O. A., Teixeira, P., and Nicolau, A. I. 2022. Raw-egg based-foods consumption and food handling practices: A recipe for foodborne diseases among Romanian and Portuguese consumers. *Food Control*. 139: 109046.

Minto, C. F., Schnider, T. W., Short, T. G., Gregg, K. M., Gentilini, A., and Shafer, S. L. 2000. Response surface model for anesthetic drug interactions. *Anesthesiology*. 92(6): 1603–1616.

Moleyar, V., and Narasimham, P. 1992. Antibacterial activity of essential oil components. *Int. J. Food Microbiol*. 16(4): 337–342.

Mourey, A., and Canillac, N. 2002. Anti-*Listeria monocytogenes* activity of essential oils components of conifers. *Food Control*. 13(4–5): 289–292.

Moussii, I. M., Nayme, K., Timinouni, M., Jamaleddine, J., Filali, H., and Hakkou, F. 2020. Synergistic antibacterial effects of Moroccan Artemisia herba alba, *Lavandula angustifolia* and *Rosmarinus officinalis* essential oils. *Synergy*. 10: 100057.

Newman, D. J., and Cragg, G. M. 2016. Natural products as sources of new drugs from 1981 to 2014. *J. Nat. Prod*. 79(3): 629–661.

Oh, J., Kim, H., Beuchat, L. R., and Ryu, J. H. 2022. Inhibition of *Staphylococcus aureus* on a laboratory medium and black peppercorns by individual and combinations of essential oil vapors. *Food Control*. 132: 108487.

Oluwatuyi, M., Kaatz, G., and Gibbons, S. 2004. Antibacterial and resistance modifying activity of *Rosmarinus officinalis*. *Phytochemistry*. 65: 3249–3254.

Parsaeimehr, M., Basti, A. A., Radmehr, B., Misaghi, A., Abbasifar, A., Karim, G., Rokni, N., Motlagh, M. S., Gandomi, H., Noori, N., and Khanjari, A. 2010. Effect of *Zataria multiflora* Boiss. essential oil, nisin, and their combination on the production of enterotoxin C and alpha-hemolysin by *Staphylococcus aureus*. *Foodborne Pathog Dis*. 7(3): 299–305.

Patrone, V., Campana, R., Vittoria, E., and Baffone, W. 2010. In vitro synergistic activities of essential oils and surfactants in combination with cosmetic preservatives against *Pseudomonas aeruginosa* and *Staphylococcus aureus*. *Curr. Microbiol*. 60(4): 237–241.

Pradhan, A. M. S., and Kim, Y. T. 2014. Relative effect method of landslide susceptibility zonation in weathered granite soil: A case study in Deokjeok-Ricreek, south Korea. *Nat. Hazards*. 72(2): 1189–1217.

Qing, S. Z., and Rui, Y. S. 1999. Quantitative design of drug compatibility by weighted modification method. *Acta Pharmacologica Sinica*. 20(11): 1043–1051.

Ramirez, M. N., and Tolmasky, M. E. 2010. Aminoglycoside modifying enzymes. *Drug Resist Updat*. 13: 151–171.

Rao, J., Chen, B., and Mcclements, D. J. 2019. Improving the efficacy of essential oils as antimicrobials in foods: Mechanisms of action. *Annu. Rev. Food Sci. Technol*. 10(1): 365–387.

Reis, D. R., Zin, G., Lemos-Senna, E., Ambrosi, A., and Di Luccio, M. 2021. A modified premix method for the emulsification of spearmint essential oil (*Mentha spicata*) by ceramic membranes. *Surfaces and Interfaces*. 26: 101328.

Reyesjurado, F., Navarrocruz, A. R., Ochoavelasco, C. E., Palou, E., Lopezmalo, A., and Avilasosa, R. 2019. Essential oils in vapor phase as alternative antimicrobials: A review. *Crit. Rev. Food Sci. Nutr*. 1–10.

Ribo, J. M., and Rogers, F. 2010. Toxicity of mixtures of aquatic contaminants using the luminescent bacteria bioassay. *Env. Toxicol*. 5(2): 135–152.

Rivas, L., McDonnell, M. J., Burgess, C. M., O'Brien, M., Navarro-Villa, A., Fanning, S., and Duffy, G. 2010. Inhibition of verocytotoxigenic *Escherichia coli* in model broth and rumen systems by carvacrol and thymol. *Int. J. Food Microbiol*. 139(1–2): 70–78.

Seow, Y. X., Yeo, C. R., Chung, H. L., and Yuk, H. 2014. Plant essential oils as active antimicrobial agents. *Crit. Rev. Food Sci. Nutr.* 54(5): 625–644.

Shah, A., Alam, S., Kabir, M., Fazal, S., Khurshid, A., Iqbal, A., and Bibi, Y. 2022. Migratory birds as the vehicle of transmission of multi drug resistant extended spectrum β lactamase producing Escherichia fergusonii, an emerging zoonotic pathogen. *Saudi J. Biol. Sci.* 29(5): 3167–3176.

Shah, S. M., Ullah, F., Ayaz, M., Sadiq, A., and Nadhman, A. 2019. Benzoic acid derivatives of Ifloga spicata (Forssk.) Sch.Bip. as potential anti-leishmanial against Leishmania tropica. *Processes.* 7(4): 208.

Shimizu, M., Shiota, S., Mizushima, T., Ito, H., Hatano, T., Yoshida, T., and Tsuchiya, T. 2001. Marked potentiation of activity of β-lactams against methicillin-resistant *Staphylococcus aureus* by Corilagin. *Antimicrob. Agents Chemother.* 45(11): 3198–3201.

Solomakos, N., Govaris, A., Koidis, P., and Botsoglou, N. 2008b. The antimicrobial effect of thyme essential oil, nisin, and their combination against *Listeria monocytogenes* in minced beef during refrigerated storage. *Food Microbiol.* 25(1): 120–127.

Somrani, M., Inglés, M. C., Debbabi, H., Abidi, F., and Palop, A. 2020. Garlic, onion, and cinnamon essential oil anti-biofilms' effect against *Listeria monocytogenes. Foods.* 9(5): 567.

Tariq, S., Wani, S., Rasool, W., Shafi, K., Bhat, M. A., Prabhakar, A., and Rather, M. A. 2019. A comprehensive review of the antibacterial, antifungal and antiviral potential of essential oils and their chemical constituents against drug-resistant microbial pathogens. *Microbial Pathogenesis.* 134: 103580.

Trzcinski, K., Cooper, B. S., Hryniewicz, W., and Dowson, C. G. 2010. Expression of resistance to tetracyclines in strains of methicillin-resistant *Staphylococcus aureus. J Antimicrob Chemother.* 45: 763–770.

Ubukata, K., Nonoguchi, R., Matsuhashi, M., and Konno, M. 1989. Expression and inducibility in *Staphylococcus aureus* of the mecA gene, which encodes a methicillin-resistant S. aureus-specific penicillin-binding protein. *J. Bacteriol.* 171(5): 2882–2885.

Wang, H., Guo, L., Liu, L., Han, B., and Niu, X. 2021. Composite chitosan films prepared using nisin and Perilla frutescense essential oil and their use to extend strawberry shelf life. *Food Biosci.* 41: 101037.

Wang, Y., Ma, E. W. M., Chow, T. W. S., and Tsui, K. L. 2013. A two-step parametric method for failure prediction in hard disk drives. *IEEE Trans. Ind. Inform.* 10(1): 419–430.

Wang, Y., Wu, C., Lu, L., Ren, G. N., Cao, X., and Shen, J. 2008. Macrolide—lincosamide-resistant phenotypes and genotypes of *Staphylococcus aureus* isolated from bovine clinical mastitis. *Vet. Microbiol.* 130(12): 118–125.

Wu, H., Zhao, F., Li, Q., Huang, J., and Ju, J. 2022. Antifungal mechanism of essential oil against foodborne fungi and its application in the preservation of baked food. *Crit. Rev. Food Sci. Nutr.* 1–13.

Zhang, J., Li, Y., Yang, X., Liu, X., Hong, H., and Luo, Y. 2021. Effects of oregano essential oil and nisin on the shelf life of modified atmosphere packed grass carp (*Ctenopharyngodon idellus*). *LWT-Food Sci. Tech.* 147: 111609.

Zhang, W., Cao, X., and Liu, S. Q. 2019. Aroma modulation of vegetable oils—A review. *Crit. Rev. Food Sci. Nutr.* 1–14.

Zheng, Q. S., and Sun, R. Y. 2000. Analysis of drug interaction in combined drug therapy by flection method. *Acta Pharmacol Sin.* 21(2): 183–187.

Zhu, Y., Li, C., Cui, H., and Lin, L. 2021. Encapsulation strategies to enhance the antibacterial properties of essential oils in food system. *Food Control.* 123: 107856.

Zolfpour-Arokhlo, M., Selamat, A., Hashim, S. Z. M., and Afkhami, H. 2014. Modeling of route planning system based on Q value-based dynamic programming with multi-agent reinforcement learning algorithms. *Eng. App. Artif. Int.* 29: 163–177.

6 Antioxidant Activity and Mechanism of Essential Oils

6.1 INTRODUCTION

Essential oils (EOs) produce a colorless liquid from all parts of the plant, including seeds, flowers, peel, stem, and bark (Ju, Xie et al., 2018; Silvestre et al., 2019). EOs have been widely used as drugs, perfumes, cosmetics, and food preservatives in many countries (Coimbra et al., 2022). The composition of EOs is very complex. They are mainly composed of terpenoids (menthol, citral, camphor, pinene, and cineole), aromatic compounds (cinnamaldehyde, eugenol, anisole, thymol, and vanillin), and aliphatic compounds (tetradecanol, nonyl alcohol, sunflower alcohol, and panaxynol) (Liu et al., 2021; Silvestre et al., 2019; Gandhi et al., 2019). EOs have a special smell and can be used as natural antioxidant and antimicrobial agents (Ju et al., 2019; Farhadi et al., 2020; Maqsoudlou et al., 2020; Mukurumbira et al., 2022; Yu et al., 2019).

At present, although the commonly used chemical synthesis antioxidants (butyl hydroxyanisole, butylated hydroxytoluene [BHT], propyl gallate, ascorbic acid, isoascorbic acid, etc.) have strong antioxidant capacity, long-term use may pose a potential threat to human health (Ju, Xu et al., 2018; Ju et al., 2020a). Therefore, the chemical synthesis of antioxidants is increasingly rejected by consumers. Because of the concerns of consumers about the safety of traditional synthetic antioxidants, the food industry tends to use more natural and safer antioxidants (Ju et al., 2020b; Shadyro et al., 2017).

As a kind of natural, green, low-toxic, and even nontoxic plant-based additive, EOs are generally considered to be relatively safe substances (generally recognized as safe [GRAS]) (Ju et al., 2022; Fabbri et al., 2020; Debonne et al., 2018). EOs contain a large number of phenolic compounds, which can protect cells from oxidative damage caused by free radicals, thus reducing the risk of cardiovascular disease, nervous system disease, and various cancers (Basavegowda and Baek, 2021; Proestos et al., 2013; Siler et al., 2014). Therefore, they are often used as an antioxidant or natural food additive in the food industry. For example, Perumal et al. (2017) found that fumigating mango with thyme EOs can increase the antioxidant enzyme activity of post-harvest mango and extend the shelf life of mango. Ju et al. (2020d) found that active packaging containing a combination of eugenol and citral can better maintain the sensory quality of bread and extend the shelf life of bread. There are also related reports showing that polysaccharide coatings containing 1–10% oregano EOs can effectively ($P < 0.05$) reduce the activities of polyphenol oxidase and peroxidase, and better maintain the appearance and color of bok choi (Vital et al., 2016). It can

DOI: 10.1201/9781003329268-6

be seen that EOs have good antioxidant activity, but in recent years there have been relatively many studies on the antibacterial activity and mechanism of EOs, while information on the antioxidant mechanism of EOs is scarce. Therefore, in this chapter, the author introduces the antioxidant mechanism of EOs in detail to provide a theoretical basis for the development of EOs as food preservatives.

6.2 OXIDATIVE STRESS

Oxidative stress refers to a state in which the body produces too many highly active molecules (active oxygen radicals [ROS] and active nitrogen radicals [RNS]) when it is subjected to various harmful stimuli, and the degree of oxidation exceeds the scope of oxide scavenging, resulting in an imbalance between the oxidation system and the antioxidant system (Jaganjac et al., 2022; Rotariu et al., 2022; Bai et al., 2022). Under the condition of oxidative stress, the polyunsaturated fatty acids that make up the cell membrane are oxidized, which leads to the oxidative damage of cell membrane and protein, thus causing cytotoxicity and genotoxicity. Oxidative stress is one of the important causes of aging and metabolic diseases (Lin et al., 2022; Yin et al., 2022; Wang et al., 2022).

6.2.1 Effect of Oxidative Stress on Protein

According to existing reports, proteins are the main cellular targets of ROS. The side chains of amino acids are easily attacked by ROS and cause oxidative damage (Wang et al., 2016). These damages mainly include (1) ROS causing the deletion of amino acid residues or peptide chain breaks. Oxidation of methionine residues or sulfhydryl groups in proteins can cause a change in the conformation of the protein, leading to protein degradation (Boguszewska-Mańkowska et al., 2015). (2) The oxidative damage of free radicals can cause the function of some important proteins to be destroyed. (3) ROS will change the charge of the protein, causing the protein to break or aggregate (Sanada et al., 2011).

6.2.2 Effect of Oxidative Stress on Lipids

It is generally believed that lipid peroxidation is the main marker of oxidative damage. However, lipid peroxidation products, such as thiobarbituric acid and isoprostaglandin, are indirect markers of oxidative stress (Collard et al., 2004; Paciorek et al., 2020). The effect of free radicals on lipids has two main aspects: (1) free radicals can induce lipid peroxidation and destroy the structure of cell membrane phospholipid bilayers, increasing tissue permeability (Hussain et al., 2008). (2) Some peroxidation products, such as malonaldehyde, can inactivate cells through protein cross-linking (Wickens et al., 1981). The process of lipid peroxidation is shown in Figure 6.1.

6.2.3 Effect of Oxidative Stress on Genetic Material

Deoxyribonucleic acid (DNA) is particularly sensitive to the attack of oxygen free radicals (ROS) (Wang et al., 2016). DNA will generate 8-oxo-guanine after oxidative damage, which is also a potential marker of cancer (Lukina et al., 2013). The

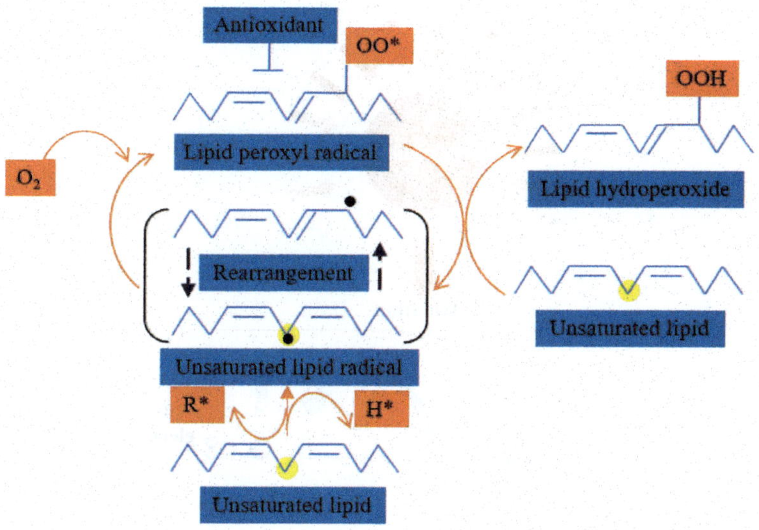

FIGURE 6.1 The process of lipid peroxidation.

damage to DNA by ROS is mainly manifested in the following aspects: (1) changes in the structure of DNA, such as mutation, rearrangement, deletion, or extension of bases, and even changes in chromosome number (Srinivas et al., 2019); (2) effect on the activity of genes or proteins; and (3) inhibition of DNA repair and exacerbation of DNA damage. Of course, some enzymatic and nonenzymatic substances in the organism can resist oxidative damage to a certain extent (Niocel et al., 2019). In addition, the ingestion of exogenous antioxidants can also play a role in antioxidative damage, for example, vitamins, trace elements, Chinese herbs, and EOs (Harun et al., 2020; Ibtisham et al., 2019; Ishkeh et al., 2019; Harun et al., 2020; Sun et al., 2018). In particular, the antioxidant capacity of some EOs is several times or even dozens of times that of ordinary antioxidants.

6.3 FREE RADICALS

Free radicals are intermediate products produced in the process of metabolism. The main pathways of free radical production include endogenous and exogenous pathways (Fasiku et al., 2020; Shadyro et al., 2019; Penabautista et al., 2019). The endogenous pathway refers to some metabolic activities and enzymatic reactions of the body. The exogenous pathway refers to the formation of free radicals by the breakdown of covalent bonds caused by stimulation of the external environment (Branduardi et al., 2013). The inducing factors of free radical production and the interaction between antioxidants and free radicals are shown in Figure 6.2.

Under normal circumstances, the free radicals in the body are constantly being eliminated while being continuously generated, and they are always maintained within a relatively stable concentration range (Bracco et al., 2018). Free radicals in the body can participate in cell proliferation, differentiation, and migration, and maintain the normal

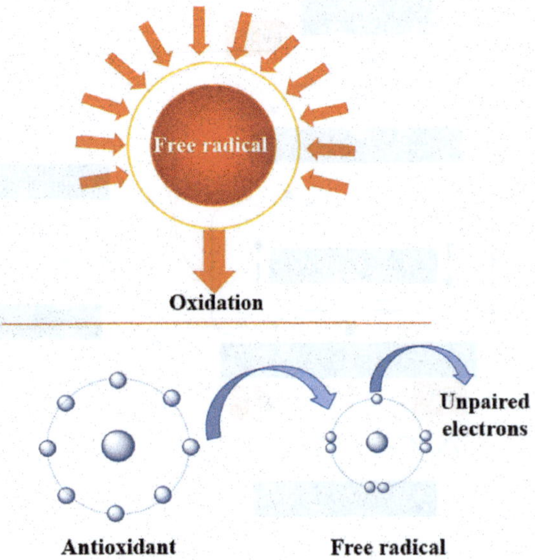

FIGURE 6.2 The inducing factors of free radical production and the interaction between antioxidants and free radicals.

metabolism of cells and tissues (Nakano et al., 2019). In contrast, they can also damage biological macromolecules in the body. However, the damaged biomolecules can be repaired and resynthesized without causing damage to the body, which depends on the antioxidant system present in the body. The types of free radicals are shown in Table 6.1.

6.4 ANTIOXIDANT SYSTEM

There are two kinds of antioxidant systems in the body: the enzyme and nonenzyme antioxidant systems (Zhang et al., 2019). The enzyme antioxidant system is mainly composed of superoxide dismutase (SOD), catalase (CTA), selenium glutathione peroxidase (SeGPx), glutathione *S*-transferase (GST), and aldehyde ketone reductase (AR). They can remove the active oxygen in cells and form the first defense line of antioxidation in vivo (Zhan et al., 2004; Cavanagh et al., 1995). Nonenzyme antioxidant systems are mainly composed of fat-soluble antioxidants (vitamin E, carotenoids, and coenzyme Q), water-soluble antioxidants (vitamin C and glutathione), and protein-based antioxidants (lactoferrin, transferrin, and thioprotein). They mainly accomplish the antioxidant effect by cutting off the reaction chain of fat oxidation (Domínguez et al., 2019; Vircheva et al., 2010).

6.5 ANTIOXIDATIVE COMPONENTS IN ESSENTIAL OILS

The types and proportions of the chemical components in EOs are the main reasons for the differences in the biological functions of EOs (Jan et al., 2020; Hamad et al.,

TABLE 6.1
Types of Free Radicals

Name	Category
ROS	Superoxide anion radical
	Hydroxyl radical
	Hydrogen peroxide
	Singlet oxygen
	Organic peroxide
	Alkoxy group
	Peroxy radical
	Hypochlorite
	ozone
RNS	Nitric oxide
	Nitrogen dioxide
	Nitrous oxide
	Nitrous oxide
	Peroxynitrite

2019). Plant species, location, growth time, growth climate, and extraction methods all affect the chemical components of EOs (Chrysargyris et al., 2021; Belhachemi et al., 2022; Fatima et al., 2019). In general, each EO may contain two to three active components, which determine the biological characteristics of the EO to a large extent. For example, pine needle EOs contain 35 compounds, the main components of which are (+)-α-pinene (20.72%), bornyl acetate (19.74%), and geranylene (13.89%) (Koutsaviti et al., 2021).

There are many kinds of EOs, most of which contain antioxidants. At present, the most studied EOs include Labiatae, *Myrtle*, Lauraceae, and Zingiberaceae. For example, the antioxidant activity of thyme, rosemary, oregano, and sage EOs has been demonstrated in the Labiatae family (Rocamora et al., 2020; Starowicz et al., 2020; Khodaei et al., 2021). The content of carvacrol and thymol in the oregano EOs accounts for about 80% of its total components. These two active components are the key substances for the antioxidant effect of oregano EOs. The content of thymol and carvacrol in thyme EOs accounts for about 50% of its total composition (Fournomiti et al., 2015). Similarly, thymol and carvacrol are the main substances by which thyme EOs exert antioxidant activity (Han et al., 2017; Pinto et al., 2009). Carnosol, rosmarinic acid, and rosmarinol are the main antioxidants in rosemary EOs (Souri and Bakhtiarizade, 2019; Abdollahi et al., 2012; Szumny et al., 2010).

Phenylpropyl esters and eugenol are the main antioxidants in *Hypericum*. Zingiberaceae mainly relies on flavonoids to play an antioxidant role. Flavonoids and anthocyanins are the main antioxidants in Umbelliferae (Daniel et al., 2009; Chakma and Konkolewicz, 2019; Mendes et al., 2015; Rakass et al., 2018). The common antioxidant components in EOs are shown in Figure 6.3.

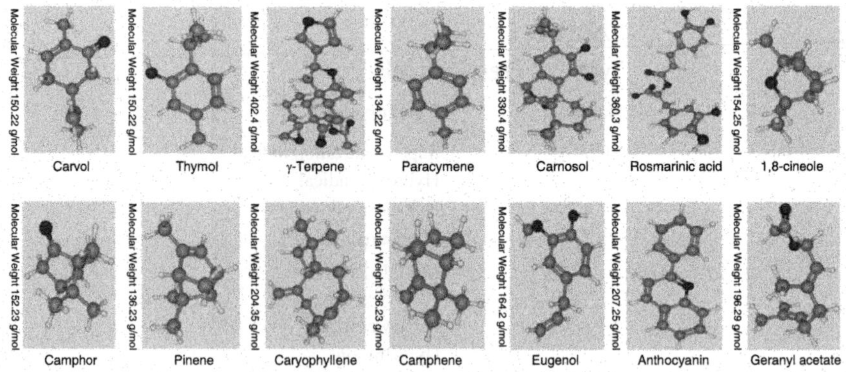

FIGURE 6.3 The common antioxidant components in essential oil.

6.6 ANTIOXIDANT MECHANISM OF ESSENTIAL OILS

6.6.1 Direct Antioxidant Effect

6.6.1.1 Free Radical Scavenging

Oxidation is the process of electron transfer from one atom to another, which is a part of aerobic activity and metabolism of the human body. This is because oxygen is the ultimate acceptor of electrons in the electron flow system (Gulcin, 2020; Brahmi et al., 2013). However, when electrons flow into an uncoupled state, free radicals will be produced.

Free radicals are usually divided into ROS and RNS, which play a dual role in the life system. ROS are the main radicals, including superoxide anion (O_2^-), hydrogen peroxide (H_2O_2), hydroxyl radicals (•OH), and singlet oxygen (RNS) (Ifeanyi et al., 2018; Uchida et al., 1988). When an organism is stimulated by the external environment, its internal balance will be broken, and a lot of free radicals and ROS will be produced. The generation of free radicals will not only affect the normal physiological functions of cells but also exacerbate membrane peroxidation, leading to an imbalance of biochemical reactions related to aerobic metabolism, and then destroy biological macromolecules such as proteins, lipids, nucleic acids, and unsaturated fatty acids (Ju et al., 2020a; Rajesh et al., 2016). If it is necessary to reduce the harm of free radicals to human health, in addition to relying on the free radical scavenger system in vivo, it is possible to block the invasion of free radicals to the human body by adding exogenous free radical scavengers (Chen et al., 2012). As a natural active substance, EOs can combine with free radicals to form stable compounds, to achieve an antioxidant effect (Bi et al., 2016). This is because the EOs contain not only reducing heteroatoms and unsaturated double bonds but also polyphenols and flavonoids. Polyphenols can quench singlet oxygen directly and prevent free radical chain reaction (Falleh et al., 2020; Fierascu et al., 2018; Leicach and Chludil, 2014). In addition, some polyphenols can also transfer active hydrogen atoms to free radicals, making them become low-activity substances and remove them. The reaction formula is as follows: R• + POH → RH + PO•. Because of the conjugation effect of the benzene ring, the phenoxy radical (PO•) is relatively stable, so the next reaction is not easy to take place. In addition, PO• can quickly react with another radical, and the reaction formula is PO• + R• → POR (Olszowy, 2019).

Previous researchers have investigated the scavenging effect of *Anoectochilus roxburghii* EOs on DPPH and ABTS free radicals. The results showed that when the concentration of EOs was in the range of 0.05–2 mg/ml, the scavenging rate of EOs to free radicals increased with the increase of concentration. When the concentration of EOs is 2 mg/ml, the clearance rate of DPPH and ABTS free radicals is 75% and 90.5%, respectively. However, under the same conditions, the scavenging rate of ascorbic acid for these two free radicals was about 90% (Shao et al., 2014). These experimental data indicated that the EOs have significant ($P < 0.05$) antioxidant activity.

Other related studies used ultrasonic-assisted extraction technology to extract *Kaempferia galanga* Linn EOs and studied their antioxidant activity through DPPH and O_2^- free-radical-scavenging experiments. The results showed that the scavenging rate of DPPH was 91.9% at the concentration of 10 mg/ml, which was significantly ($P < 0.05$) higher than that of BHT in the same conditions. At the concentration of 1.0 mg/ml, the removal rate of ABTS was 82.7%, which was significantly ($P < 0.05$) higher than that of ascorbic acid under the same conditions (Qin et al., 2015).

6.6.1.2 Chelation With Metal Ions

The phenolic substances in the EOs have good chelating properties. For example, eugenol can chelate with copper, iron, and calcium ions, which can not only inhibit the auto-oxidation of fats but also prevent the generation of free radicals induced by metal ions (Jia et al., 2021). In addition, some transition metal ions (Fe^{2+} and Cu^{2+}) can also undergo a Fenton reaction with H_2O_2 to generate hydroxyl radicals (•OH). The reaction process is shown in Figure 6.4.

Hydroxyl radical is one of the most harmful oxygen radicals. It can start a free radical chain reaction related to almost all biomacromolecules. Polyphenols in EOs can be chelated with metal ions to form relatively stable complexes, thus inhibiting the generation of free radicals (Bi et al., 2015). The reaction process is shown in Figure 6.5.

Tiwary et al. (2007) analyzed the chemical components of Rutaceae EOs using GC-MS and evaluated its antioxidant activity by a free-radical-scavenging

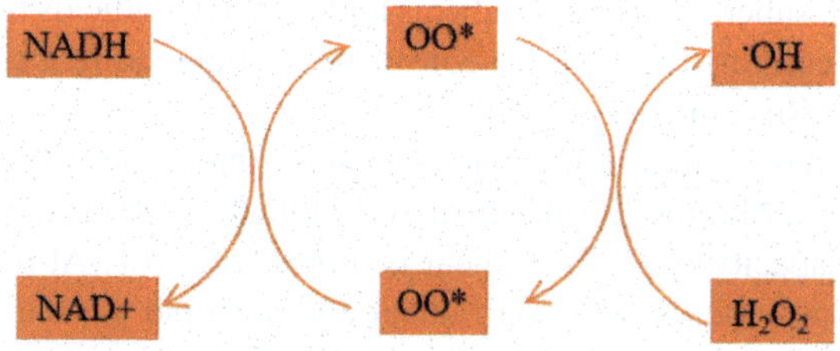

FIGURE 6.4 Schematic diagram of OH produced by Fenton reaction.

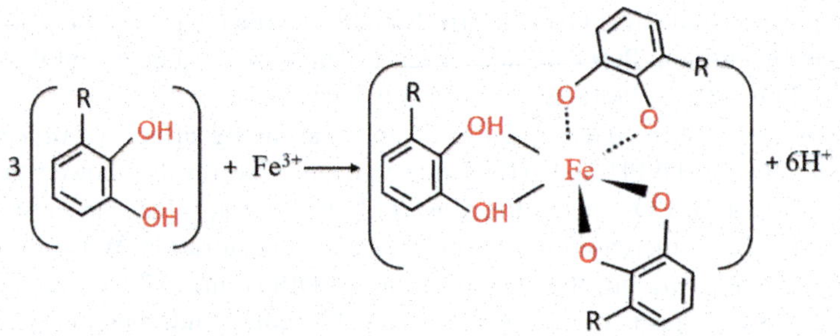

FIGURE 6.5 Chelation of phenols with metal ions.

experiment and a metal ion chelating experiment. The results showed that the main chemical component of the EOs was limonene (62.288%), and it had good free-radical-scavenging ability and metal ion chelating ability.

There are also related studies on free-radical-scavenging experiments, metal ion chelation experiments, and total reduction experiments to explore the antioxidant activity of *Nandina domestica* EOs. The results showed that the IC50 values of DPPH radical-scavenging rate, ABTS radical-scavenging rate, O_2^- radical-scavenging rate, metal ion chelating, and total reducing power were 28.39, 20.61, 53.22, 92.52, and 145.35 μg/ml, respectively. It can be seen that the EOs also have significant ($P < 0.05$) antioxidant activity (Gholivand et al., 2010).

6.6.2 Indirect Antioxidant Effect

6.6.2.1 Inhibit Lipid Peroxidation

For food, lipid oxidation will not only give an unpleasant smell to food but also poses a potential threat to human health. At present, lipid oxidation is one of the main factors that cause a decline in food quality. More and more researchers are investigating the possibility of using natural antioxidants to prevent food lipid peroxidation (Wu et al., 2022; Lampi et al., 2019).

The antioxidant activity of EOs may be mainly related to phenolic components (Jummes et al., 2020; Das et al., 2020; Cagol et al., 2020). Because the hydroxyl carried by phenolic components can be used as the donor of free radicals in the process of fat oxidation, it can prevent the formation of the peroxidation reaction, thus protecting the lipid from oxidation. For example, *Pogostemon cablin* Benth EOs can not only scavenge superoxide anion radicals and hydroxyl radicals but also inhibit the peroxidation of lecithin. The EOs are considered to be good free radical scavengers and antioxidants (Roshan et al., 2022).

In addition, it has been reported that the scavenging capacity of *Alliumfistulosum* L. var. *caespitosum* Makino EOs for hydroxyl radicals (•OH) is more than 90%, and the scavenging capacity to DPPH radical is more than 70%. At the same time, the ability to scavenge superoxide anion radical and antilipid peroxidation of the EOs was more than 50% (Liu et al., 2018). This indicated that the EOs could inhibit lipid peroxidation in addition to its strong free-radical-scavenging ability.

6.6.2.2 Regulate the Level of Antioxidant Enzymes

It is reported that some active components in EOs can bind to cell surface receptors. Then, through signal transduction to improve the level of antioxidant enzyme secretion, they enhance the body's antioxidant defense ability (Yu et al., 2022; Li et al., 2022). However, the specific regulatory mechanism is not fully determined, but it can be inferred that terpenes in EOs can modify the sulfhydryl group of Keapl. Keapl is a multidomain repressor protein in the cytoplasm, which participates in the redox reaction of cells. Keapl was modified to reduce the ubiquitination level of Nrf2. Keapl and Nrf2 have an uncoupling effect. After activation, Nrf2 can enter the nucleus and bind to the Maf protein to form a dimer. The substance can participate in the antioxidant response, activate the expression of target genes, and increase the expression level of antioxidant enzymes (glutathione peroxidase, glutathione, protein reductase, and quinone oxidoreductase). In addition, some EOs containing phenylpropane can also participate in the redox regulation of cells through the Keapl signaling pathway. For example, *Lavandula angustifolia* Mill EOs can improve the levels of catalase, SOD, and glutathione peroxidase in mouse cells. Similarly, *Coriandrum sativum* L. EOs can regulate the activities of glutathione transferase and cytochrome p450A1 in mouse cells (Zaid et al., 2017). The mechanism of EOs regulating antioxidant enzymes is shown in Figure 6.6.

6.7 FACTORS AFFECTING THE ANTIOXIDANT ACTIVITY OF ESSENTIAL OILS

The antioxidant activity of EOs is affected by many factors, such as the type, origin, physical and chemical properties, extraction methods, storage conditions, and

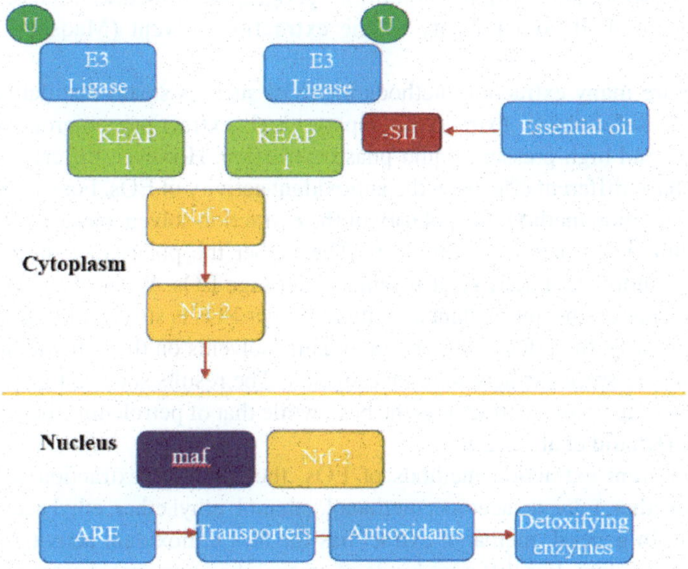

FIGURE 6.6 The mechanism of plant essential oil regulating peroxidase.

initiators and receptors of oxidation. In addition, the antioxidant effect of the same EOs is also different under different conditions. The antioxidant activities of different EOs or their active components have been introduced previously and will not be described here.

6.7.1 Solubility

EOs are oil-soluble antioxidants. It is generally believed that oil-soluble antioxidants have a good antioxidant effect on oil-based products. Similarly, water-soluble antioxidants have a good antioxidant effect on water-based products. However, practice has proved that this is not entirely the case. Musakhanian et al. (2022) found that water-soluble antioxidants have a better antioxidant effect than oil-soluble antioxidants in oil-based systems. In an emulsion system, however, the result is the opposite. This may be because water-soluble antioxidants can be evenly distributed in the interface between air and oil molecules in the oil-soluble system, thus preventing oil oxidation; when in the water-based emulsion system, water-soluble antioxidants will be diluted by water to reduce their effective concentration; in contrast, oil-soluble antioxidants can maintain their effective concentration in the emulsion system, so they have a better antioxidant effect than water-soluble antioxidants. This is why, in many cases, we often combine EOs with emulsifiers in different kinds of food systems.

6.7.2 Extraction Method

The activity and content of antioxidant components in EOs are affected by plant variety, position, age, climatic condition, geographical location, harvest season, extraction method, and polarity of the extraction solvent (Maqsoudlou et al., 2020).

There are many extraction methods for EOs, such as steam distillation, solvent extraction, solid-phase extraction, supercritical extraction, microwave-assisted extraction, and high-pressure liquid-phase extraction. However, different extraction methods have different effects on the antioxidant activity of EOs. For example, in the solvent extraction method, the polarity of the extraction solvent has a great influence on the antioxidant activity of the EOs. The higher the polarity of the extract, the higher the antioxidant activity of rosemary and sage EOs. When the polarity of the extract is smaller, the antioxidant activity of ginger EOs is stronger (Trojakova et al., 2001). Similarly, the effects of different organic solvents on the antioxidant capacity of six kinds of spice extracts were investigated. The results showed that the antioxidant effect of methanol extract was the best, while that of petroleum ether extract was the worst (Khalid et al., 2020).

For different extraction methods of EOs, the common extraction solvents are petroleum ether, toluene, acetone, methanol, ethanol, ethyl ether, ethyl acetate, water, and so on. In general, methanol extract has the best antioxidant activity; of course, this is not absolute. For different kinds of spices, the appropriate extraction solvent should be selected.

6.7.3 Initiator of Oxidation

The antioxidant activity of EOs was affected by the initiator of oxidation, showing different degrees of antioxidant effect. Chlorophyll is a co-oxidant that causes oil oxidation in olive oil. The EOs will show the effect of promoting oxidation or antioxidation depending on the experimental method and the concentration used.

Similarly, it has been reported that when the concentration of α-tocopherol, γ-tocopherol, and δ-tocopherol in soybean oil exceeds a certain range, they all change from antioxidants to oxidants (Player et al., 2006). Similar reports also showed that with the increase of the amount of cinnamon extract, the antioxidant activity of the extract on lard oxidation increases, but when the addition reaches a certain concentration, the extract itself will also undergo an oxidation reaction, causing oil oxidation to be activated (Dang et al., 2000).

6.7.4 Receptor Type

The evaluation of the antioxidant effect of EOs can not only use oil as an evaluation index because the antioxidant effect of EOs measured under the condition of a single oil is not necessarily suitable for the final processed products. For example, biscuits and chocolates contain not only oil ingredients but also non-oil components such as sodium ions, iron ions, and emulsifiers, which can affect the antioxidant effect of EOs. Therefore, to use natural antioxidants, various receptors and components should be considered as far as possible.

6.7.5 Storage Condition

The antioxidant effect of spice extract or EOs also depends on its storage conditions. First of all, spice extracts or EOs must be stored in sealed, light-shielded, and low-temperature conditions; otherwise, spice extracts or EOs themselves are very easy to oxidize, resulting in weakening of their antioxidant effect and even promoting oxidation. Second, when the spice extract or EO is added to the food, the storage conditions of the food will also affect its antioxidant activity. If it is reported that the storage temperature has a great effect on the antioxidation of garlic EOs, the food with garlic EOs should be stored at as low a temperature as possible (Bakkali et al., 2008).

6.8 CONCLUDING REMARKS

EOs have been studied by many researchers because of their unique antioxidation and antimicrobial effects. In recent years, there have been many studies on the antimicrobial and antioxidation effects of EOs. However, current research mainly focuses on the antimicrobial properties of EOs. Related research on the antioxidation mechanism of EOs is still very scarce. Therefore, the antioxidative mechanism of EOs was here analyzed in detail.

Compared with the traditional synthetic antioxidants, EOs are safer. As natural antioxidant and antimicrobial agents, they have broad application prospects and research value with regard to food preservation and processing. However, the composition of EOs is very complex, which leads to a lack of research on the antioxidant activities of various monomers in EOs. Therefore, we should pay more attention to the study of the antioxidation and antibacterial mechanism of the active components in the EOs.

ACKNOWLEDGEMENTS

This work was supported by the National Natural Science Foundation of China (32202192), Special fund for Taishan Scholars Project.

REFERENCES

Abdollahi, M., Rezaei, M., and Farzi, G. 2012. A novel active bionanocomposite film incorporating rosemary essential oil and nanoclay into chitosan. *J. Food Eng.* 111(2): 343–350.

Bai, R., Guo, J., Ye, X. Y., Xie, Y., and Xie, T. 2022. Oxidative stress: The core pathogenesis and mechanism of Alzheimer's disease. *Ageing Res. Rev.* 101619.

Bakkali, F., Averbeck, S., Averbeck, D., and Idaomar, M. 2008. Biological effects of essential oils--a review. *Food Chem. Toxicol.* 46: 446–475.

Basavegowda, N., and Baek, K. H. 2021. Synergistic antioxidant and antibacterial advantages of essential oils for food packaging applications. *Biomolecules.* 11(9): 1267.

Belhachemi, A., Maatoug, M. H., and Canela-Garayoa, R. 2022. GC-MS and GC-FID analyses of the essential oil of *Eucalyptus camaldulensis* grown under greenhouses differentiated by the LDPE cover-films. *Ind. Crops Prod.* 178: 114606.

Bi, S. F., Zhu, G. Q., Wu, J., Li, Z. K., Lv, Y. Z., and Fang, L. 2015. Chemical composition and antioxidant activities of the essential oil from nandina domestica fruits. *Nat. Prod. Chem. Res.* 30(3): 1–4.

Bi, S. F., Zhu, G. Q., Wu, J., Li, Z., Lv, Y. Z., and Fang, L. 2016. Chemical composition and antioxidant activities of the essential oil from Nandina domestica fruits. *Nat. Prod. Chem. Res.* 30(3): 362–365.

Boguszewska-Mańkowska, D., Nykiel, M., and Zagdańska, B. 2015. Protein oxidation and redox regulation of proteolysis. *IntechOpen.* 17–35.

Bracco, P., Costa, L., Luda, M. P., and Billingham, N. C. 2018. A review of experimental studies of the role of free-radicals in polyethylene oxidation. *Polym. Degrad. Stab.* 67–83.

Brahmi, F., Mechri, B., Flamini, G., and Madiha, D. 2013. Antioxidant activities of the volatile oils and methanol extracts from olive stems. *Acta Physiol. Plant.* 35(4): 1061–1070.

Branduardi, P., Longo, V., Berterame, N. M., Rossi, G., and Porro, D. 2013. A novel pathway to produce butanol and isobutanol in *Saccharomyces cerevisiae*. *Biotechnol. Biofuels.* 6(1): 68–68.

Cagol, L., Baldisserotto, B., Becker, A. G., Souza, C. D. F., and Ballester, E. L. C. 2020. Essential oil of *Lippia* alba in the diet of Macrobrachium rosenbergii: Effects on antioxidant enzymes and growth parameters. *Aquaculture Res.* 51(6): 2243–2251.

Cavanagh, E. M. V. D., Inserra, F., León Ferder, Romano, L., Ercole, L., and César G. Fraga. 1995. Superoxide dismutase and glutathione peroxidase activities are increased by enalapril and captopril in mouse liver. *FEBS Letters.* 361(1): 0–24.

Chakma, P., and Konkolewicz, D. 2019. Dynamic covalent bonds in polymeric materials. *Angewandte Chemie.* 131(29): 9784–9797.

Chen, A. F., Chen, D., Daiber, A., Faraci, F. M., Li, H., Rembold, C. M., and Laher, I. 2012. Free radical biology of the cardiovascular system. *Clin. Sci.* 123(2): 73–91.

Chrysargyris, A., Evangelides, E., and Tzortzakis, N. 2021. Seasonal variation of antioxidant capacity, phenols, minerals and essential oil components of sage, spearmint and sideritis plants grown at different altitudes. *Agronomy.* 11(9): 1766.

Coimbra, A., Ferreira, S., and Duarte, A. P. 2022. Biological properties of *Thymus zygis* essential oil with emphasis on antimicrobial activity and food application. *Food Chem.* 133370.

Collard, K. J., Godeck, S., Holley, J. E., and Quinn, M. W. 2004. Pulmonary antioxidant concentrations and oxidative damage in ventilated premature babies. *Arch. Dis. Child.* 89(5): 241–242.

Dang, M. N., Takacsova, M., and Nguyen, D. V. 2000. The influence of extracts and essential oils from various spices on the oxidation stability of lard. *Czech J. Food Sci. UZPI (Czech Republic).* 23(2): 421–435.

Daniel, A. N., Sartoretto, S. M., Schmidt, G., Caparrozassef, S. M., Bersaniamado, C. A., and Cuman, R. K. 2009. Anti-inflammatory and antinociceptive activities A of eugenol essential oil in experimental animal models. *Res. Rev.: J. Pharmacogn. Phytochem.* 212–217.

Das, M., Roy, S., Asit, K. S., Chandan, G., and Singh, M. 2020. In vitro evaluation of antioxidant and antibacterial properties of supercritical co 2 extracted essential oil from clove bud (*Syzygium aromaticum*). *J. Plant Biochem. Biotechnol.* 1–5.

Debonne, E., Van Bockstaele, F., De Leyn, I., Devlieghere, F., and Eeckhout, M. 2018. Validation of in-vitro antifungal activity of thyme essential oil on *Aspergillus niger* and Penicillium paneum through application in par-baked wheat and sourdough bread. *LWT-Food Sci. Technol.* 87: 368–378.

Domínguez, R., Pateiro, M., Gagaoua, M., Barba, F. J., Zhang, W., and Lorenzo, J. M. 2019. A comprehensive review on lipid oxidation in meat and meat products. *Antioxidants.* 8(10): 429.

Fabbri, J., Maggiore, M. A., Pensel, P. E., Guillermo, M. D., and María, C. E. 2020. In vitro efficacy study of *Cinnamomum zeylanicum* essential oil and cinnamaldehyde against the larval stage of echinococcus granulosus. *Exp. Parasitol.* 214: 107904.

Falleh, H., Jemaa, M. B., Saada, M., and Ksouri, R. 2020. Essential oils: A promising eco-friendly food preservative. *Food Chem.* 330: 127268.

Farhadi, N., Babaei, K., Farsaraei, S., Moghaddam, M., and Pirbaloti, A. G. 2020. Changes in essential oil compositions, total phenol, flavonoids and antioxidant capacity of achillea millefolium at different growth stages. *Ind. Crops Prod.* 112570.

Fasiku, V., Andeve, C., and Govender, T. 2020. Free radical-releasing systems for targeting biofilms. *J. Controlled Release.* 322(10): 248–273.

Fatima, R., Addí, R., Navarro-Cruz, C. E., and Ochoa, V. 2019. Essential oils in vapor phase as alternative antimicrobials: A review. *Crit. Rev. Food Sci. Nutr.* 140–155.

Fierascu, R. C., Ortan, A., Fierascu, I. C., and Fierascu, I. 2018. In vitro and in vivo evaluation of antioxidant properties of wild-growing plants. A short review. *Curr. Opin. Food Sci.* 1–8.

Fournomiti, M., Kimbaris, A., Mantzourani, I., Plessas, S., Theodoridou, I., Papaemmanouil, V., and Alexopoulos, A. 2015. Antimicrobial activity of essential oils of cultivated oregano (*Origanum vulgare*), sage (Salvia officinalis), and thyme (*Thymus vulgaris*) against clinical isolates of *Escherichia coli*, Klebsiella oxytoca, and *Klebsiella pneumoniae.*" *Microb. Ecol. Health Dis.* 23289–23289.

Gandhi, G. R., Vasconcelos, A. B., Haran, G. H., Calisto, V. K., Jothi, G., Quintans, J. S., and Gurgel, R. Q. 2019. Essential oils and its bioactive compounds modulating cytokines: A systematic review on anti-asthmatic and immunomodulatory properties. *Phytomedicine.* 73: 152854.

Gholivand, M. B., Rahimi-Nasrabadi, M., Batooli, H., and Ebrahimabadi, A. H. 2010. Chemical composition and antioxidant activities of the essential oil and methanol extracts of Psammogeton canescens. *Food Chem. Toxicol.* 48(1): 01–28.

Gulcin, İ. 2020. Antioxidants and antioxidant methods: An updated overview. *Arch. Toxicol.* 94(3): 651–715.

Hamad, Y. K., Abobakr, Y., Salem, M. Z. M., Ali, H. M., and Alzabib, A. A. 2019. Activity of plant extracts/essential oils against three plant pathogenic fungi and mosquito larvae: GC/MS analysis of bioactive compounds. *Bioresources.* 14(2): 4489–4511.

Han, F., Ma, G., Yang, M., Yan, L., Xiong, W., Shu, J., and Xu, H. 2017. Chemical composition and antioxidant activities of essential oils from different parts of the oregano. *J. Agric. Sci. Technol. B*. 18(1): 79–84.

Harun, H., Daud, A., Hadju, V., Arief, C. P. P., and Mallongi, A. 2020. Antioxidant effect of moringa oleifera leaves in hemoglobin oxidation compare with vitamin c. *Enfermería Clínica*. 30: 18–21.

Hussain, A. I., Anwar, F., Sherazi, S. T., and Przybylski, R. 2008. Chemical composition, antioxidant and antimicrobial activities of basil (*Ocimum basilicum*) essential oils depends on seasonal variations. *Food Chem*. 108(3): 986–995.

Ibtisham, F., Nawab, A., Niu, Y., Wang, Z., Wu, J., and Xiao, M. 2019. The effect of ginger powder and Chinese herbal medicine on production performance, serum metabolites and antioxidant status of laying hens under heat-stress condition. *J. Therm. Biol*. 81: 20–24.

Ifeanyi, O. E. 2018. A review on free radicals and antioxidants. *Int. J. Curr. Res. Med. Sci*. 4(2): 123–133.

Ishkeh, S. R., Asghari, M., Shirzad, H., Alirezalu, A., and Ghasemi, G. 2019. Lemon verbena (*Lippia citrodora*) essential oil effects on antioxidant capacity and phytochemical content of raspberry (Rubus ulmifolius subsp. sanctus). *Sci. Hortic*. 297–304.

Jaganjac, M., Milkovic, L., Zarkovic, N., and Zarkovic, K. 2022. Oxidative stress and regeneration. *Free Radical Biol. Med*. 181: 154–165.

Jan, S., Rashid, M., Abd_Allah, E. F., and Ahmad, P. 2020. Biological efficacy of essential oils and plant extracts of cultivated and wild ecotypes of *Origanum vulgare* l. *BioMed Res. Int*. (1): 1–16.

Jia, Z., Wen, M., Cheng, Y., and Zheng, Y. 2021. Strategic advances in spatiotemporal control of bioinspired phenolic chemistries in materials science. *Adv. Funct. Mater*. 31(14): 2008821.

Ju, J., Guo, Y., Cheng, Y., and Yaoc, W. 2022. Analysis of the synergistic antifungal mechanism of small molecular combinations of essential oils at the molecular level. *Ind. Crops Prod*. 188: 115612.

Ju, J., Xie, Y., Guo, Y., Cheng, Y., Qian, H., and Yao, W. 2018. Application of starch microcapsules containing essential oil in food preservation. *Crit. Rev. Food Sci. Nutr*. 1–12.

Ju, J., Xie, Y., Guo, Y., Cheng, Y., Qian, H., and Yao, W. 2019. Application of edible coating with essential oil in food preservation. *Crit. Rev. Food Sci. Nutr*. 59(15): 2467–2480.

Ju, J., Xie, Y., Yu, H., Guo, Y., Cheng, Y., Chen, Y., and Yao, W. 2020a. Synergistic properties of citral and eugenol for the inactivation of foodborne molds in vitro and on bread. *LWT-Food Sci. Technol*. 122: 109063.

Ju, J., Xie, Y., Yu, H., Guo, Y., Cheng, Y., Qian, H., and Yao, W. 2020b. Analysis of the synergistic antifungal mechanism of eugenol and citral. *LWT-Food Sci. Technol*. 123: 109128.

Ju, J., Xie, Y., Yu, H., Guo, Y., Cheng, Y., Zhang, R., and Yao, W. 2020c. Major components in Lilac and *Litsea cubeba* essential oils kill *Penicillium roqueforti* through mitochondrial apoptosis pathway. *Ind. Crops Prod*. 149: 112349.

Ju, J., Xie, Y., Yu, H., Guo, Y., Cheng, Y., Zhang, R., and Yao, W. 2020d. A novel method to prolong bread shelf life: Sachets containing essential oils components. *LWT-Food Sci. Technol*. 23: 109744.

Ju, J., Xu, X., Xie, Y., Guo, Y., Cheng, Y., Qian, H., and Yao, W. 2018. Inhibitory effects of cinnamon and clove essential oils on mold growth on baked foods. *Food Chem*. 850–855.

Jummes, B., Sganzerla, W. G., Cleonice Gonçalves da Rosa, Noronha, C. M., and Barreto, P. L. M. 2020. Antioxidant and antimicrobial poly-ε-caprolactone nanoparticles loaded with *Cymbopogon martinii* essential oil. *Biocatal. Agric. Biotechnol*. 23: 101499.

Khalid, S., Shaheen, S., Hussain, K., Shahid, M. N., and Sarwar, S. 2020. Pharmacological analysis of obnoxious water weed: Eichhornia crassipes (Mart.) Solms. *JAPS: J. Anim. Plant Sci*. 30(6): 167–180.

Khodaei, N., Nguyen, M. M., Mdimagh, A., Bayen, S., and Karboune, S. 2021. Compositional diversity and antioxidant properties of essential oils: Predictive models. *LWT-Food Sci. Technol.* 138: 110684.

Koutsaviti, A., Toutoungy, S., Saliba, R., Loupassaki, S., Tzakou, O., Roussis, V., and Ioannou, E. 2021. Antioxidant potential of pine needles: A systematic study on the essential oils and extracts of 46 species of the genus Pinus. *Foods.* 10(1): 142.

Lampi, A. M., Yang, Z., Mustonen, O., and Piironen, V. 2019. Potential of faba bean lipase and lipoxygenase to promote formation of volatile lipid oxidation products in food models. *Food Chem.* 311: 125982.

Leicach, S. R., and Chludil, H. D. 2014. Plant secondary metabolites: Structure–activity relationships in human health prevention and treatment of common diseases. *Stud. Nat. Prod. Chem.* 42: 267–304.

Li, G., Xiang, S., Pan, Y., Long, X., Cheng, Y., Han, L., and Zhao, X. 2021. Effects of cold-pressing and hydrodistillation on the active non-volatile components in lemon essential oil and the effects of the resulting oils on aging-related oxidative stress in mice. *Front. Nutr.* 8: 329.

Lin, J., Niu, Z., Xue, Y., Gao, J., Zhang, M., Li, M., and Li, X. 2022. Chronic vitamin D3 supplementation alleviates cognition impairment via inhibition of oxidative stress regulated by PI3K/AKT/Nrf2 in APP/PS1 transgenic mice. *Neurosci. Lett.* 136725.

Liu, Y. C., He, Y., Wang, F., Xu, R., Yang, M., Ci, Z., and Lin, J. 2021. From longevity grass to contemporary soft gold: Explore the chemical constituents, pharmacology, and toxicology of Artemisia argyi H. Lév. & vaniot essential oil. *J. Ethnopharmacol.* 279: 114404.

Liu, Y. C., Yuan, Y, Weng, A. H. 2018. Study on the antioxidant and antibacterial effects of the essential oil of red onion in vitro. *Int. J. Food Sci.* 18(11): 246–252.

Lukina, Maria, V., and Popov, A. 2013. Dna damage processing by human 8-oxoguanine-DNA glycosylase mutants with the occluded active site. *J. Biol. Chem.* 288: 28936–28947.

Maqsoudlou, A., Assadpour, E., Mohebodini, H., and Jafari, S. M. 2020. Improving the efficiency of natural antioxidant compounds via different nanocarriers. *Adv. Colloid Interface Sci.* 119(17): 173–191.

Mendes, G. M., Rodriguesdasdores, R. G., Campideli, L. C., Mendes, G. M., Rodriguesdasdores, R. G., and Campideli, L. C. 2015. Evaluation of the content of antioxidants, flavonoids and phenolic compounds in culinary preparations. *Rev. Bras. Plantas Med.* 17: 636–649.

Mukurumbira, A. R., Shellie, R. A., Keast, R., Palombo, E. A., and Jadhav, S. R. 2022. Encapsulation of essential oils and their application in antimicrobial active packaging. *Food Control.* 108883.

Musakhanian, J., Rodier, J. D., and Dave, M. 2022. Oxidative stability in lipid formulations: A review of the mechanisms, drivers, and inhibitors of oxidation. *AAPS PharmSciTech.* 23(5): 1–30.

Nakano, Y., Biegasiewicz, K. F., and Hyster, T. K. 2019. Biocatalytic hydrogen atom transfer: An invigorating approach to free-radical reactions. *Curr. Opin. Chem. Biol.* 16–24.

Niocel, M., Appourchaux, R., Nguyen, X., Delpeuch, M., and Cimarelli, A. 2019. The DNA damage induced by the cytosine deaminase APOBEC3A leads to the production of ROS. *Sci. Rep.* 9(1): 4714–4728.

Olszowy, M. 2019. What is responsible for antioxidant properties of polyphenolic compounds from plants? *Plant Physiol. Biochem.* 144: 135–143.

Paciorek, P., Uberek, M., and Grzelak, A. 2020. Products of lipid peroxidation as a factor in the toxic effect of silver nanoparticles. *Materials.* 13(11): 2460.

Penabautista, C., Baquero, M., Vento, M., and Chaferpericas, C. 2019. Free radicals in Alzheimer's disease: Lipid peroxidation biomarkers. *Clinica Chimica Acta.* 85–90.

Perumal, A. B., Sellamuthu, P. S., Nambiar, R. B., and Sadiku, E. R. 2017. Effects of essential oil vapour treatment on the postharvest disease control and different defence responses in two mango (*Mangifera indica* L.) cultivars. *Food Bioprocess Technol.* 10(6): 1131–1141.

Pinto, E., Valesilva, L. A., Cavaleiro, C., and Salgueiro, L. 2009. Antifungal activity of the clove essential oil from *Syzygium aromaticum* on *Candida*, *Aspergillus* and dermatophyte species. *J. Med. Microbiol.* 58(11): 1454–1462.

Player, M. E., Kim, H. J., Lee, H. O., and Min, D. B. 2006. Stability of α-, γ-, or δ-Tocopherol during soybean oil oxidation. *J. Food Sci.* 71(8): C456–460.

Proestos, C., Lytoudi, K., Mavromelanidou, O. K., Zoumpoulakis, P., and Sinanoglou, V. J. 2013. Antioxidant capacity of selected plant extracts and their essential oils. *Antioxidants.* 2(1): 11–22.

Qin Ma, XiaoDan Fan, XiaoCao Liu, TaiQiu Qiu, and JianGuo Jiang. 2015. Ultrasound-enhanced subcritical water extraction of essential oils from Kaempferia galangal L. and their comparative antioxidant activities. *Sep. Purif. Technol.* 150: 73–79.

Rajesh, M., Cai, L., Mukhopadhyay, P., and Vedantham, S. 2016. Redox signaling and myocardial cell death: Molecular mechanisms and drug targets. *Oxid. Med. Cell. Longevity.* 3190753–3190753.

Rakass, S., Babiker, H. A., and Oudghirihassani, H. 2018. Comparative evaluation of total phenolic content, total flavonoids content and antioxidants activity in Skin & Pulp extracts of Cucurbita maxima. *Moroccan J. Chem.* 6(2): 26–46.

Rocamora, C. R., Ramasamy, K., Lim, S. M., Majeed, A. B. A., and Agatonovic-Kustrin, S. 2020. HPTLC based approach for bioassay-guided evaluation of antidiabetic and neuroprotective effects of eight essential oils of the Lamiaceae family plants. *J. Pharm. Biomed. Anal.* 178: 112909.

Roshan, A. B., Dubey, N. K., and Mohana, D. C. 2022. Chitosan nanoencapsulation of Pogostemon cablin (Blanco) Benth. essential oil and its novel preservative effect for enhanced shelf life of stored Maize kernels during storage: Evaluation of its enhanced antifungal, antimycotoxin, antioxidant activities and possible mode of action. *Int. J. Food Sci. Technol.* 57(4): 2195–2202.

Rotariu, D., Babes, E. E., Tit, D. M., Moisi, M., Bustea, C., Stoicescu, M., and Bungau, S. G. 2022. Oxidative stress—Complex pathological issues concerning the hallmark of cardiovascular and metabolic disorders. *Biomed. Pharmacother.* 152: 113238.

Sanada, M., Kuroda, K., and Ueda, M. 2011. Ros production and apoptosis induction by formation of gts1p-mediated protein aggregates. *Biocatal. Pharm. Biotechnol. Ind.* 75(8): 1546–1553.

Shadyro, O. I., Samovich, S. N., and Edimecheva, I. P. 2019. Free-radical and biochemical reactions involving polar part of glycerophospholipids. *Free Radical Biol. Med.* 6–15.

Shadyro, O. I., Sosnovskaya, A. A., and Edimecheva, I. P. 2017. Flaxseed oil stabilization using natural and synthetic antioxidants. *Eur. J. Lipid Sci. Technol.* 119(10): 1700079.

Shao, Q., Deng, Y., Liu, H., Zhang, A., Huang, Y., Xu, G., and Li, M. 2014. Essential oils extraction from Anoectochilus roxburghii using supercritical carbon dioxide and their antioxidant activity. *Ind. Crops Prod.* 104–112.

Siler, B., Živkovic, S., Banjanac, T., Cvetkovic, J., Živkovic, J. N., Ciric, A., and Misic, D. 2014. Centauries as underestimated food additives: Antioxidant and antimicrobial potential. *Food Chem.* 367–376.

Silvestre, W. P., Livinalli, N. F., Baldasso, C., and Tessaro, I. C. 2019. Pervaporation in the separation of essential oil components: A review. *Trends Food Sci. Technol.* 93: 42–52.

Souri, M. K., and Bakhtiarizade, M. 2019. Biostimulation effects of rosemary essential oil on growth and nutrient uptake of tomato seedlings. *Scientia Horticulturae.* 243: 472–476.

Srinivas, U. S., Tan, B. W., Vellayappan, B. A., and Jeyasekharan, A. D. 2019. ROS and the DNA damage response in cancer. *Redox Biology.* 25: 101084.

Starowicz, M., Lelujka, E., Ciska, E., Lamparski, G., Sawicki, T., and Wronkowska, M. 2020. The application of Lamiaceae Lindl. promotes aroma compounds formation, sensory properties, and antioxidant activity of oat and buckwheat-based cookies. *Molecules*. 25(23): 5626.

Szumny, A., Figiel, A., Gutierrezortiz, A., and Carbonellbarrachina, A. A. 2010. Composition of rosemary essential oil (*Rosmarinus officinalis*) as affected by drying method. *J. Food Eng*. 97(2): 253–260.

Tiwary, M., Naik, S. N., Tewary, D. K., Mittal, P. K., and Yadav, S. 2007. Chemical composition and larvicidal activities of the essential oil of *Zanthoxylum armatum* DC (Rutaceae) against three mosquito vectors. *J Vector Borne Dis*. 44(3): 198–204.

Trojakova, L., Reblova, Z., Nguyen, H. T., and Pokornya, J. 2001. Antioxidant activity of rosemary and sage extracts in rapeseed oil. *J. Food Lipids*. 8(1): 1–13.

Uchida, S., Ohta, H., Edamatsu, R., Hiramatsu, M., and Ozaki, M. 1988. Active oxygen free radicals are scavenged by condensed tannins. *Adv. Res. Pharm. Biol.* 280: 135–138.

Vircheva, S., Alexandrova, A., Georgieva, A., Mateeva, P., Zamfirova, R., Kubera, M., and Kirkova, M. 2010. In vivo effects of pentoxifylline on enzyme and non-enzyme antioxidant levels in rat liver after carrageenan-induced paw inflammation. *Cell Biochem. Funct*. 28(8): 668–672.

Vital, A. C. P., Guerrero, A., Monteschio, J. D. O., Valero, M. V., Carvalho, C. B., de Abreu Filho, B. A., and Do Prado, I. N. 2016. Effect of edible and active coating (with rosemary and oregano essential oils) on beef characteristics and consumer acceptability. *PloS One*. 11(8): e0160535.

Wang, W. T., Fan, M. L., Hu, J. N., Sha, J. Y., Zhang, H., Wang, Z., and Li, W. 2022. Maltol, a naturally occurring flavor enhancer, ameliorates cisplatin-induced apoptosis by inhibiting NLRP3 inflammasome activation by modulating ROS-mediated oxidative stress. *J. Funct. Foods*. 94: 105127.

Wang, W. T., Wang, W., and Azadzoi, K. M. 2016. Alu RNA accumulation in hyperglycemia augments oxidative stress and impairs e NOS and SOD2 expression in endothelial cells. *Mol. Cell. Endocrinol*. 426: 91–100.

Wickens, D. G., Wilkins, M. H., Lunec, J., Ball, G., and Dormandy, T. L. 1981. Free-radical oxidation (peroxidation) products in plasma in normal and abnormal pregnancy. *Ann. Clin. Biochem*. 18(3): 158–162.

Wu, H., Richards, M. P., and Undeland, I. 2022. Lipid oxidation and antioxidant delivery systems in muscle food. *Compr. Rev. Food Sci. Food Saf*. 21(2): 1275–1299.

Yin, L., Xu, L., Chen, B., Zheng, X., Chu, J., Niu, Y., and Ma, T. 2022. SRT1720 plays a role in oxidative stress and the senescence of human trophoblast HTR8/SVneo cells induced by D-galactose through the SIRT1/FOXO3a/ROS signalling pathway. *Reprod. Toxicol*. 111: 1–10.

Yu, S., Long, Y., Li, D., Shi, A., Deng, J., Ma, Y., and Guo, J. 2022. Natural essential oils efficacious in internal organs fibrosis treatment: Mechanisms of action and application perspectives. *Pharmacol. Res*. 106339.

Yu, Z., Tang, J., Khare, T., and Kumar, V. 2019. The alarming antimicrobial resistance in eskapee pathogens: Can essential oils come to the rescue? *Fitoterapia*. 140: 104433.

Zaid, H., Silbermann, M., Amash, A., Gincel, D., Abdelsattar, E., and Sarikahya, N. B. 2017. Medicinal plants and natural active compounds for cancer chemoprevention/chemotherapy. *Evid.-based Complement. Altern. Med*. 7952417–7952417.

Zhan, C. D., Sindhu, R. K., Pang, J., Ehdaie, A., and Vaziri, N. D. 2004. Superoxide dismutase, catalase and glutathione peroxidase in the spontaneously hypertensive rat kidney: Effect of antioxidant-rich diet. *J. Hypertension*. 22(10): 2025–2033.

Zhang, S., Xu, B., and Gan, Y. 2019. Seed treatment with Trichoderma longibrachiatum T6 promotes wheat seedling growth under NaCl stress through activating the enzymatic and nonenzymatic antioxidant defense systems. *Int. J. Mol. Sci*. 20(15): 3729.

7 Antiviral Activity and Mechanism of Essential Oils

7.1 INTRODUCTION

In the process of the development of civilization, human beings are constantly plagued by viral infectious diseases. Virus infection has the characteristics of large quantity, strong infectivity, wide spread, high fatality rate, and lack of effective prevention (Gorbalenya and Lauber, 2022; Pezzi et al., 2020). From the large-scale outbreak of smallpox virus in mid-20th century to the Middle East respiratory syndrome (MERS) caused by coronavirus to the global pneumonia epidemic caused by novel coronavirus, it has posed a great threat to human health.

In 2020, SARS-CoV-2 swept through most countries around the world. It has dealt a major blow to world trade, tourism, and economic development, while posing a serious threat to human life and health. Coronaviruses are a large family of viruses and are known to cause the common cold, as well as more serious diseases such as Middle East respiratory syndrome and severe acute respiratory syndrome (SARS) (Gharebaghi et al., 2020). SARS-CoV-2 is a new coronavirus that has never previously been found in humans (Figure 7.1) (Naqvi et al., 2020). The virus mainly infects cells in the lungs and upper respiratory tract, resulting in severe pneumonia, respiratory distress, and in some cases death. At present, there is no specific treatment for the virus and most therapies are mainly aimed at symptomatic relief (Pawar). According to Worldometers real-time statistics, SARS-CoV-2 caused a total of 6 million deaths and 460 million infections by April 14, 2022, and these totals are increasing rapidly because of the high cost and long time required to develop new drugs and the seriousness of the pandemic.

Therefore, it is valuable to propose simple, effective, and available treatments to slow down the spread of the disease. This could have a significant impact on the control of SARS-CoV-2 pandemic. Traditional medicine is a reliable source for the development of drugs to treat new diseases. It seems reasonable to focus on long-term traditional medicines and medicinal plants (Bahramsoltani and Rahimi, 2020; Chinsembu, 2020). Traditional medicine can provide advice on the prevention and treatment of new diseases such as the SARS-CoV-2 pandemic (Mahroozade et al., 2020; Mi et al., 2020). Although the epidemic occurred in the 21st century, the manifestation of coronary disease is similar to that of obstruction described in traditional Persian medicine. Several medicinal plants have appeared in TPM textbooks as anti-obstruction drugs. The combination of herbal medicine and modern medicine represents the effectiveness of herbal medicine in SARS-CoV-2's management (Lin et al., 2020; My et al., 2021).

DOI: 10.1201/9781003329268-7

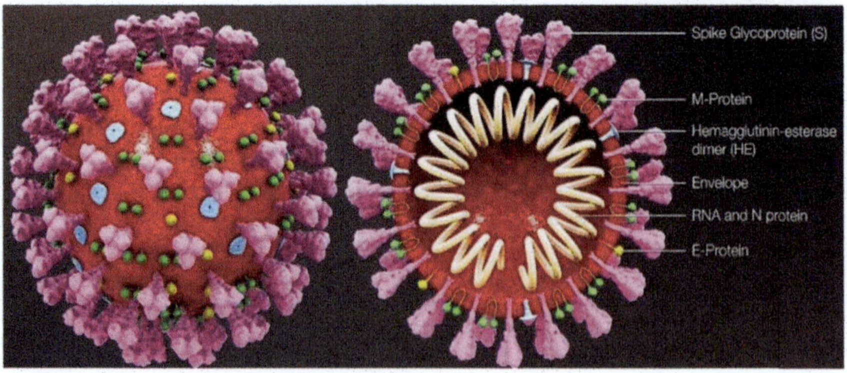

FIGURE 7.1 Schematic diagram of SARS- CoV-2 structure.

Plant essential oil is a natural mixture containing a variety of bioactive components. Many plant essential oils have been classified as generally recognized safe substances by the Food and Drug Administration (FDA) (generally regarded as safe [GRAS]). The biological characteristics and aroma of EOs are derived from the terpenoid and phenylpropanoid classes of natural products, which are their main bioactive components (Figure 7.2) (Jian et al., 2018; Ju et al., 2020).

At present, studies have confirmed that plant essential oils or their active components have antiviral activity against herpes simplex virus type 1 (HSV-1), herpes simplex virus type 2 (HSV-2), influenza A/PR/8 virus (H1N1 subtype), infectious bronchitis virus (IBV), and severe acute respiratory syndrome coronavirus (SARS-CoV) (Yang et al., 2022; Wani et al., 2021; Mendoza et al., 2022). These studies showed that plant essential oils have very important research value and application potential in antivirus.

7.2 AN OVERVIEW OF VIRUSES

A virus is a small, simple, acellular organism that contains a single nucleic acid (DNA or RNA) and must be parasitic in living cells and replicate (Ma et al., 2021). The virus is mainly composed of internal nucleic acid and external protein shell. Generally speaking, according to whether the virus has envelope or not, the virus can be divided into two types: enveloped virus (envelope virus) and unenveloped virus (nonenveloped virus). Among them, the enveloped virus is lipophilic. Similarly, plant essential oils are also lipophilic. Therefore, the effect of essential oil on envelope virus is stronger (Tang et al., 2020). Because the unenveloped virus is hydrophilic, the drug preparation which is soluble in water has a better killing effect on the unenveloped virus. At the same time, existing studies have shown that plant essential oil can effectively resist the virus before the adsorption stage, but not after the adsorption stage, which indicates that the essential oil may interact with the envelope virus before its adsorption, thus playing an antiviral role (Koch et al., 2008).

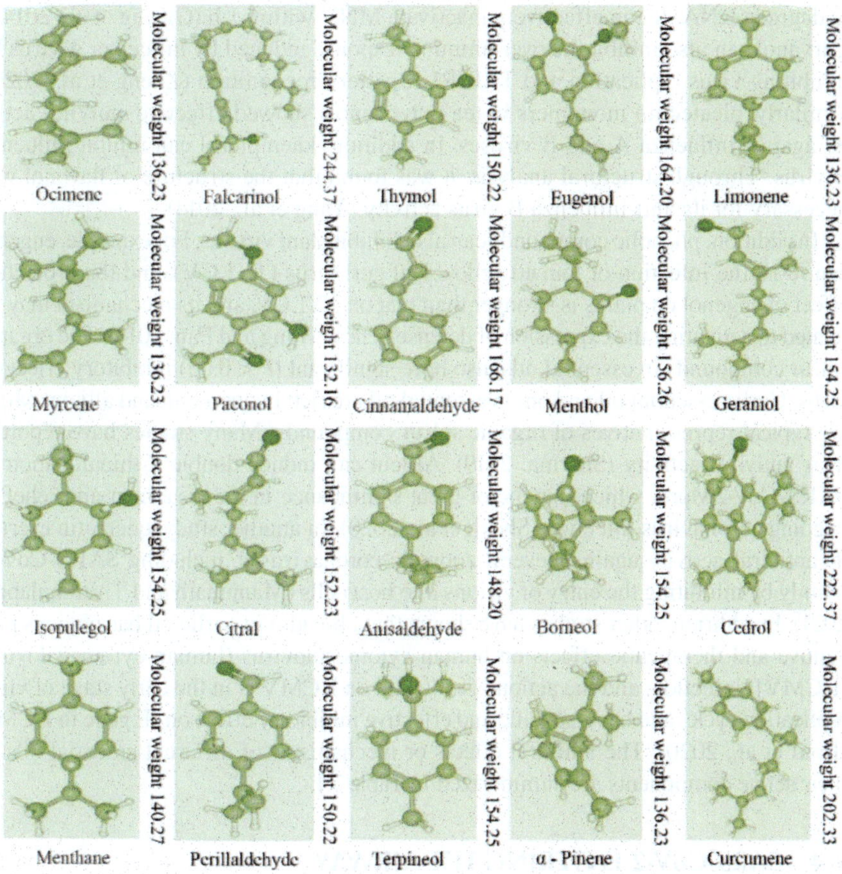

FIGURE 7.2 Structures of the main active components of essential oils.

7.3 PLANT ESSENTIAL OILS WITH ANTIVIRAL ACTIVITY AND THEIR ACTIVE COMPONENTS

Monoterpenes in plant essential oils have strong antiviral activity and can act directly on viruses, including monoterpene alcohols and monoterpene aldehydes. Eucalyptus alcohol is a monoterpene alcohol, which treats respiratory virus infection by synergism with other components in plant essential oils (Li et al., 2017). Linalool can be used to treat lower respiratory tract infection caused by virus (Junior et al., 2021). Sesquiterpenoids (*trans*-anisole, β-caryophyllene, and β-epoxy caryophyllene) from the fennel essential oil can directly act on HSV, the inhibition rate is 40–99%, and the selectivity index (SI) is 140 (Pathak et al., 2017). At the same time, phenylpropanoid and sesquiterpene in essential oil also contribute to antivirus, and its mechanism may be to interfere with the envelope structure of virion or mask the structure of virus (Saddi et al., 2007). Carvanol can directly act on the capsid of MNV, which increases the diameter of capsid, expands, and disintegrates. After the capsid broke, it began to degrade RNA. It can effectively inactivate MNV within 1 h (Gilling et al., 2014a). Carvanol can also inhibit the over immune response induced by influenza A virus by inhibiting virus replication and TLR/RLR pattern recognition (Zheng et al., 2020). Similarly, bicatechin monomers in tea polyphenols showed effective antiviral activity against influenza A and B viruses. In addition, kaempferol can inhibit influenza B virus. Through structural analysis, it was found that the structure of flavonol was necessary for its anti-influenza B virus activity (Yang et al., 2014).

In addition, phenolic compounds can also inhibit plant viruses. For example, eugenol can resist the infection of tomato yellow leaf curl virus (TYLCV), and the protective effect of eugenol on plants is stronger than that on TYLCV, and its mechanism may be related to activating the expression of defense genes (Wang and Fan, 2014). The organic sulfur compounds in essential oils also have significant ($P < 0.05$) inhibitory effect on virus. Garlic essential oil and onion essential oil are rich in quercetin and allicin, which are typical representatives of organic sulfur compounds. Many studies have reported their antiviral activity (Sharma, 2019). Allicin can induce double *S*-thioallylation of SARS-CoV-2Mpro, which may be of great significance in the treatment and relief of ongoing coronavirus infection (Shekh et al., 2020). In another study, quercetin exerted its antiviral activity against several zoonotic coronaviruses, including SARS-CoV-2, mainly by inhibiting the entry of virions into host cells (Manjunath and Thimmulappa, 2021). In addition, other studies have shown that allitridin in garlic oil has obvious preventive and therapeutic effects on human cytomegalovirus (human cytomegalovirus [HCMV]) infection, and the action of allitridin on HCMV is in the early stage of virus replication cycle, and has a specific and effective inhibitory effect on IE gene in HCMV (Rouf et al., 2020). The antiviral effects or mechanisms of different essential oils or their active components are summarized in Table 7.1.

7.4 SARS-CoV-2 ENTERING THE AIRWAY

The binding of SARS-CoV-2 to host cell receptor is the first step of virus infection, which determines the severity and pathogenic mechanism of virus infection. A better understanding of viral structural proteins and targeted receptors on host cells is helpful for the treatment of infection. SARS-CoV-2 infection begins when virions attach to the host cell surface through glycosylated S protein, a trimer class I fusion protein composed of two main subunits: receptor binding domain S1 and S2. When S1 binds to the host cell receptor, the fusion process between the viral membrane and the host cell membrane is triggered, such as angiotensin-converting enzyme 2 (ACE2) of SARS-CoV-2 (Kirtipal et al., 2020).

ACE2 receptors are widely distributed in trachea, alveolar epithelial cells, and macrophages (Kuba et al., 2005). When the virus attacks these cells, mature virions are released to other target cells, forming a new infection. ACE2 is also widely expressed in arteriovenous endothelial cells, brain neurons, immune cells, renal tubular epithelial cells, intestinal mucosal cells, and renal tubular epithelial cells. It is a variety of susceptible targets for SARS-CoV-2 infection. Unlike SARS-CoV-2, however, MERS-CoV targets the respiratory tract, liver, small intestine, kidneys, prostate, and activated white blood cells (Widagdo et al., 2016). When the lymphovirus is attached to the surface of the host cell, the virus enters the cell in two different

TABLE 7.1
Antiviral Effect or Mechanism of Different Essential Oils or Their Active Components

Source of EOs	Major active component	Virus type	Inhibition rate or IC50	Way or mode of action	Reference
Zanthoxylum myriacanthum var. *pubescens*	L-Limonene	Influenza A and B viruses	EC_{50} of 1.87 ± 1.04 μg/ml	Inhibition of influenza virus during intracellular replication phase	Yang et al. (2022)
L. alba	β-Caryophyllene	Zika virus	IC = 32.2 μg/ml; LC = 73.4 ± 1.68 μg/ml	Inhibit virus replication in the early stages of the virus cycle	Sobrinho et al. (2021)
Hornstedtia bella Škorničk	–	Vaccinia virus	EC values 80 μg/ml	Inhibit virus replication	Sanna et al. (2021)
Thymus vulgaris	–	Feline coronavirus	27 μg/ml	Inhibit virus replication	Catella et al. (2021)
Lavandula angustifolia L. and *Salvia officinalis* L.	–	H5N1 virus	IC50 0.11 ± 0.01 μg/ml and 0.41 ± 0.02 μg/ml	Inhibit virus replication	Abou Baker et al. (2021)
Lippia multiflora, *Zingiber officinale*		PV-1 and enterovirus type 1	9 and 9.077 μg/ml	Inhibition of protein formation and viral replication	José-Rita et al. (2022)
Oregano essential oil	Carvacrol	MNV	Exposure for 15 min reduced by 0.95-log10	It acts directly on the viral capsid and then on RNA	Gilling et al. (2014b)
Fennel essential oil	Eugenol	TYLCTHV	The incidence rate was reduced by 75%	Inhibit the accumulation of DNA-A, so as to limit the infection of virus in seedlings	Weng et al. (2019)
Marigold essential oil	Z-β-Basilene dihydropentanone	Ca VMV	More than 90%	Unclear mechanism of action	Singh et al. (2002)
Orange peel essential oil	Limonene	TMV	Inhibition rate 76%	TMV is directly inactivated and may interfere with coat protein or inhibit capsid	Lu et al. (2013)
Lemon essential oil	Limonene	TMV	Inhibition rate 63.5%	The protective effect of inhibiting protein formation is weaker than that of treatment and inactivation	Lu et al. (2013)

(*Continued*)

TABLE 7.1
(Continued)

Source of EOs	Major active component	Virus type	Inhibition rate or IC50	Way or mode of action	Reference
Oregano essential oil	Carvacrol	BOHV-2	IC50 = 58.4 μg/ml	It can inhibit the virus before and after virus inoculation	Pilau et al. (2011)
Eucalyptus essential oil	–	AIV	Killing rate 100%	Unclear mode of action	Barbour et al. (2008)
Zinnia essential oil	Carvacrol	NDV	–	It mainly destroys the infectivity of the virus or inhibits the early proliferation cycle of the virus phase	Mohammadi et al. (2015)
Star anise essential oil	Phenylpropanoid	HSV-1	IC50 = 1 μg/ml	HSV is directly inactivated and may interfere with the envelope structure of virus particles or mask the virus structure	Allahverdiyev et al. (2013)

ways according to the availability of the host cell protease to activate the spike protein attached to the receptor (Figure 7.3).

In the first pathway, the virus invades the host cell as an endosome, and endocytosis is mediated by cage protein-dependent and independent endocytosis (Figure 7.3). This phenomenon led to the conformational change of the virus particles, and then the virus envelope fused with the inner body wall. In the second pathway, the direct invasion of virions into host cells is mediated by hydrolytic cleavage of receptor-attached spike proteins by transmembrane serine protease 2 (TMPRSS2) or transmembrane serine protease 11D (TMPRSS11D) on the surface of host cells. As first observed in SARS-Co-2, the S2 domain of spike protein completes the direct fusion between the virus and the plasma membrane (Benton et al., 2020).

7.5 THE MAIN ANTIVIRAL PATHWAY AND MECHANISM OF ESSENTIAL OIL AND ITS ACTIVE COMPONENTS

7.5.1 The Main Antiviral Ways of Essential Oils and Their Active Components

Virus infection is essentially a molecular infection, including not only the replication and proliferation of the virus in the host cell, but also the physiological process of

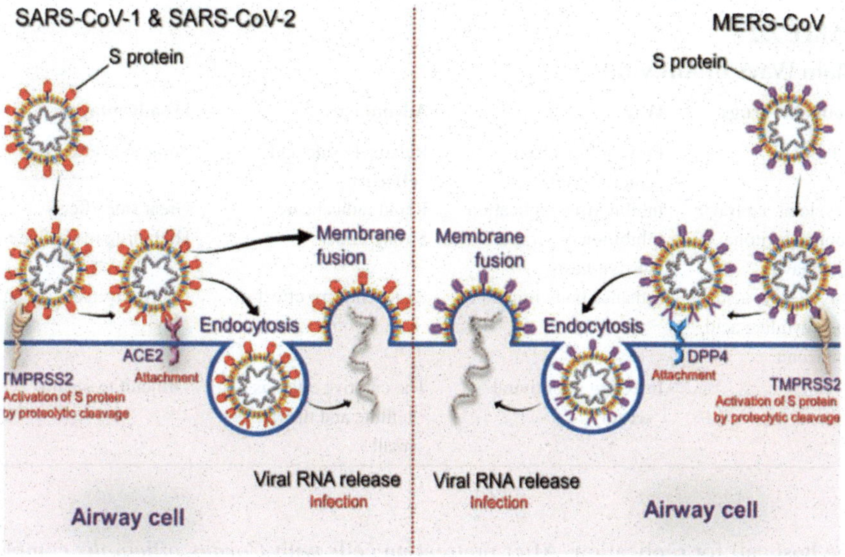

FIGURE 7.3 Schematic diagram of CoV attachment and entry into airway cells.

Source: Kirtipal et al. (2020).

the virus and the host producing immunity (Reichling et al., 2009). Based on the infection process and characteristics of the virus, the main ways of antivirus and the advantages and disadvantages of each approach are shown in Table 7.2.

7.5.2 Antiviral Mechanism of Essential Oil and Its Active Components

At present, the virus plaque reduction method is usually used in the laboratory to detect the stage and location at which plant essential oils may inhibit virus infection (Reichling et al., 2009). Generally, there are four stages of pretreatment (pretreatment of host cells with essential oil before virus infection), pretreatment of virus (pretreatment of virus with essential oil before infection), adsorption stage (addition of essential oil during adsorption), and intracellular replication (host cells are cultured with essential oil after virus invasion and replication). At the same time, different plant essential oils have different mechanisms of resistance to different viruses, but there are also some commonalities. For example, most of the antiviral effects of plant essential oils are before or during the adsorption of the virus, and a few after the virus has penetrated into the host cell. The main mode of action of plant essential oil is direct interaction with virus. We will describe the possible antiviral mechanism of essential oils from the following five aspects. The first is to protect the host cells. In the early stage of virus infection, after the cells were pretreated with essential oil, the small molecules of essential oil entered the host cells and took the lead in preempting various receptors on the cells. When the receptor is filled with essential oil molecules, the genetic information of the virus cannot be released into

TABLE 7.2
Main Ways of Antiviral

Common drugs	Way	Advantages	Disadvantages
Arbidol	Prevent virus from entering cells	Relatively safe and effective	Some side effects
Acyclovir, fabiravir	Inhibit virus replication	Rapid radical cure	Great side effects
Lopinavir ritonavir, darunavir	Inhibition of viral biosynthesis	Strong effect	High drug resistance rate
Polyinosinic acid, polycytidine acid, nitazonit	Enhance body immunity	Exact curative effect	Many adverse reactions
–	Injection of antiviral serum	The curative effect is definite and the risk is small	Difficult to obtain

the host cell for replication. After pretreating cells with *Cornus officinalis* complex for 24 h, the plaque decreased to less than 60% oregano and sage before respiratory syncytial virus (RSV) infection, but the treatment with compound essential oil for a short time or after the virus entered the cells did not show the effect of anti-RSV virus (Melpomeni et al., 2019). The second is to inactivate the virus. This is the mechanism of most plant essential oils inhibiting viruses, mainly because the lipophilic compounds in plant essential oils can interfere with the envelope function of the virus, disintegrate the envelope, and inactivate the virus. However, this kind of mechanism is only effective for enveloped viruses and ineffective for nonenveloped viruses. Thyme essential oil was mixed with human adenovirus (HAdV3) suspension for 2 h, and then monolayer cell culture was added. The results showed that the damage degree of the cells was significantly ($P < 0.05$) less than that of the positive control and the other three modes (protection, treatment, adsorption), and the virus titer decreased from $10^{5.33}$ $TCID_{50}$/ml to 10^4 $TCID_{50}$/ml (Abdelhakim et al., 2020). The third is to inhibit the growth of the virus. Inhibiting the replication and proliferation of the virus in the body or preventing the spread of the virus between cells is the main way to treat viral diseases because the virus can maintain its physiological activity only by continuously replicating itself and spreading to new host cells after infection. Citronellol, the main active component of citronella essential oil, can inhibit the activity of human immunodeficiency virus reverse transcriptase (HIV-1RT) in a dose-dependent manner, thus inhibiting the replication and reproduction of HIV virus. In addition, clove essential oil can inhibit HSV-1 replication in vitro. The *Sandalwood* essential oil can inhibit the replication of HSV virus in cells (Mohammadi et al., 2015). Similarly, *Ridolfia segetum* and *Oenanthecrocata* essential oils can affect HIV reverse transcriptase activity (Barbour et al., 2008; Pilau et al., 2011). The fourth is to inhibit the adsorption capacity of the virus. A small number of plant essential oils can interfere with the structure of virus particle film, thus inhibiting the adsorption of virus to host cells. Oregano essential oil had the most obvious effect on inhibiting virus adsorption to host cells, and the number of virus copies decreased by 10 orders of magnitude after 48 h (Melpomeni et al., 2019). After the action of lemon grass essential oil, the murine norovirus capsid remains intact and the genome is intact, but the infectivity of the virus is greatly reduced, which may be through the essential oil molecules to wrap the virus capsid to prevent the specific adsorption of the virus to the host cells, thus giving play to the antiviral effect (Gilling et al., 2014b). Finally, the essential oil can enhance the host's own immunity. After entering the human body, small molecules such as alcohols, phenols, and aldehydes contained in essential oils can promote bone marrow cells to divide neutrophils and increase the number of neutrophils, thus enhancing human immunity. There have been many reports on the regulation of the immune system by essential oils. For example, lavender and perilla EOs can enhance the immunity of rats through olfactory administration. After inhaling oil vapor, the rats' white blood cell count, lymphoid percentage, and red blood cell count all increased (Chen, 2019). The Cypress EOs significantly ($P < 0.05$) increases the activity of peritoneal macrophages and enhances nonspecific immunity in mice (Cook et al., 1987). Similarly, cinnamon EOs significantly ($P < 0.05$) improves the immune response of mice and the phagocytic function of their macrophages (El-Hack et al., 2020). The Qiai EOs can significantly ($P < 0.05$) increase the weight of the spleen and thymus, thicken the sebum, and increase lymphocyte counts in mice with low functional immunity induced by phosphoramide, as well as promoting humoral and cellular immunity (Huang et al., 2020). Perilla EOs significantly ($P < 0.05$) increases the spleen and thymus indices of mice and also increases the activity of macrophage acid phosphatase, thus enhancing immunity (Liang et al., 2018).

7.6 CONCLUDING REMARKS

In this chapter, we mainly introduced the plant essential oils with antiviral effects and their active components. The main antiviral pathway and mechanism of essential oil and its active components were analyzed. It has important guiding significance for understanding the antiviral activity of plant essential oils and promoting the application of plant essential oils or plant active components in the research and development of antiviral drugs.

At present, the inhibitory effect of plant essential oils on virus has been confirmed by a large number of studies. However, due to the complex chemical composition of essential oil, the antiviral mechanism of essential oil has not been systematically revealed. In particular, the research on essential oils against unenveloped viruses is still limited. In addition, new essential oil resources and more clinical trials are needed to further evaluate the antiviral mechanism of essential oils and their cost and applicability in practical applications. Of course, the application of plant essential oils in the antiviral field and the combined use of essential oils and antibiotics are also worthy of further study in the future.

ACKNOWLEDGEMENTS

This work was supported by the National Natural Science Foundation of China (32202192), Special fund for Taishan Scholars Project.

REFERENCES

Abdelhakim, B., Imane, C., Fatima, E. G., Taoufiq, B., Abdelaali, B., Nasreddine, El. O., Douae, T., Mohamed, El. S., and Naoual, El. M. 2020. Ethnomedicinal use, phytochemistry, pharmacology, and food benefits of thymus capitatus. *J. Ethnopharmacol.* 259: 112925.

Abou Baker, D. H., Amarowicz, R., Kandeil, A., Ali, M. A., and Ibrahim, E. A. 2021. Antiviral activity of *Lavandula angustifolia* L. and Salvia officinalis L. essential oils against avian influenza H5N1 virus. *J. Agric. Food Res.* 4: 100135.

Allahverdiyev, A. M., Bagirova, M., Yaman, S., Koc, R. C., Abamor, E. S., Ates, S. C., and Oztel, O. N. 2013. Development of new antiherpetic drugs based on plant compounds. In *Fighting multidrug resistance with herbal extracts, essential oils and their components* (pp. 245–259). Cambridge, MA: Academic Press.

Bahramsoltani, R., and Rahimi, R. 2020. An evaluation of traditional Persian medicine for the management of SARS-CoV-2. *Front. Pharmacol.* 11: 571434.

Barbour, E. K., Yaghi, R. H., Jaber, L. S., Shaib, H. A., and Harakeh, S. 2008. Safety and antiviral activity of essential oil against Avian Influenza and New Castle Disease viruses. *Int. J. Appl. Res. Vet. Med.* 8(1): 60–64.

Benton, D. J., Wrobel, A. G., Xu, P., Roustan, C., and Gamblin, S. J. 2020. Receptor binding and priming of the spike protein of SARS-CoV-2 for membrane fusion. *Nature.* 588: 327–330.

Catella, C., Camero, M., Lucente, M. S., Fracchiolla, G., Sblano, S., Tempesta, M., and Lanave, G. 2021. Virucidal and antiviral effects of *Thymus vulgaris* essential oil on feline coronavirus. *Res. Vet. Sci.* 137: 44–47.

Chen, S. 2019. Research progress on pathogenesis of osteoarthritis. *World Latest Medicine Information.* 46: 560–571.

Chinsembu, K. C. 2020. Coronaviruses and Nature's pharmacy for the relief of coronavirus disease 2019. *Revista Brasileira De Farmacognosia.* 30(5): 603–621.

Cook, R. M., Musgrove, N., and Ashworth, R. F. 1987. Activity of rat peritoneal eosinophils following induction by different methods. *Int. Arch. Allergy Immunol.* 83(4): 423–427.

El-Hack, M., Alagawany, M., Abdel-Moneim, A., Mohammed, N. G., and Elnesr, S. S. 2020. Cinnamon (*Cinnamomum zeylanicum*) oil as a potential alternative to antibiotics in poultry. *Antibiotics.* 9(5): 210–221.

Gharebaghi, R., Heidary, F., Moradi, M., and Parvizi, M. 2020. Metronidazole a potential novel addition to the COVID-19 treatment regimen. *SSRN Elect. J.* 8(1): e40.

Gilling, D. H., Kitajima, M., Torrey, J. R., and Bright, K. R. 2014a. Antiviral efficacy and mechanisms of action of oregano essential oil and its primary component carvacrol against murine norovirus. *J. Appl. Microbiol.* 116(5): 1149–1163.

Gilling, D. H., Kitajima, M., Torrey, J. R., and Bright, K. R. 2014b. Mechanisms of antiviral action of plant antimicrobials against murine norovirus. *J. Appl. Environ. Microbiol.* 80(16): 4898–4910.

Gorbalenya, A. E., and Lauber, C. 2022. Bioinformatics of virus taxonomy: Foundations and tools for developing sequence-based hierarchical classification. *Opinion in Virology.* 52: 48–56.

Huang, C., Wang, Y., Li, X., Ren, L., and Cao, B. 2020. Clinical features of patients infected with 2019 novel coronavirus in Wuhan, China. *The Lancet.* 395: 10223.

Jian, J., Yunfei, X., Yahui, G., Yuliang, C., He, Q., and Weirong, Y. 2018. The inhibitory effect of plant essential oils on foodborne pathogenic bacteria in food. *Crit. Rev. Food Sci. Nutr.* 59: 3281–3292.

José-Rita, B. J., Bertin, G. K., Ibrahime, S. K., Yannick, K., Erick-Kévin, B. G., Riphin, K. L., and Mireille, D. 2022. Study of the chemical and in vitro cytotoxic activities of essential oils (EOs) of two plants from the Ivorian flora (*Lippia multiflora*, *Zingiber officinale*) and their antiviral activities against non-enveloped viruses. *S. Afr. J. Bot.* 172: 110984.

Ju, J., Xie, Y., Yu, H., Guo, Y., and Yao, W. 2020. Synergistic interactions of plant essential oils with antimicrobial agents: A new antimicrobial therapy. *Crit. Rev. Food Sci. Nutr.* 2: 1–12.

Junior, G. B., Souza, C., Silva, H., Bianchini, A. E., and Baldisserotto, B. 2021. Combined effect of florfenicol with linalool via bath in combating Aeromonas hydrophila infection in silver catfish (Rhamdia quelen). *Aquaculture.* 545: 737247.

Kirtipal, N., Bharadwaj, S., and Kang, S. G. 2020. From SARS to SARS-CoV-2, insights on structure, pathogenicity and immunity aspects of pandemic human coronaviruses. *Infect., Genet. Evol.* 85: 104502.

Koch, C., Reichling, J., and Schneele, J. 2008. Inhibitory effect of essential oils against herpes simplex virus type 2. *Phytomedicine.* 15(1–2): 71–78.

Kuba, K., Imai, Y., Rao, S., Gao, H., Guo, F., Guan, B., Huan, Y., Yang, P., Zhang, Y., and Deng, W. 2005. A crucial role of angiotensin converting enzyme 2 (ACE2) in SARS coronavirus-induced lung injury. *Nat. Med.* 11(8): 875.

Li, Y., Xu, Y. L., Lai, Y. N., Liao, S. H., Liu, N., and Xu, P. P. 2017. Intranasal co-administration of 1,8-cineole with influenza vaccine provide cross-protection against influenza virus infection. *Phytomedicine.* 127–135.

Liang, Y. T., He, X. D., Guo, J. T., and Jing, H. J. 2018. Effects of N:P ratio of Artemisia ordosica on growth influenced by soil calcium carbonate. *Sci. Cold and Arid Regions.* 10(04): 333–340.

Lin, A., Song, E., Lee, H. W., and Lee, M. S. 2020. Herbal medicine for the treatment of coronavirus disease 2019 (COVID-19): A systematic review and meta-analysis of randomized controlled trials. *J. Clin. Med.* 9(5): 1583.

Lu, M., Han, Z., Xu, Y., and Yao, L. 2013. In vitro and in vivo anti-tobacco mosaic virus activities of essential oils and individual compounds. *J. Microbiol. Biotechnol.* 23(6): 771–778.

Ma, X. Y., Hill, B. D., Hoang, T., and Wen, F. 2021. Virus-inspired strategies for cancer therapy. *Semin. Cancer Biol.* 26: 10610.

Mahroozade, S., Kenari, H. M., Eghbalian, F., Ghobadi, A., and Yousefsani, B. S. 2020. Avicenna's points of view in epidemics: Some advice on coronavirus 2 (COVID-19). *J. Complementary Med. Res.* 28(2): 1–12.

Manjunath, S. H., and Thimmulappa, R. K. 2021. Antiviral, immunomodulatory, and anticoagulant effects of quercetin and its derivatives: Potential role in prevention and management of COVID-19. *J. Pharm. Anal.* 12(1): 29–34.

Melpomeni, T., Stergios, A. P., Christos, L., Elias, C., and George, S. 2019. Antiviral effect of an essential oil combination derived from three aromatic plants (Coridothymus capitatus (L.) Rchb. f., *Origanum dictamnus* L. and *Salvia fruticosa* Mill.) against viruses causing infections of the upper respiratory tract. *J. Herb. Med. Toxicol.* 17–23.

Mendoza, J. P. I. M., Ubillús, B. P. T., Bolívar, G. T. S., Palacios, R. D. P. C., Lopez, P. S. G. H., Rodríguez, D. A. P., and Koecklin, K. H. U. 2022. Antiviral effect of mouthwashes against SARS-COV-2: A systematic review. *The Saudi Dental J.* 34(3): 167–193.

Mi, A., Yk, A., Fm, A., and Mp, B. 2020. Persian medicine recommendations for the prevention of pandemics related to the respiratory system: A narrative literature review. *J. HerbMed Pharmacol.* 10(1): 100483.

Mohammadi, A., Mosleh, N., Shomali, T., Ahmadi, M., and Sabetghadam, S. 2015. In vitro evaluation of antiviral activity of essential oil from *Zataria multiflora* Boiss. against Newcastle disease virus. *J. HerbMed Pharmacol.* 4: 71–74.

My, A., Zt, A., Mma, A., Mav, B., Mv, C., and Ss, D. 2021. Cinnamon and its possible impact on COVID-19: The viewpoint of traditional and conventional medicine. *Biomed. Pharmacother.* 143: 112221.

Naqvi, A., Fatima, K., Mohammad, T., Fatima, U., and Hassan, M. I. 2020. Insights into SARS-CoV-2 genome, structure, evolution, pathogenesis and therapies: Structural genomics approach. *Biochimica et Biophysica Acta (BBA)- Molecular Basis of Disease*. 165878.

Pathak, M., Mandal, G. P., Patra, A. K., Samanta, I., Pradhan, S., and Haldar, S. 2017. Effects of dietary supplementation of cinnamaldehyde and formic acid on growth performance, intestinal microbiota and immune response in broiler chickens. *Anim. Prod. Sci.* 57: 15816.

Pezzi, L., Diallo, M., Rosa-Freitas, M. G., Vega-Rua, A., and Siqueira, A. M. 2020. GloPID-R report on chikungunya, o'nyong-nyong and Mayaro virus, part 5: Entomological aspects. *Antiviral Res.* 174: 104670.

Pilau, M. R., Alves, S. H., Weiblen, R., Arenhart, S., Cueto, A. P., and Lovato, L. T. 2011. Antiviral activity of the *Lippia graveolens* (*Mexican oregano*) essential oil and its main compound carvacrol against human and animal viruses. *Braz. J. Microbiol.* 42(4): 1616–1624.

Reichling, J., Schnitzler, P., Suschke, U., and Saller, R. 2009. Essential oils of aromatic plants with antibacterial, antifungal, antiviral, and cytotoxic properties--an overview. *Forschende Komplementrmedizin*. 16(2): 79–90.

Rouf, R., Uddin, S. J., Sarker, D. K., Islam, M. T., and Sarker, S. D. 2020. Anti-viral potential of garlic (*Allium sativum*) and it's organosulfur compounds: A systematic update of pre-clinical and clinical data. *Trends Food Sci. Technol.* 104: 219–234.

Saddi, M., Sanna, A., Cottiglia, F., Chisu, L., Casu, L., Bonsignore, L., and Logu, A. D. 2007. Antiherpevirus activity of *Artemisia arborescens* essential oil and inhibition of lateral diffusion in Vero cells. *Ann. Clin. Microbiol. Antimicrob.* 6(1): 10–18.

Sanna, G., Madeddu, S., Serreli, G., Nguyen, H. T., Le, N. T., Usai, D., and Donadu, M. G. 2021. Antiviral effect of Hornstedtia bella Škorničk essential oil from the whole plant against vaccinia virus (VV). *Nat. Prod. Res.* 35(24): 5674–5680.

Sharma, N. 2019. Efficacy of Garlic and Onion against virus. *Int. J. Res. Pharm. Sci. Technol.* 10(4): 3578–3586.

Shekh, S., Reddy, K. K. A., and Gowd, K. H. 2020. In silico allicin induced S-thioallylation of SARS-CoV-2 main protease. *J. Sulfur Chem.* (6489): 1–12.

Singh, B., Joshi, V. P., Ram, R., Sharma, A., and Zaidi, A. A. 2002. *U.S. Patent No. 6,444,458*. Washington, DC: U.S. Patent and Trademark Office.

Sobrinho, A. C. N., de Morais, S. M., Marinho, M. M., de Souza, N. V., and Lima, D. M. 2021. Antiviral activity on the Zika virus and larvicidal activity on the Aedes spp. of *Lippia alba* essential oil and β-caryophyllene. *Ind. Crops Prod.* 162: 113281.

Tang, B., Wang, X., Li, Q., Bragazzi, N. L., Tang, S., Xiao, Y., and Wu, J. 2020. Estimation of the transmission Risk of the 2019-nCoV and its implication for public health interventions. *J. Clin. Med.* 9(2): 462.

Wang, C., and Fan, Y. 2014. Eugenol enhances the resistance of tomato against tomato yellow leaf curl virus. *J. Sci. Food Agric.* 94(4): 677–682.

Wani, A. R., Yadav, K., Khursheed, A., and Rather, M. A. 2021. An updated and comprehensive review of the antiviral potential of essential oils and their chemical constituents with special focus on their mechanism of action against various influenza and coronaviruses. *Microb. Pathog.* 152: 104620.

Weng, S. H., Tsai, W. A., Lin, J. S., Tsai, W. S., and Chen, M. C. 2019. Priming of plant resistance to heat stress and tomato yellow leaf curl Thailand virus with plant-derived materials. *Front. Recent Dev. Plant Sci.* 10: 00906.

Widagdo, W., Raj, V. S., Schipper, D., Kolijn, K., Gjlh, V. L., Bosch, B. J., Bensaid, A., Segalés, J., Baumgärtner, W., and Adme, O. 2016. Differential expression of the middle east respiratory syndrome coronavirus receptor in the upper respiratory tracts of humans and dromedary camels. *J. Virol. Res.* 90(9): 4838–4842.

Yang, J., Zhao, L., Li, R., Yan, Y., Yin, J., Dai, Q., and Li, S. 2022. In vitro and in vivo antiviral activity of Maqian (*Zanthoxylum myriacanthum* var. pubescens) essential oil and its major constituents against strains of influenza virus. *Ind. Crops Prod.* 177: 114524.

Yang, Z. F., Bai, L. P., Huang, W. B., Li, X. Z., Zhao, S. S., Zhong, N. S., and Jiang, Z. H. 2014. Comparison of in vitro antiviral activity of tea polyphenols against influenza A and B viruses and structure-activity relationship analysis. *Fitoterapia.* 93: 47–53.

Zheng, K., Wu, S. Z., Lv, Y. W., Pang, P., and Chen, X. Y. 2020. Carvacrol inhibits the excessive immune response induced by influenza virus A via suppressing viral replication and TLR/RLR pattern recognition. *J. Ethnopharmacol.* 268: 113555.

8 The Repellent or Insecticidal Effect of Essential Oils Against Insects and Its Mechanism

8.1 INTRODUCTION

Botanical pesticides have been used against pests (Sousa et al., 2021; Pandey et al., 2021; Camilo et al., 2022) for centuries. For example, during the Roman Empire, the use of plant extracts as pesticides was common. However, after World War II, some widely used plant extracts were replaced by synthetic chemical pesticides. With the advent of synthetic pesticides, the use of botanical pesticides began to decline sharply. Later, people gradually recognized the adverse effects of chemical synthetic pesticides, such as environmental pollution, residues in food and feed, and insect resistance. This has led to growing interest in the use of plant extracts or essential oils. It is reported that more than 1,500 species of plants have been found to have insecticidal value (Rattan, 2010). Although compared with modern synthetic chemical insecticides, the efficacy of plant extracts is relatively low, but its relatively safe properties open up new prospects for the research of plant pesticides. Some common plants with deworming or insecticidal effects are shown in Figure 8.1.

Plants have evolved a variety of defense mechanisms to reduce insect attacks, including structural and induced ones. Unfortunately, insects have also evolved new systems to overcome these plant defenses. Chemical ecology is a discipline that studies how specific chemicals participate in the interaction between organisms and their surroundings. This chemical acts as a chemical defense and affects the molecular targets of herbivores or microbes. Mixtures of secondary metabolites may have a longer deterrent effect on insects and herbivores than single compounds. Different physical properties may allow for longer defense persistence. In addition, different authors have studied the mode and site of insecticidal activity of plant extracts or essential oils (Ibrahim et al., 2022; Boutjagualt et al., 2022; Johnson et al., 2022; Mattar et al., 2022; Narayanankutty et al., 2021).

Even though these plant extracts or essential oils represent an overall low risk to humans and the environment, strict regulations and low persistence still need to be overcome because of their volatility (Antonelli and Donelli, 2020). Therefore, new formulation and nanotechnology are being developed to improve the physical and

DOI: 10.1201/9781003329268-8

FIGURE 8.1 Plants with insect repellent or insecticidal effects. (a) *Artemisia argyi* Lév. & Vaniot. (b) *Lavandula angustifolia* Mill. (c) *Pelargonium hortorum.* (d) *Pyrethrum cinerariifolium* Trev. (e) *Murraya paniculata* L. Jack. (f) *Telosma cordata* (Burm. f.) Merr.

chemical stability and biological activity of essential oils to promote the development of these natural products as modern pesticides. The main purpose of this study is to analyze the active components with repellent or insecticidal activity in plant essential oils. On this basis, the repellent or insecticidal effects of plant essential oils were reviewed. Finally, the insecticidal mechanism of plant essential oil was explored. There are a lot of chemical defenses in nature, but little is known about how they act at the molecular level.

8.2 ACTIVE COMPONENTS IN PLANT ESSENTIAL OILS THAT HAVE INSECTICIDAL OR INSECTICIDAL EFFECTS

Before the development of modern pesticides, plant extracts containing pyrethroids and nicotine were widely used as pesticides in agriculture. With the increasing awareness of safety and environmental protection, chemical synthetic pesticides are required to be replaced by natural substitutes. Most secondary metabolites, such as terpenes and alkaloids, have been reported as candidates for insecticidal compounds and may be effective substitutes for pest control. In addition, linalool, camphor, safrole, borneol, citral, 1,8-cineole, neroli tertiary aldehyde, isoneroli tertiary aldehyde, farnesol, methyl eugenol, and α-terpineol in Lauraceae are of great economic value. All kinds of plant essential oils have toxic effects on different kinds of pests (Figure 8.2). For example, *C. austroindica* leaf oil and *trans*-anethole demonstrated superior contact and fumigant toxicities against adult beetles of *S. oryzae* and *T. castaneum*. *C. austroindica* leaf oil and *trans*-anethole showed potential fumigant toxicity against *S. oryzae* (lethal concentration 50 [LC50] 38.80 and 76.98 μl/l) and

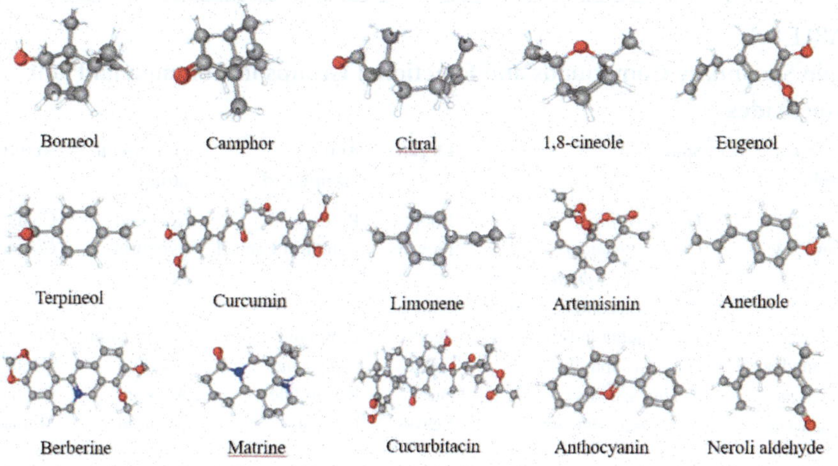

FIGURE 8.2 Active components in essential oil that can avoid or kill insects.

T. castaneum (LC50 35.65 and 29.10 μl/l) (Johnson et al., 2022). Fucus spiralis brown algae essential oil has significant toxicity to *Drosophila melanogaster*. The LD50 and LD90 are 0.239 and 2.467 ppm, respectively (Boutjagualt et al., 2022). Similarly, fennel essential oil has a significant ($P < 0.05$) killing effect on currant-lettuce aphid (Cantó-Tejero et al., 2022). At the same time, field experiments showed that the essential oil had no plant toxicity to any crops.

At present, *Cinnamomum camphora* L., *Cinnamomum burmannii*, *Cinnamomum cassia* Presl., *Laurus nobilis*, *Litsea cubeba*, *Lindera glauca*, *Lindera communis* Hemsl., *Alseodaphne hainanensis* Merr., and *Sassafras tzumu* (Hemsl.) Hemsl. are the main plant species studied more in insect repellent and insecticidal. In a review about plant essential oil, Naranjo et al. (2015) emphasized the complexity of pest management and its impact on ecology and economy. In addition to their role in chemical defense, plant extracts or essential oils can also be used as chemical signals to attract beneficial insects for pollination and seed transmission. For example, anthocyanins and monoterpenes can be used as insect attractants to attract insects for pollination, but may have insecticidal and antibacterial effects when anthocyanins and monoterpenes are present in leaves. However, in nontarget species, many compounds exhibit some useful biological activities. These compounds can be divided into three categories: terpenes, alkaloids, and phenols. The main secondary compounds and functional groups in commercial botanical pesticides are listed in Table 8.1.

8.2.1 Terpene Compound

Terpenes have been widely studied in the field of pesticides. Many terpenoids are secondary metabolites produced by plants and insects in the process of coevolution. They can not only attract pollinators to assist in plant pollination, but also resist some insects and help plants resist pests and pathogenic microorganisms (Pichersky

TABLE 8.1
Main Secondary Compounds and Functional Groups in Commercial Plant Insecticides

Botanical family	Species	Organ	Active compound	Functional group	Action
Fabaceae	*Derris elliptica*	Roots	Rotenone	Rotenoid	Contact
Asteraceae	*Chrysanthemum cinerariaefolium*, *C. roseus*, and *C. carreum*	Flowers	Pyrethrin I and II	Pyrethrin	Contact
Meliaceae	Azadirachta indica	Seeds	Limonoid	Azadirachtin	Contact
Amaranthaceae	Anabasis aphylla	Leaves	Anabasine	Alkaloid	Contact
Salicaceae	Ryania speciosa	Bark	Ryanodine	Alkaloid	Contact
Solanaceae	*Nicotiana tabacum*	Leaves	Nicotine	Alkaloid	Stomach poison
Piperaceae	*Piper nigrum*	Seeds	Piperine	Alkaloid	Stomach poison

and Gershenzon, 2002). Some monoterpenes such as limonene and geraniene have good insecticidal activity. In addition, some monoterpene derivatives, such as pyrethroids, are very efficient botanical pesticides. Some triterpenes such as limonins are also strong insect defenses and are often used as insecticides for crops. Other triterpenoids, such as cucurbitacin C and diterpene polygonum dialdehyde (polygodial), are also involved in plant insect resistance. However, terpene mixtures with different physical properties may be more toxic to insects and have longer defense ability. For example, mixtures of monoterpenes and diterpenes have synergistic inhibitory effects on some insects and pathogens. In addition, many terpenoids have certain insect repellent, antimalarial, and antibacterial effects, such as camphor, Ascaris driving hormone, artemisinin, andrographolide, etc. (Naimi et al., 2022; Zhou et al., 2022; Lim et al., 2021).

8.2.2 Phenolic Compounds

Phenolic compounds are characterized by the existence of a hydroxyl group (OH) on the benzene ring or other complex aromatic ring structure, for example, catechol, resorcinol, hydroquinone, pyrogallol, and so on. They range from simple phenols to polyphenols, for example, anthocyanins and tannins. Plant phenols have been proved to be one of the important defenses against insects in many studies (Puri et al., 2022; Naimi et al., 2022; Vachon et al., 2020). For example, different phenolic compounds were purified from the bark of the medicinally important plant *Acacia nilotica*, which is rich in polyphenols and were evaluated for their insecticidal potential against a polyphagous pest, *Spodoptera litura*. The results showed that various concentrations of these pure compounds had adverse effects on larval growth, survival rate, adult emergence rate, pupa weight, and different nutritional indexes. The hydrolyzable tannin in oak is a well-known phenolic substance that has a negative effect on the growth of gypsy moths. However, their specific mode of action is not clear.

8.2.3 Alkaloid

Alkaloids are alkaline, naturally occurring organic compounds that contain at least one nitrogen atom. These compounds also have insecticidal effects at low concentrations. Their modes of action vary, but many affect acetylcholine receptors (such as nicotine) or neuronal sodium channels (such as resveratrol) in the nervous system. Examples of insecticidal activities include *Nicotinia* spp., *Anabasine*, *Schoenocaulon officinale*, and *Ryania speciosa*. In addition, some studies have confirmed that matrine extracted from legumes has significant insecticidal and acaricidal activity, in which R as a long straight-chain aliphatic alkane is necessary for the insecticidal and acaricidal activity of *Sophora flavescens* (Huang et al., 2020). Quinolidine alkaloids isolated from *Sophora secundiflora* have mosquito killing effect. The median lethal concentration was 3.11 ppm after 24 h and 0.66 ppm after 48 h (Li et al., 2019). The lentil isolated from adzuki bean is used as a model compound for the development of carbamate insecticides. Alkaloids are abundant in Berberidaceae, Leguminosae, Solanaceae, and Maomaceae, all of which are widely used as traditional insect repellents (Pavela et al., 2019; Xiong et al., 2020).

8.3 INHIBITORY EFFECT OF PLANT ESSENTIAL OILS ON THE REPRODUCTIVE ABILITY OF INSECT PESTS

It has been confirmed that essential oils as aromatic substances have negative effects on the behavior of insects during oviposition (Lazarevi et al., 2020; Papanastasiou et al., 2020; Stepanycheva et al., 2019; Wu et al., 2020). Some experiments have been carried out and this phenomenon has been confirmed. Ostadi et al. (2020) evaluated the oviposition deterrence and ovicidal activity of peppermint essential oil to Chinese chelate. The results showed that the oviposition of adults was completely inhibited at the concentration of 1 μl/l, while the emergence of adults was completely inhibited at higher concentrations (200 μl/l). In another study, Bounoua-Fraoucene et al. (2019) tested *M. piperita* L. oil against *A. obtectus* and observed that the mean oviposition in female was reduced to 39% while mean number of adult emergence was decreased to 32%. Ansari et al. (2000) studied the oviposition and reproductive ability of female mosquitoes treated with different concentrations of pepper essential oil. The results showed that when the concentration was 2 ml/m^2, the reproduction and fertility of eclosion female adults decreased significantly.

8.4 REPELLENT OR INSECTICIDAL ACTIVITY OF PLANT ESSENTIAL OILS

Insect repellents work by providing a vapor barrier to prevent arthropods and other insects by touching the required surfaces. The repellent properties of plant essential oils and extracts have been well documented. However, most of these studies have focused on Coleoptera and Diptera pests. Gbolade et al. (2000) studied the repellency of five different concentrations (1%, 5%, 10%, 15%, 20%, and 25% v/v) of Nigerian clove basil and Thai sweet basil essential oils to adult *Aedes aegypti* in the laboratory. The results showed that the five different concentrations of essential oil solutions showed obvious anti-mosquito biting activity (anti-biting time 2–3 h). Among them, clove Basil essential oil showed stronger mosquito-repellent activity. In addition, based on the remarkable anti-mosquito bite activity of clove basil essential oil, the essential oil has been commercially developed as an insect repellent. Eleonora et al. (2021) determined the insecticidal activity of clove basil volatile oil against black fly, and the results showed that local use of 20% (v/w) essential oil could reduce the bite rate of black fly by more than 80%. Similarly, prenone and menthone also have significant insecticidal activity. Their lethal doses to rape and *Drosophila melanogaster* were 0.13 and 1.29 l/ml, respectively (Pavlidou et al., 2004). Among them, the killing effect of prenone on *D. melanogaster* larvae was better than that of menthol (Karpouhtsis et al., 1998). Other researchers studied the mosquito-repellent and insecticidal activity of basil essential oil at three different sites (World Bank Estate, Ihitte, and Umuekunne) in Imo state in eastern Nigeria. The volatile oil of basil was prepared with olive oil and palm oil into four different emulsions with concentrations of 20% and 30%, respectively, and then the mosquito-repellent and mosquito-killing activity of basil essential oil was evaluated. The results showed that local application of four different emulsions could significantly reduce the bite rate of mosquitoes in World Bank Estate area. Thirty percent (v/w) olive oil and palm kernel oil were found in three experimental centers to provide overnight protection and mosquito bites. Essential oils were found at the World Bank Estate experimental center to quickly knock down and paralyze some mosquitoes (Oparaocha et al., 2010). Therefore, basil, which grows in eastern Nigeria, has mosquito-repellent and mosquito-killing activity. The use of this essential oil formula can reduce the spread of human diseases caused by mosquito bites.

8.5 INSECTICIDAL MECHANISM OF PLANT ESSENTIAL OILS

The mode of action of essential oils or their components as pesticides is unclear. However, due to the observed repellent, antifeedant, and growth regulation, essential oils can obviously affect the physiology of insects in different ways. Essential oils and their components can affect biochemical processes, thus destroying the endocrine balance of insects (Rattan, 2010). It has been reported that essential oils play a role by reversibly competing to inhibit acetylcholinesterase (AChE). Inhibition of AChE can lead to the accumulation of acetylcholine at the synapse, which makes the synapse in a state of permanent stimulation, resulting in ataxia. That is, there is a general lack of coordination in the neuromuscular system, which eventually leads to the death of insects (Aygun et al., 2002). Ginger essential oil has been found to alter the behavior and memory of the cholinergic system (dos Reis Barbosa et al., 2019), and linalool has been identified as an inhibitor of acetylcholinesterase (Ryan and Byrne, 1988). Other alkaloid compounds berberine, palmatine, and sanguinarine are also toxic to insects and vertebrates. The main way of their toxic effect is to affect acetylcholinesterase, butylcholinesterase, choline acetyltransferase, α1 and α2 adrenergic, nicotinic, toxic and 5-hydroxytryptamine-2 receptors. However, there are different views at present. For example, Lee et al. (2001) found no direct link between insect toxicity and AChE inhibition. Similarly, menthone had strong toxicity to *Rhizoctonia solani*, but had relatively little inhibitory effect on AChE activity. However, the less toxic β-pinene showed a strong inhibitory effect on AChE (K_i 0.0028 mM). Therefore, it is suspected that in addition to the inhibition of AChE, monoterpenes may also act on other vulnerable sites (Lee et al., 2001).

Some active components in plant essential oils can also exert insecticidal effects by affecting protein phosphorylation (H-ATP: proton pump) or the activity of some enzyme systems (such as ATPase). Rotenone is a well-known and powerful mitochondrial poison, which acts by inhibiting the activity of mitochondria. Breviscapine showed insecticidal and bactericidal activity by inhibiting mitochondrial complex III (Khambay et al., 2003). Terpenoids are the main components of essential oils, which have been proved to have neurotoxic effects on houseflies and cockroaches. They also act as pheromones in insects, complicating the interaction between plants and insects (Ramadan et al., 2020; Kuriyama et al., 2002). Oxygen-containing monoterpenes (such as carvanol, linalool, and terpineol) are more toxic to adult longicorn beetles than nonoxidizing compounds (cymene, cinnamaldehyde, anethole), which may be due to the different modes of action of different components (Regnault-Roger and Hamraoui, 1994). In Table 8.2, we summarize the action mechanism of some botanical insecticides.

8.6 DEVELOPMENT OF RESISTANCE TO PLANT ESSENTIAL OILS OR PHYTOCHEMICALS

At present, the drug resistance of plant pests and microorganisms is still a worldwide problem. Although it has been suggested that plant extracts contain many compounds, they are more complex than synthetic pesticides, thus delaying the establishment of microbial drug resistance (Chamkhi et al., 2021; Zehra et al., 2021). In laboratory studies, it was found that *Plutea xylostella* L. developed resistance to all major synthetic pesticides quite quickly. However, after 42 generations of renewal iteration, it failed to develop resistance to Indonesian wood essential oil, which is due to the composition and complex mode of action of Indonesian wood essential oil. The combination of behavioral and physiological effects of botanical insecticides prevents the development of drug resistance. However, farmers often use large amounts of synthetic pesticides to control herbivorous insects, which leads to drug resistance. If the problem of insect resistance continues in the future, more pesticides and funds are needed to manage these agricultural pests. Therefore, people have been looking for novel insecticides as a means to deal with the problem of drug resistance. However, the existence of a variety of resistance mechanisms indicates that any kind of insecticides may produce resistance within a certain period of time. Of course, new plant-derived molecules with specific or multiple targets are less likely to develop drug resistance. The botanical insecticides developed in the future should be inactivated to insects and safe to human beings and the environment.

TABLE 8.2
Action Mechanism of Botanical Insecticides

S. No.	System	Mechanism of action	Plant source	Compound	Reference
1	GABA system	GABA-gated chloride channel	*Thymus vulgaris*	Thymol, silphinenes	Tavares et al. (2022)
2	Cholinergic system	Inhibition of acetylecholinestrase (AChE)	*Azadirachtina indica, Mentha* spp., *Lavendula* spp.	Essential oils	Alanazi et al. (2022)
		Cholinergic acetylcholine nicotinic receptor agonist/antagonist	*Nicotiana* spp., *Delphinium* spp., *Haloxylon salicornicum, Stemona japonicum*	Nicotine	Lloyd and Williams (2020)
3	Octopaminergic system	Octopaminergic receptors	*Cedrus* spp., *Pinus* spp., *Citronella* spp., *Eucalyptus* spp.	Essential oils	Isman (2020)
		Block octopamine receptors by working through tyramine receptors cascade	*Thymus vulgaris*	Thymol	Foudah et al. (2022)
	Miscellaneous	Hormonal balance disruption	*Azadiractina indica*	Azadirachtin	Kulkarni (2020)
4	Mitochondrial system	Sodium and potassium ion exchange disruption	*Crysanthemum cinerariaefolium*	Pyrethrin	Dabiri et al. (2021)
		Inhibitor of cellular respiration (mitochondrial complex I electron transport inhibitor or METI)	*Lonchocarpus* spp.	Rotenone	Khambay et al. (2003)
		Affect calcium channels	*Ryania* spp.	Ryanodine	Copping and Menn (2000)
		Affect nerve cell membrane action	*Schoenocaulon officinale*	Sabadilla	Weinzierl and Henn (2020)

8.7 CONCLUDING REMARKS

In this chapter, we describe in detail the efficacy of plant essential oils as insect repellents or insecticides. In particular, the insecticidal mechanism of plant essential oils was analyzed, which has important guiding significance for the future development of botanical insecticides. Sustainable growth in agriculture is essential for most developing countries to maintain a growing population. Due to the problems of resistance and residue of chemical synthetic insecticides, it is therefore necessary to develop safe alternatives. More and more researchers are reconsidering plant drugs containing active ingredients in an effort to solve these problems. Therefore, more targeted and biodegradable compounds are needed to replace chemicals with broad-spectrum toxicity.

A variety of mechanisms give unique properties to plant extracts or plant essential oils, making them very useful in today's agriculture and industry. For example, the development of imidacloprid and a series of new nicotine may be one of the greatest milestones in the history of pesticide chemistry. Understanding the mode of action of existing pesticides will help to determine the chemical properties of new compounds, which may be the molecular basis for future research and development of plant pesticides. Because botanical insecticides are unlikely to cause ecological damage, a large number of plants have been developed because of their insecticidal properties against various pests. In addition to all the advantages and safety of plant preparations compared with artificial synthesis, it will also bring necessary psychological satisfaction to farmers. At the same time, it will help to reduce the current overuse of synthetic pesticides. Plant active substances can be used as model compounds for the development of chemically synthetic derivatives with enhanced activity or environment-friendliness. Therefore, it will be particularly important to explore the insecticidal mechanism of plant active components in the future. Botanical insecticides can be used to develop new molecules with high-precision targets, which is of great significance for the sustainable development of ecological agriculture.

ACKNOWLEDGEMENTS

This work was supported by the National Natural Science Foundation of China (32202192), Special fund for Taishan Scholars Project.

REFERENCES

Alanazi, A. D., Ben Said, M., Shater, A. F., and Al-Sabi, M. N. S. 2022. Acaricidal, larvacidal, and repellent activity of *Elettaria cardamomum* essential oil against Hyalomma anatolicum ticks infesting Saudi Arabian cattle. *Plants*. 11(9): 1221.

Ansari, M. A., Vasudevan, P., Tandon, M., and Razdan, R. K. 2000. Larvicidal and mosquito repellent action of peppermint (*Mentha piperita*) oil. *Bioresour. Technol.* 71(3): 267–271.

Antonelli, M., and Donelli, D. 2020. Efficacy, safety and tolerability of aroma massage with lavender essential oil: An overview. *Int. J. Ther. Massage Bodyw.* 13(1): 32.

Aygun, D., Doganay, Z., Altintop, L., Guven, H., Onar, M., Deniz, T., and Sunter, T. 2002. Serum acetylcholinesterase and prognosis of acute organophosphate poisoning. *J. Toxicol., Clin. Toxicol.* 40(7): 903.

Bounoua-Fraoucene, S., Kellouche, A., and Debras, J. F. 2019. Toxicity of four essential oils against two insect pests of stored grains, *Rhyzopertha dominica* (Coleoptera: Bostrychidae) and *Sitophilus oryzae* (Coleoptera: Curculionidae). *African Entomol.* 27(2): 344–359.

Boutjagualt, I., Hmimid, F., Errami, A., Bouharroud, R., Qessaoui, R., Etahiri, S., and Benba, J. 2022. Chemical composition and insecticidal effects of brown algae (Fucus spiralis) essential oil against *Ceratitis capitata* Wiedemann (Diptera: Tephritidae) pupae and adults. *Biocatalysis and Agricultural Biotechnology*. 40: 102308.

Camilo, C. J., Leite, D. O. D., Nonato, C. D. F. A., de Carvalho, N. K. G., Ribeiro, D. A., and da Costa, J. G. M. 2022. Traditional use of the genus *Lippia* sp. and pesticidal potential: A review. *Biocatal. Agric. Biotechnol.* 102296.

Cantó-Tejero, M., Pascual-Villalobos, M. J., and Guirao, P. 2022. Aniseed essential oil botanical insecticides for the management of the currant-lettuce aphid. *Ind. Crops Prod.* 181: 114804.

Chamkhi, I., Benali, T., Aanniz, T., El Menyiy, N., Guaouguaou, F. E., El Omari, N., and Bouyahya, A. 2021. Plant-microbial interaction: The mechanism and the application of microbial elicitor induced secondary metabolites biosynthesis in medicinal plants. *Plant Physiology and Biochemistry*. 167: 269–295.

Copping, L. G., and Menn, J. J. 2000. Biopesticides: A review of their action, applications and efficacy. *Pest Manage. Sci.* 56: 651–676.

Dabiri, M., Majdi, M., and Bahramnejad, B. 2021. Spatial and developmental regulation of putative genes associated with the biosynthesis of sesquiterpenes and pyrethrin I in Chrysanthemum cinerariaefolium. *Biologia*. 76(5): 1603–1616.

dos Reis Barbosa, A. L., Bezerr, J. N. S., Neto, M. A., a Fonteles, M. M. D. F., and de Barros Viana, G. S. 2019. Alterations in behavior and memory induced by the essential oil of *Zingiber officinale* Roscoe (ginger) in mice are cholinergic-dependent. *Int. J. Med. Plants Res.* 8(12): 1–8.

Eleonora, S., Filippo, Maggi., Giulia, B., Roman, P., Maria, C. B., Nickolas, G. K., Angelo, C., Donato, R., Nicolas, D. A., Wilkei, John, C., and Beieri, G. B. 2021. Apiaceae essential oils and their constituents as insecticides against mosquitoes—A review. *Ind. Crops Prod.* 171: 113892.

Foudah, A. I., Shakeel, F., Alqarni, M. H., Ali, A., Alshehri, S., Ghoneim, M. M., and Alam, P. 2022. Determination of thymol in commercial formulation, essential oils, traditional, and ultrasound-based extracts of *Thymus vulgaris* and *Origanum vulgare* using a greener HPTLC approach. *Molecules*. 27(4): 1164.

Gbolade, A. A., Oyedele, A. O., Sosan, M. B., Adewayin, F. B., and Soyela, O. L. 2000. Mosquito repellent activities of essential oils from two Nigerian Ocimum species. *J. Trop. Med. Plants*. 67(3): 789–799.

Huang, J., Li, S., Lv, M., Li, T., and Xu, H. 2020. Non-food bioactive products for insecticides (II): Insights into agricultural activities of matrine-type alkaloid analogs as botanical pesticides. *Ind. Crops Prod.* 154: 112759.

Ibrahim, S. S., Abou-Elseoud, W. S., Elbehery, H. H., and Hassan, M. L. 2022. Chitosan-cellulose nanoencapsulation systems for enhancing the insecticidal activity of citronella essential oil against the cotton leafworm *Spodoptera littoralis*. *Ind. Crops Prod.* 184: 115089.

Isman, M. B. 2020. Commercial development of plant essential oils and their constituents as active ingredients in bioinsecticides. *Phytochem. Rev.* 19(2): 235–241.

Johnson, A. J., Venukumar, V., Varghese, T. S., Viswanathan, G., Leeladevi, P. S., Remadevi, R. K. S., and Baby, S. 2022. Insecticidal properties of *Clausena austroindica* leaf essential oil and its major constituent, trans-anethole, against *Sitophilus oryzae* and *Tribolium castaneum*. *Ind. Crops Prod.* 182: 114854.

Karpouhtsis, I., Pardali, E., Feggou, E., Kokkini, S., and Mavragani-Tsipidou, P. 1998. Insecticidal and genotoxic activities of oregano essential oils. *J. Agric. Food Chem.* 46(4): 1111–1115.

Khambay, B., Batty, D., Jewess, P. J., Bateman, G. L., and Hollomon, D. W. 2003. Mode of action and pesticidal activity of the natural product dunnione and of some analogues. *Pest Manage. Sci.* 59(2): 174–182.

Kulkarni, D. S. 2020. Gonadal histoarchitecture and reproductive hormones studies of female albino rats treated with azadirachtin. *Editorial Board*. 9(11): 199.

Kuriyama, T., Schmidt, T. J., Okuyama, E., and Ozoe, Y. 2002. Structure—activity relationships of seco-prezizaane terpenoids in γ-aminobutyric acid receptors of houseflies and rats. *Bioorg. Med. Chem.* 10(6): 1873–1881.

Lazarevi, J., Jevremovi, S., Kosti, I., Kosti, M., and Jovanovi, D. E. 2020. Toxic, oviposition deterrent and oxidative stress effects of *Thymus vulgaris* essential oil against *Acanthoscelides obtectus*. *Insects*. 11(9): 563–574.

Lee, S. E., Lee, B. H., Choi, W. S., Park, B. S., Kim, J. G., and Campbell, B. C. 2001. Fumigant toxicity of volatile natural products from Korean spices and medicinal plants towards the rice weevil, *Sitophilus oryzae* (L). *Pest Manage. Sci.* 57(6): 548–553.

Li, Y. F., Liu, W. Z., Fan, J. W., Huang, C. L., Deng, L. H., Zhuang, H. F., and Guan, Y. X. 2019. Quinoline alkaloids isolated from *Scolopendra subspinipes* mutilans. *Chin. Herb. Med.* 11(3): 344–346.

Lim, A. M. T., Oyong, G. G., Tan, M. C. S., Shen, C. C., Ragasa, C. Y., and Cabrera, E. C. 2021. Quorum quenching activity of *Andrographis paniculata* (Burm f.) Nees andrographolide compounds on metallo-β-lactamase-producing clinical isolates of *Pseudomonas aeruginosa* PA22 and PA247 and their effect on lasR gene expression. *Heliyon*. 7(5): e07002.

Lloyd, G. K., and Williams, M. 2000. Neuronal nicotinic acetylcholine receptors as novel drug targets. *J. Pharmacol. Exp. Ther*. 292(2): 461–467.

Mattar, V. T., Borioni, J. L., Hollman, A., and Rodriguez, S. A. 2022. Insecticidal activity of the essential oil of *Schinus* areira against *Rhipibruchus picturatus* (F.) (Coleoptera: Bruchinae), and its inhibitory effects on acetylcholinesterase. *Pestic. Biochem. Physiol.* 105134.

Naimi, I., Zefzoufi, M., Bouamama, H., and M'hamed, T. B. 2022. Chemical composition and repellent effects of powders and essential oils of *Artemisia absinthium, Melia azedarach, Trigonella foenum-graecum*, and *Peganum harmala* on *Tribolium castaneum* (Herbst) (Coleoptera: Tenebrionidae). *Ind. Crops Prod.* 182: 114817.

Naranjo, S. E., Ellsworth, P. C., and Frisvold, G. B. 2015. Economic value of biological control in integrated pest management of managed plant systems. *Annu. Rev. Entomol.* 60(1): 621–645.

Narayanankutty, A., Sasidharan, A., Job, J. T., Rajagopal, R., Alfarhan, A., Kim, Y. O., and Kim, H. J. 2021. Mango ginger (*Curcuma amada* Roxb.) rhizome essential oils as source of environmental friendly biocides: Comparison of the chemical composition, antibacterial, insecticidal and larvicidal properties of essential oils extracted by different methods. *Environ. Res.* 202: 111718.

Oparaocha, E. T., Iwu, I., and Ahanakuc, J. E. 2010. Preliminary study on mosquito repellent and mosquitocidal activities of *Ocimum gratissimum* (L.) grown in eastern nigeria. *J. Vector Borne Dis.* 47(1): 45–50.

Ostadi, A., Javanmard, A., Machiani, M. A., Morshedloo, M. R., Nouraein, M., Rasouli, F., and Maggi, F. 2020. Effect of different fertilizer sources and harvesting time on the growth characteristics, nutrient uptakes, essential oil productivity and composition of *Mentha x piperita* L. *Ind. Crops Prod.* 148: 112290.

Pandey, A. K., Silva, A. S., Varshney, R., Chávez-González, M. L., and Singh, P. 2021. Curcuma-based botanicals as crop protectors: From knowledge to application in food crops. *Curr. Res. Biotechnol.* 3: 235–248.

Papanastasiou, S. A., Ioannou, C. S., and Papadopoulos, N. T. 2020. Oviposition-deterrent effect of linalool-a compound of citrus essential oils-on female Mediterranean fruit flies, *Ceratitis capitata* (Diptera: Tephritidae). *Pest Manage. Sci.* 5858.

Pavela, R., Maggi, F., Iannarelli, R., and Benelli, G. 2019. Plant extracts for developing mosquito larvicides: From laboratory to the field, with insights on the modes of action. *Acta tropica*. 193: 236–271.

Pavlidou, V., Karpouhtsis, I., Franzios, G., Zambetaki, A., Scouras, Z., and Mavragani-Tsipidou, P. 2004. Insecticidal and genotoxic effects of essential oils of Greek sage, *Salvia fruticosa*, and mint, *Mentha pulegium*, on *Drosophila melanogaster* and *Bactrocera oleae* (Diptera: Tephritidae). *J. Agric. Urban Entomol.* 21(1): 39–49.

Pichersky, E., and Gershenzon, J. 2002. The formation and function of plant volatiles: Perfumes for pollinator attraction and defense. *Curr. Opin. Plant Biol.* 5(3): 237–243.

Puri, S., Singh, S., and Sohal, S. K. 2022. Oviposition behaviour and biochemical response of an insect pest, *Zeugodacus cucurbitae* (Coquillett) (Diptera: Tephritidae) to plant phenolic compound phloroglucinol. *Comp. Biochem. Physiol. Part - C: Toxicol.* 255: 109291.

Ramadan, G. R., Abdelgaleil, S. A., Shawir, M. S., El-bakary, A. S., Zhu, K. Y., and Phillips, T. W. 2020. Terpenoids, DEET and short chain fatty acids as toxicants and repellents for *Rhyzopertha dominica* (Coleoptera: Bostrichidae) and *Lasioderma serricorne* (Coleoptera: Ptinidae). *J. Stored Prod. Res.* 87: 101610.

Rattan, R. S. 2010. Mechanism of action of insecticidal secondary metabolites of plant origin. *Crop Protect.* 29(9): 913–920.

Regnault-Roger, C., and Hamraoui, A. 1994. Comparison of the insecticidal effects of water extracted and intact aromatic plants on *Acanthoscelides obtectus*, a bruchid beetle pest of kidney beans. *Chemoecology*. 6(1): 1–5.

Ryan, M. F., and Byrne, O. 1988. Plant-insect coevolution and inhibition of acetylcholinesterase. *J. Chem. Ecol.* 14(10): 1965–1975.

Sousa, R. M. O., Cunha, A. C., and Fernandes-Ferreira, M. 2021. The potential of Apiaceae species as sources of singular phytochemicals and plant-based pesticides. *Phytochem.* 187: 112714.

Stepanycheva, E., Petrova, M., Chermenskaya, T., and Pavela, R. 2019. Fumigant effect of essential oils on mortality and fertility of thrips Frankliniella occidentalis Perg. *Environ. Sci. Pollut. Res.* 06239.

Tavares, C. P., Sousa, I. C., Gomes, M. N., Miró, V., Virkel, G., Lifschitz, A., and Costa-Junior, L. M. 2022. Combination of cypermethrin and thymol for control of *Rhipicephalus microplus*: Efficacy evaluation and description of an action mechanism. *Ticks and Tick-borne Diseases*. 13(1): 101874.

Vachon, J., Assad-Alkhateb, D., Baumberger, S., Van Haveren, J., Gosselink, R. J., Monedero, M., and Bermudez, J. M. 2020. Use of lignin as additive in polyethylene for food protection: Insect repelling effect of an ethyl acetate phenolic extract. *Composites Part C: Open Access*. 2: 100044.

Weinzierl, R., and Henn, T. 2020. Botanical insecticides and insecticidal soaps. In *Handbook of integrated pest management for turf and ornamentals* (pp. 541–555). Boca Raton, FL: CRC Press.

Wu, M., Xiong, Y., Han, R., Dong, W., and Xiao, C. 2020. Fumigant toxicity and oviposition deterrent activity of volatile constituents from Asari radix et rhizoma against *Phthorimaea operculella* (Lepidoptera: Gelechiidae). *Int. J. Insect Sci.* 20(6): 1–6.

Xiong, Y., Sui, X., Ahmed, S., Wang, Z., and Long, C. 2020. Ethnobotany and diversity of medicinal plants used by the Buyi in eastern Yunnan, China. *Plant Diversity*. 42(6): 401–414.

Zehra, A., Raytekar, N. A., Meena, M., and Swapnil, P. 2021. Efficiency of microbial bioagents as elicitors in plant defense mechanism under biotic stress: A review. *Curr. Res. Microb. Sci.* 2: 100054.

Zhou, Y., Qiu, T. X., Hu, Y., Ji, J., Liu, L., and Chen, J. 2022. Evaluation on the antiviral activity of artemisinin against rhabdovirus infection in common carp. *Aquaculture*. 738410.

9 Preparation Strategy and Application of Porous Starch Microcapsule and Cyclodextrin Containing Essential Oils

9.1 INTRODUCTION

Consumers' pursuit to a healthy lifestyle has promoted people to develop new technologies that can prolong the shelf life of food without the use of preservatives. Compared with other types of preservation, edible microcapsules containing essential oils are becoming more and more popular, especially the starch microcapsules containing essential oil (EOs–starch microcapsules) because of their environment-friendly, healthier characteristics and the ability to carry active ingredients. In addition, the EOs–starch microcapsules can also reduce the flavor influence and prolong the action time of essential oil on food through its slow-release effect, which can promote the use of EOs in food (Ju et al., 2018; Reis et al., 2022). Understanding the different collocation of edible starch microcapsules and EOs and the related antibacterial mechanism will be more effective and targeted to promote the application of EOs in the real food system.

9.2 THE CONCEPT AND DEVELOPMENT OF MICROENCAPSULATION

Microencapsulation technology originated in the 1940s and was used to prepare microcapsules of cod liver oil. Wurster et al. had successfully prepared drug-coated microcapsules by air suspension method in the late 1940s, so this method is also called Wurster method. From the 1950s to 1960s, the theory of interfacial polymerization has attracted great attention. Researchers have gradually applied the method of synthesizing macromolecules to the preparation of microcapsules. From the 1970s to now, with the increasingly mature technology and the application and development of more high-tech, the preparation technology of microcapsules is no longer limited to the field of medicine research, but gradually expanded to food, pesticides, textiles, coatings, adhesives, cosmetics, and other industries (Zhou et al., 2021). Microencapsulation technology uses film-forming materials to encapsulate liquid or solid to form microparticles which are called microcapsules, the diameter of which

DOI: 10.1201/9781003329268-9

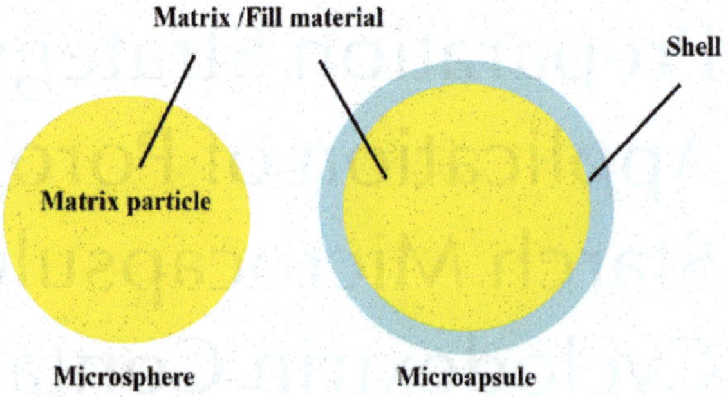

FIGURE 9.1 The structure of microcapsules.

is generally 1–1,000 nm. Film-forming materials for encapsulation are called wall materials. Many natural and synthetic materials can be used as wall materials. The encapsulated materials are called core materials, which are generally liquid, solid, or gas with high reactivity, sensitivity, or volatility. Microencapsulation technology can separate the core material from the surrounding environment, thus effectively reducing the response of the core material to environmental factors such as water, light, oxygen, and temperature, thereby improving the appearance and performance of the core material. Figure 9.1 shows the structure of microcapsules.

9.3 WALL AND CORE MATERIALS OF MICROCAPSULES

9.3.1 Wall Materials

The wall materials of microcapsules can be divided into natural macromolecule materials, derivatives, and synthetic macromolecule materials. Natural macromolecule materials mainly include plant gum, starch, sodium alginate, paraffin, and agar. Their advantages are low toxicity, high viscosity, and easy film formation, whereas the disadvantage is their low strength. Synthetic materials mainly include polyester, polyether, and polyamide. Their advantages are good mechanical properties and easy to be controlled by chemical or physical modification. Their disadvantages are poor biocompatibility and limited application in the field of medicine and food (Dajic Stevanovic et al., 2020).

Although Arabic gum is considered as the most ideal wall material by scholars, due to its insufficient supply and high price in recent years, the search for new microcapsule wall materials has become a research hotspot. Starch is a green, environment-friendly material (which can be reused and degraded repeatedly). Besides being edible, it can also be used as wall material of microcapsules to make up for the shortcomings of traditional materials. Starch-based materials are colorless, tasteless, and processable. They not only do not affect the texture and flavor of core materials, but also play an active role in promoting gastrointestinal health and mineral absorption and controlling blood lipid, sugar, etc. At present, in the process of starch deep processing, it not only can be used independently, but also mixed with other materials. The main starch materials of microcapsule wall materials are porous starch, resistant starch, modified starch, maltodextrin, beta-cyclodextrin and alkenyl succinate starch ester, etc. (Huang et al., 2018).

9.3.2 Core Materials

Core material is the target material encapsulated in microcapsules. There are many kinds of core materials, such as drugs, food additives, antifouling agents, and catalysts. Generally speaking, the core material must be released from the capsule in order to play its role. Its release rate can be divided into instantaneous release and slow release. Instantaneous release is the release of core material from wall materials broken by external forces such as crushing and deformation, while slow release is the release of core materials through wall diffusion or wall melting degradation. This slow-release technology has a good effect in medicine, food, paint, and agriculture (Sotelo-Boyás et al., 2017).

The release rate of core material is mainly affected by wall thickness, wall hole, wall deformation mode, crystallinity, crosslinking degree, etc. In addition, the solubility and diffusion coefficient of the core material itself will also have some influence on its release rate. Generally, the release process of the core material from microcapsules will follow zero-order or first-order release rate equation.

9.4 APPLICATION OF MICROENCAPSULATION TECHNOLOGY IN FOOD INDUSTRY

The application of microencapsulation technology in food industry has simplified many traditional processes and solved many problems that cannot be solved by conventional technology. It has greatly promoted the transformation of food industry from low-level primary processing to high-level deep processing industry. At present, many microencapsulated foods have been developed by microencapsulation technology, such as powdered oil, powdered wine, and capsule beverage. Flavoring agents (flavor oils, spices, condiments), natural pigments, nutrient fortifiers (vitamins, amino acids, minerals), sweeteners, acids, preservatives, antioxidants, etc. have also been widely used in the production of microencapsulated food additives (Misra et al., 2021; Coelho et al., 2022; Mamusa et al., 2021). Generally speaking, the application of microcapsule technology in food industry can play the following roles.

(1) It can prevent some unstable food raw materials and accessories from volatilizing, oxidizing, and deteriorating. Many essence and spice essential oil are unstable, volatile, and oxidizable. VE, VC, highly unsaturated oils (such as DHA and EPA), etc. are easy to be oxidized, thus losing their functions. However, these components need to be highly dispersed in production. Microencapsulation is the best way to solve this contradiction.

(2) It can reduce or conceal bad taste and volatility of core material. Some food ingredients have unpleasant odors or tastes, such as stink, pungency, bitterness, odor, etc., which can be reduced or masked by microcapsule technology. The microencapsulated products are not dissolved in the mouth but releasing the contents and playing its nutritional role in the digestive tract.

(3) It can change the shape of the material, that is, solidify the liquid material into fine flowable powder, which can not only be easy to use, transport, and preserve, but also simplify the food production process and develop new products.
(4) It can control the release rate of core material. There are many examples that active ingredients in food need to be controlled release. For example, microencapsulated ethanol preservatives slowly release ethanol vapor in sealed packaging to prevent contamination of food by molds. The acid–base expanded microcapsules will release gas after reaching the required temperature.

9.5 MODIFIED STARCH

Natural starch extracted from plants could not always meet the requirements of special products, such as biomaterials with certain carrying capacity, or could not always subject to extreme process conditions, such as high temperature, freeze-drying, strong acids, and strong bases. Hence, the natural starch is limited and unacceptable in many industrial applications. Therefore, it is very important for natural starch to be modified by different technologies to enhance or inhibit its inherent properties or make its properties meet the requirements of industrial applications (Zhao et al., 2022; Das and Sit, 2021; Jaymand, 2021). Fortunately, a large number of reactive hydroxyl groups in molecular chain of starch provide a structural basis for the modification of starch.

Usually, suitable methods are used to introduce new functional groups on starch molecules or alter the size of starch molecules and properties of starch granules, thereby changing the natural characteristics of starch and expanding its application range. This is the preparation of modified starch. And the usual modification methods include physics (e.g., high-pressure autoclave, osmotic pressure treatment), chemistry (e.g., oxidation, esterification, etherification, hydroxypropylation), and enzyme modification (Lin et al., 2017). Nowadays, the modified starch has more and more varieties and more and more applications, which has brought great convenience to our life. However, in the modified starch with different functions, the application of porous starch in food has attracted more and more attention in recent years due to the high loading capacity of porous starch. For example, Qiu et al. (2017) and Souza et al. (2013) made use of porous corn starch and cassava starch to prepare starch microcapsules and antibacterial films which can carry essential oil, respectively. Marefati et al. (2017) prepared a particle emulsifier which could carry the emulsifier with modified starch successfully. Some other researchers have even added curcumin into porous starch to improve the solubility of curcumin.

9.6 STRUCTURAL CHARACTERISTICS AND PROPERTIES OF POROUS STARCH

Porous starch, as a new type of modified starch, has two characteristics: first, it is different from the original starch in structure; second, porous starch breaking through the limitation of the original starch's own characteristics, has special properties different from natural starch, and its application scope is widened, so it is more suitable for food and pharmaceutical industry. In structure, porous starch is characterized by the formation of hollow structure in the whole starch tissue due to its abundant voids, which greatly improves the porosity and specific surface area of porous starch. Based on different preparation methods and mechanisms, porous structures are also quite different. For example, the porous starch prepared by ball milling or ultrasonic method only shows some pits or cracks on the surface of granules through physical action. Compared with the original starch, the increase of porosity and specific surface area is limited. Solvent exchange method utilizes the gelatinization and retrogradation properties of starch to form a reticulated hydrocolloid, and then through solvent replacement, porous starch with three-dimensional reticulation can be prepared. The porous starch skeleton is similar to the spongy structure. The voids penetrate the whole starch tissue and have very high specific surface area and porosity. The principle of acid hydrolysis or enzymatic hydrolysis is to prepare porous starch by controlling the degree of hydrolysis of raw starch. Acids or enzymes first attach to the surface of starch, preferentially hydrolyze the irregular and amorphous areas of starch from outside to inside and then form channels, thus forming holes from the surface to the core on starch granules. The porous starch prepared by this hydrolysis method has a honeycomb-like hollow structure. Small holes about 1 μm in diameter covered the whole surface of starch granules and goes deep into the center. The volume of the holes accounts for about 50% of the total granular volume (Figure 9.2).

The structural characteristics of porous starch make it more advantageous in the application of food or medicine than the original starch. These advantages are mainly as follows:

(1) Enhance the stability and avoid the effects of light, heat, air, and chemical environment on target substances (such as DHA, EPA, vitamin E, vitamin A, vitamin D, beta-carotene, and lycopene).
(2) Prevent volatilization and retain aroma and other components (such as spices, asparagus sweeteners, acidic agents, spices, enzymes, and spices).

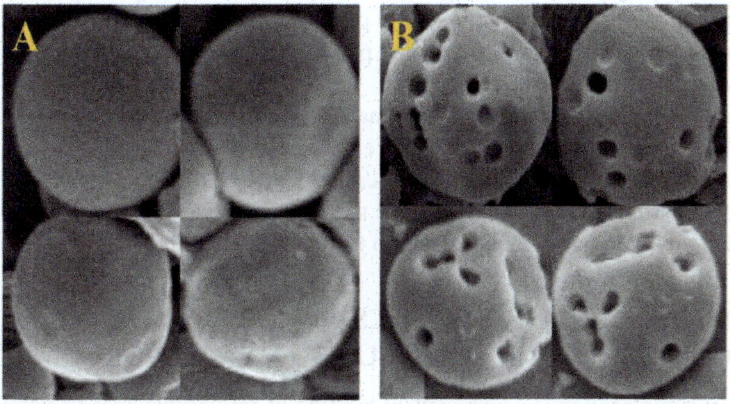

FIGURE 9.2 Schematic diagrams of porous starch. (a) Original starch. (b) Porous starch.

(3) Improve the solubility and solubility of the target substance.
(4) Cover up the irritating odor and the bitterness and stink in food and medicine (such as the bitterness of peptides, extracts of traditional Chinese medicine, Ganoderma lucidum, ginseng and aloe, the beany smell of soybean products, and the fishy smell of seafood).
(5) Ability to combine with drugs to achieve slow release.
(6) Other functions (powdering of medicines and foodstuffs which is easy to be transported and preserved, etc.)

9.7 PREPARATION OF MICROENCAPSULATED POROUS STARCH

Because of the wide application of microcapsules, the preparation process or method of microcapsules has greatly attracted attention of relevant scholars since its advent. According to incomplete statistics, there are more than 200 preparation methods of microcapsules (Glenn et al., 2010). According to the formation mechanism and properties of microcapsules, the preparation methods of microcapsules can be divided into three categories: physical chemical method, physical method, and chemical method. Physicochemical methods include phase separation method and spray-drying method. Physical methods include electrostatic deposition and air suspension. Chemical methods include complex coacervation, single coacervation, and emulsification (Jadhav et al., 2017).

At present, the preparation of microencapsulated porous starch mainly adopts the following two methods:

1. *Direct Adsorption Method:* Using porous starch as core material to absorb and embed the target substance by direct adsorption, and then the appropriate wall material was selected and combined with core material in appropriate proportion to prepare microcapsules. For example, cassava porous starch was used as core material by some scholars to absorb silkworm chrysalis oil to make powdered silkworm chrysalis oil, then zein was used as wall material and combined with powdered silkworm chrysalis oil in a certain proportion, and then the microcapsules containing silkworm chrysalis oil were prepared. Others used porous starch as wall material to adsorb and encapsulate grape seed oil and prepare microcapsules containing grape seed oil.
2. *Spray Drying Method:* The core material is dispersed in the wall solution to form a uniform emulsion. The emulsion is atomized into tiny droplets through airflow and evenly dispersed in the hot gas stream, so that the wall material is quickly evaporated and the wall material is solidified, and then the core material is encapsulated to form a microcapsule.

9.8 BINDING CHARACTERISTICS OF ESSENTIAL OILS–STARCH MICROCAPSULES

Inclusion complexation is a noncovalent binding of small molecules to starch where the guest molecule is bound by inclusion into the helical cavity of amylose. The molecules bound to starch are also termed ligands, since the theories developed in biochemistry for describing the binding of small molecules to macromolecules are also valid for this type of interaction (Condepetit et al., 2006). The main ligands include alcohols, aldehydes, phenols, and lactones. Besides the endogenous starch–lipid complexes, the properties of starch complexes with added lipids have been studied in detail in the past. This type of interaction is applied to tailor the textural properties of starch-based foods. In contrast, less literature is available on EOs–starch microcapsules, and the findings on starch–lipid complexes have a limited validity for EOs–starch microcapsules, since the properties of the complexes vary according to the character of the ligands.

Compared to lipids, EOs are smaller ligands with shorter carbon chains and frequently present a ring structure as in the case of terpenes and lactones. A common property of lipids and EOs is their hydrophobic character which is accompanied by poor water solubility. However, while most complexing lipids like monoglycerides are surface active and therefore well dispersible in water, EOs generally possess poor water dispersibility which may impair the interaction with starch in aqueous systems.

Regarding the complexation ability of amylopectin, Rutschmann and Solms observed a nonspecific binding at high ligand concentrations and concluded that the long external branches of amylopectin permit the formation of complexes with very low stability, but with very high binding capacity values. The finding that potato amylopectin solutions form soft gels at high ligand concentrations is a further indication that amylopectin takes part in the complexation process (Heinemann et al., 2003). Likewise, EOs retention measurements suggest that both polymers, amylose and amylopectin, are able to bind EOs (Arvisenet et al., 2002). However, the exact nature of EOs binding to amylopectin remains unknown and the scarce literature on this topic reflects the difficulty in seizing amylopectin complexes rather than the inexistence of this type of interaction. Mixed complexes can be formed in systems where more than one type of ligand is present at the same time. This aspect is of relevance in real food systems since it is likely that a food contains lipids as well as different flavor compounds that are able to complex with starch. For chemically modified starches, which are often used in the food industry, Wulff et al. (1998) showed that the complexation ability decreases with increasing degree of substitution. On the other hand, a slight hydroxypropylation of starch provides a route for obtaining soluble amylose complexes since this chemical modification improves the interaction with water and prevents the aggregation of amylose.

9.9 PREPARATION OF ESSENTIAL OILS–STARCH MICROCAPSULES

At present, there are many materials for preparing EOs–microcapsules, and according to our investigation, they are mainly divided into three categories which are starch, chitosan, and colloid. Table 9.1 makes a statistical analysis of the main materials, preparation methods, and related evaluation indexes for the preparation of EOs–microcapsules. From Table 9.1 it is found that the material used to prepare EOs–microcapsules is mainly amyloid, which mainly includes porous starch and dextrin. The reason for this phenomenon may be that the preparation of EOs–starch microcapsules is relatively simple. The porous starches are first mixed with essential

TABLE 9.1
The Main Materials, Preparation Methods, and Related Evaluation Indexes for the Preparation of EOs–Microcapsules

Material classification	Starch types	Essential oil	Production method	Evaluation index	Results or conclusions	Reference
Starch	Porous starch	Clove essential oil	Spray-drying	Heat and corrosion resistance	The modified microcapsule has a strong heat resistance and has a significant inhibitory effect on the mold spore	Qiu et al. (2022)
	Porous starch	Oregano essential oil	Embedding method	Entrapment rate and yield	The encapsulation rate and yield of microcapsules prepared by this method was 88.25% and 44.27%, respectively	Wang et al. (2016)
	Jackfruit seed starch	Vanilla essential oil	Spray-drying	Entrapment rate, yield, and sustained release	The encapsulation rate and yield of microcapsules prepared by this method was 74.49% and 92.06%, respectively. It has good sustained release	Zhu et al. (2017)
	Jackfruit seed starch	Vanilla essential oil	Freeze-drying in vacuum	Stability and sustained release	The microcapsule has good storage stability and sustained release potential, and the shelf life can reach 250 days	Zhu et al. (2018)
	Succinylated taro starch	Pomegranate seed oil	Spray-drying	Embedding rate	A Central composite design (CCD) was applied and the treatment with the highest pomegranate seed oil encapsulation efficiency (61.09% ± 0.41%) was selected PSO-loaded microparticles obtained with 15% feed solids using 190°C inlet air temperature showed low Aw (0.08 ± 0.01), moisture (1.26 ± 0.05%), hygroscopicity (11.69% ± 0.57%), and water solubility (9.81% ± 0.24%)	Cortez-Trejo et al. (2021)
	Starch nanoparticles	Peppermint essential oil, oregano essential oil, cinnamon essential oil, and lavender essential oils	In situ polymerization	Embedding rate	As the complexation temperature increases, the embedding rate increases significantly. Microcapsules can significantly extend the antioxidant and antimicrobial time of essential oils	Qiu et al. (2017)

	β-Cyclodextrin and beet pectin	Garlic essential oil	Compound cohesion method	Entrapment rate	The highest encapsulation and production efficiencies were obtained for the samples with 3:6–1:2 and 1:6–1:2 of core:wall material and sugar beet pectin:β-CD ratios, respectively	Emadzadeh et al. (2021)
	β-Cyclodextrin	Oregano essential oil	Embedding method	Embedding rate	Microcapsules delayed the sustained release of essential oils	Shi et al. (2022)
	β-Cyclodextrin	Cinnamon and oregano essential oils	Compound cohesion method	Sustained release and antifungal activity	The microcapsule can effectively delay the release rate of essential oil	Zhou et al. (2017)
	β-Cyclodextrin	Willow leaf essential oil	Freeze-drying in vacuum	Embedding rate	The encapsulation rate of microcapsule produced by this method is 85.28% and has significant bacteriostasis	Sheng et al. (2017)
	Chitosan and gallic acid	Garlic essential oil	Compound cohesion method	Embedding rate	The microcapsules prepared using GA-CS (at a mass ratio of 0.5:1) presented the best physicochemical properties, including antioxidant activity, encapsulation efficiency, sustained release, etc.	Teng et al. (2022)
	Maltodextrin and whey protein	Lime essential oil	Spray-drying	Embedding rate and antioxidation	The encapsulation rate of microcapsules produced by this method can reach 83.3%, and has significant antioxidant activity	Campelo et al. (2017)
Chitosan	Chitosan	Thyme essential oil	Compound cohesion method	Embedding rate and bacteriostasis activity	The embedding rate of the microcapsules was more than 68%, and has a strong inhibitory effect on *Bacillus cereus*	Sotelo-Boyás et al. (2017)
	Chitosan	*Gaultheria procumbens* L. essential oil	Spray-drying	Appearance and antibacterial activity	The appearance of the microcapsule was uniform spherical and it was able to significantly inhibit *Aspergillus flavus*	Kujur et al. (2017)
	Chitosan and sodium alginate	Thyme essential oil	Layer self-assembly method	Embedding rate	The embedding rate of the microcapsule was up to 80.23%	Huang et al. (2016)

(Continued)

TABLE 9.1
(Continued)

Material classification	Starch types	Essential oil	Production method	Evaluation index	Results or conclusions	Reference
	Chitosan and gelatin	Rosemary essential oil	Spray-drying	Antibacterial activity	The microcapsules with an encapsulation efficiency of more than 70% were used to impart durable mosquito-repellency (more than 90%) against Anopheles mosquitoes, antibacterial properties against *E. coli* (>93%) and *S. aureus* (>95%) bacteria, significant (>91%) antioxidant activity and a pleasant aroma	Singh and Sheikh (2022)
	Chitosan, Tween 20, Tween 40, and Tween 60	Citrus essential oil	Emulsion polymerization	Entrapment rate and rheology	The microcapsule produced by the combination of chitosan and Tween 60 has better rheology and the minimum embedding rate was 68.1%	Song et al. (2021)
	Pectin and whey protein	*Ziziphora clinopodiodes* essential oil	Electrostatic binding method	Appearance	The microcapsule produced by this method has a regular spherical shape	Hosseinnia et al. (2017)
Colloid	Gelatin and acacia	Oregano essential oil	Compound cohesion method	Embedding rate	The microcapsule produced by this method has an embedding rate of up to 83.1% and markedly increases its bacteriostasis durability	Tian et al. (2016)
	Gum Arabic, maltodextrin, sodium alginate, and whey protein	Juniper berry essential oil	Spray-drying	Embedding rate	Best results were obtained when JBEO was encapsulated using GA/MD (1:1) as a carrier producing microcapsule with the highest encapsulation efficiency (70.07%) and the best results in density properties, porosity, dissolution time, and thermal properties	Bajac et al. (2022)

Gelatin and polyvinyl alcohol	Pine tree essential oil	Solvent evaporation method	Corrosion resistance	The microcapsules showed excellent corrosion resistance	Li et al. (2017)
Sodium alginate	Coriander essential oil	Spray-drying	Morphology and sustained release	The microcapsule is spherical with smooth surface and the maximum release rate at pH 6.5	Dima et al. (2016)
Gum Arabic and maltodextrin	Peppermint essential oil	Spray-drying	Average particle size	The size of the microcapsules ranged from 10 to 125 μm, with a Sauter mean diameter of 37.8 ± 1.4 μm and shell thickness of 0.91 ± 0.15 μm. The microcapsules were spherical with a relatively smooth surface	Baiocco et al. (2021)
Gelatin	Dandelion root extract	Solvent evaporation method	Antibacterial activity	Microcapsules enhance the antibacterial activity of the films	Al-Maqtari et al. (2022)

oils in proper proportion at room temperature and the essential oils will be entrapped in porous starch because of the larger specific surface area and honeycomb hole of porous starch. Studies have shown that porous starch can adsorb the essential oil of 80–100% of its own weight to form solid granules (Sun et al., 2021). Besides, the use of porous starches for preparing EOs–microcapsules does not require complex processing such as emulsification and homogenization of essential oils. This does not affect the activity of the essential oil, nor will it have an adverse effect on the components and flavor of essential oil. Moreover, the cost is low and the loading capacity is high (Huang et al., 2017). Whereas other materials such as chitosan or colloid may require more complex preparation processes when used to prepare EOs–microcapsules. And in most cases, they themselves have no ability to form microcapsules and need to be combined with other materials when needed. This increases the complexity of the preparation process to a certain extent and may also cause the loss of essential oil. Since the starch-based EOs–microcapsules play an important role in food preservation, we further summarize the preparation formulas and research purposes of starch based EOs–microcapsules in Table 9.2.

9.10 ACTION MECHANISM

It is EO that plays an important role in antimicrobial in EOs–starch microcapsules. There is a close relationship between the chemical composition and structure of EO and its bacteriostasis. The inhibitory mechanisms of EOs vary from components to components, which indicate that the bacteriostasis mechanism of general EO is not a single one and is the result of multipoint inhibition approach (Wu et al., 2022). The antimicrobial effect of EOs–starch microcapsules is mainly completed in two steps. First of all, EOs spread out through the micropores, fracture, or semipermeable membrane in starch microcapsule. According to the physicochemical properties of the microcapsule material, the process is mainly divided into three cases. First, the antibacterial agent is diffused and released from the microcapsule. Second, the microcapsule swells to promote the release of the antibacterial agent. Third, the degradation or corrosion of microcapsule leads to the release of the antibacterial agent, and then the released EO work on the microbial cell membrane, so as to exert bacteriostatic action. Because of the hydrophilicity of the porous starch, the EO has better contact with the microbial cell membrane. According to relative studies, the interaction between the released EO and the microbial cell membrane may be the following. First, EO can destroy the cell membrane by altering the integrity of the cell phospholipid bimolecular layer or by interfering with the active transport protein (Moghimi et al., 2016). Second, the positively charged capsule material can interact with the negatively charged microorganism cell wall, which will result in the increase of the EO concentration. However, this hypothesis is still controversial. For example, cationic surfactant (Yue Guisuan) also has strong antimicrobial activity. It is also reported that the charge of the emulsion does not affect the antimicrobial activity of the EO (Majeed et al., 2016). Third, the edible coating may delay the release of essential oils and prolong its action time to microbes. Consistent with this hypothesis, some studies found that EOs–microcapsules could significantly ($P < 0.05$) prolong their antimicrobial activity compared with free essential oils.

TABLE 9.2
The Preparation Formulas and Research Purposes of Starch-Based EOs–Microcapsules

Starch type	Essential oil	Method	Preparation conditions	Research purposes	Function	Reference
S. edule fruit starch	Cinnamon essential oil	Spray-drying	The drying conditions were inlet/outlet air drying temperatures of 150/85°C, using a parallel arrangement of the nozzle concerning the drying airflow, and a nozzle pressure of 2.5 kPa. Throughout the drying process, the emulsion was continuously agitated with a magnetic stirrer	To evaluate the effects of starch, whey protein concentrate, and gum Arabic on the physical properties of microcapsules	Protective effect on active ingredients	Porras-Saavedra et al. (2021)
Octenyl succinic anhydride modified starches	Clove oil	Spray-drying	The emulsions were fed (10 ml/min) into the spray-dryer operating at inlet and outlet temperatures of 170°C ± 5°C and 80°C ± 5°C, respectively	Explore the oxidative stability of microcapsules	Antioxidant and lipophilic bioactive agents	Sharif et al. (2017)
Sweet potato starch	Thyme essential oil	Emulsion polymerization	Glycerol was added as a plasticizer at a concentration of 2 g/100 g (w/w, on dry basis of the weight of starch); 0.25 g/g of essential oil	Extends the shelf life of shrimp meat during cold storage	Antibacterial and antioxidation	Alotaibi et al. (2018), Cruz-Tirado et al. (2020)
Modified corn starch	Thyme essential oil	Spray-drying	The input and output temperatures of the spray-dryer were 190 and 85°C, respectively	Extend the action time of essential oil through microcapsule release technology	Extended antifungal time	Kalateh-Seifari et al. (2021)
Succinylated taro starch	Pomegranate seed oil	Spray-drying	A Central composite design (CCD) with a central point of three factors (feed solids concentration, S: 15%, 25%; oil to wall material ratio, *r*: 1:3, 1:4, and inlet temperature, *T*: 170°C, 190°C	Avoid oxidative degradation of pomegranate seed oil	Preparation of functional food	Cortez-Trejo et al. (2021)

(Continued)

TABLE 9.2
(Continued)

Starch type	Essential oil	Method	Preparation conditions	Research purposes	Function	Reference
Cassava starch	Cinnamon essential oil	Emulsion polymerization	2% starch + 0.01% cinnamon essential oil	The effect of the essential oil-containing microcapsules on "Pedro Sato" storage quality was explored	Antifungal	Zhao et al. (2022)
Porous starch (StarrierR®)	High-oleic sunflower oil	Plating	Temperature and light intensity were 32.5°C and 600 Klux, respectively	The study evaluates the oxidation level of high-oleic sunflower oil (HOSO) plated onto porous starch as an alternative to spray-drying	Antioxidation	Belingheri et al. (2015)
Native waxy corn starch	Cinnamon essential oil	In situ polymerization	The complexation temperature is 90°C; the volumes of ethanol solution and essential oil are 20 and 0.250 ml, respectively	Explored the carrying capacity of starch granules to essential oils	Antioxidant and antimicrobial	Qiu et al. (2017)
Corn starch	Garlic essential oil	Emulsion polymerization	2% starch + 0.01% cinnamon essential oil	The optimum conditions for the production of microcapsules containing garlic essential oil were explored	It provides the basis for the industrialized production of microcapsule products containing garlic oil	Li et al. (2015)
Cassava starch	*S. guianensis* essential oil	Freeze-drying	The microencapsulation consisted of heating 2 g of starch in 30 ml of water to its gelatinization temperature (68°C). Then polysorbate was added at a ratio of 2% with dropwise addition of *S. guianensis* essential oil under constant stirring, and finally, the preparation was cooled to room temperature	Controlling pests by wrapping plant essential oil with starch.	The cassava starch-based microparticles exhibit promising functionality as carriers for essential oils with mosquitocidal activities	Moura et al. (2021)

Corn starch	Oregano essential oil	Sharp hole method	Emulsifier: Tween 80; Time: 30 min	In order to improve the embedding effect of natural plant essential oils	The encapsulation rate of the essential oil reached 88.25%	Wang et al. (2016)
Potato starch (St)	*Zataria multiflora* essential oil	Emulsion polymerization	First, the whey protein isolate powder (2 g) was dissolved in distilled water (50 ml) by magnetic stirring, and then *Zataria multiflora* essential oil (0.5 g) was added to the solution, and then Tween 80 (0.2 ml) was added as an emulsifier to dissolve the essential oil in solution	In this study, a composite film made of potato starch (St)/apple peel pectin (Pec) was prepared. The St/Pec film was modified with microencapsulated *Zataria multiflora* essential oil (MEO) and zirconium oxide (ZrO_2) nanoparticles (St/Pec/MEO/ZrO_2)	Improve the antioxidant and antibacterial activities of the film and prolong the shelf life of quail meat	Sani et al. (2021)
Porous starch	Cumin essential oil	Complex coacervation process	PS was weighed and mixed with FEO at 1:1 ratio and the mixture was placed in a magnetic agitator and stirred at room temperature for 30 min, followed by 30 min of ultrasound. Then the mixture was transferred to a centrifugal tube for 30 min at 12,000 r/min	Delay the action time of fennel essential oil and improve its antibacterial activity	Extending the shelf life of pork	Sun et al. (2021)
Corn starch	Clove oil	Spray-drying	Inlet temperature: 170°C, Feed flow rate: 40 ml/min, dry air flow: 60 m^3/h	The antibacterial effect of microcapsules containing clove oil on meat products was studied	Antibacterial	Hasan et al. (2022)
β-Cyclodextrin	Cinnamon and oregano essential oil	Solvent evaporation method	50 g sample dissolved in ethanol (10 g/l), oil bath temperature: 55°C ± 2°C, vacuum filtration, air drying temperature: 25°C	The antibacterial activity of the microcapsules on the fruit pathogens was analyzed	Antibacterial	Munhuweyi et al. (2018)

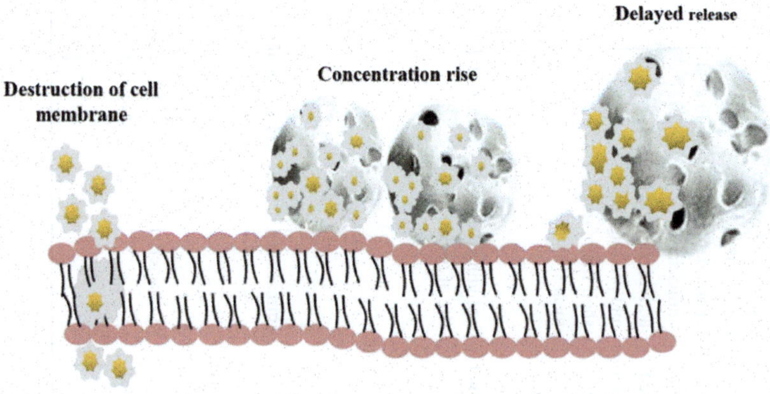

FIGURE 9.3 Schematic diagram of the mechanism of different routes.

For example, microcapsules containing pine nut essential oil were prepared by spray-drying using Arabic gum, maltodextrin, sodium alginate, and whey protein concentrate as wall materials. The results showed that the microcapsules could significantly ($P < 0.05$) prolong the release time of pine nut essential oil (Bajac et al., 2022). Similarly, Singh et al. (2022) used chitosan–gelatin composite as the shell material and rosemary essential oil as the core material to prepare functional microcapsules for the development of multifunctional flax fabric. The results show that microcapsules can effectively avoid the loss of essential oil caused by repeated washing of flax fabric. Similarly, chitosan microcapsules containing garlic essential oil can significantly ($P < 0.05$) prolong the release time of garlic essential oil (Teng et al., 2022). In recent years, there were related literatures which introduced that the use of macromolecules may help to improve the antimicrobial activity of EO. For example, the combination of sodium caseinate and baili EO, or the combination of sodium alginate and citronella EO can play a synergistic bacteriostasis effect (Xue et al., 2015). Therefore, further exploration is needed to improve the understanding of the basic mechanism of EOs–starch microcapsules for improving the specific antibacterial effect of antibacterial agents on target microorganisms. The mechanisms described in Figure 9.3. may coexist and are difficult to appear alone.

9.11 APPLICATION OF ESSENTIAL OILS–STARCH MICROCAPSULES IN FOOD PRESERVATION

With the continuous improvement of living standard, people not only expect to be able to eat local food, but also expect to be able to taste exotic food. However, it is unfortunate that the long storage and transport conditions have had a great negative impact on the taste and quality of these products. Although the plastic packaging had reduced these negative effects to a certain extent, with the problem of white pollution caused by plastic packaging waste and the improvement of people on food safety, developing degradable and edible, environment-friendly natural green preservation

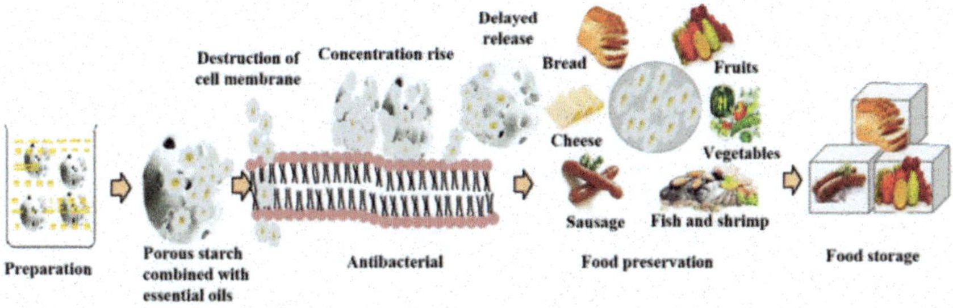

FIGURE 9.4 Potential food to be considered for EOs–starch microcapsules application.

material has become a research focus in the field of food. Among them, edible microcapsule technology has great potential for application and popularization.

Edible microcapsule technology, as a method of food preservation, has been recognized by international community and has been listed as the key technology developed in 21st century. Although edible microcapsule technology has a short history, its appearance has solved many food problems and showed great superiority, especially in the food preservation industry. Many products are wasted because of the defects of fresh-keeping technology, but it can be better solved by the technology of edible microcapsule. In particular, the microcapsules made of starch not only have a wide range of materials, but also are safe and pollution-free. The main functions of EOs–starch microcapsules in food application are antimicrobial and antioxidant effects, which may be mainly related to the effect of EO. In addition, from the type of food we can see that the application of EOs–starch microcapsules mainly concentrated in cereals and meat products, which may be related to the structural characteristics of the starch microcapsule itself. The appearance of starch microcapsules is generally granular, and so its mechanical and permeable properties may be worse than the ordinary synthetic films, which limit its application in some foods. But with the continuous improvement and exploration of processing technology, these limitations will be overcome. Figure 9.4 shows potential food to be considered for EOs–starch microcapsules application.

9.12 APPLICATION OF THE COMBINATION OF CYCLODEXTRIN AND ESSENTIAL OILS IN FOOD PRESERVATION

In addition to porous starch, cyclodextrin (CDs) is also one of the most commonly used materials for the formation of inclusion complexes. CDs are obtained by starch degradation. They are cyclic oligosaccharides consisting of 6 (α-CD), 7 (β-CD), or 8 (γ-CD) glucose units, which are bound together by α-(1–4) bond to form a cyclic structure (Figure 9.5) (Lima et al., 2016). Figure 9.5 also showed a schematic diagram of the cavity of CDs and its main properties. There is a nonpolar pore in the structure of CDs and some hydroxyl groups on the surface. The inclusion of hydrophobic compounds is mainly carried out through the hydrophobic interaction between guest molecules and the inner wall of CDs cavity. However, other forces, such as van der

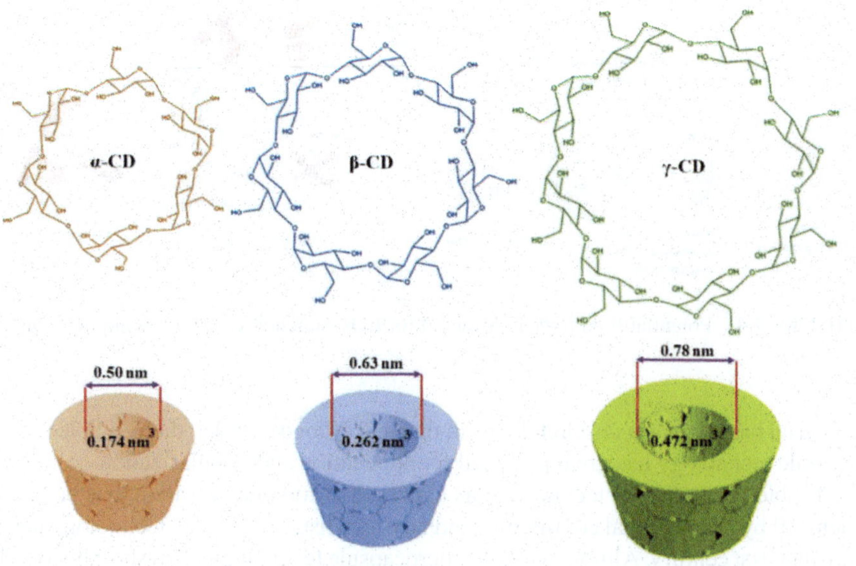

FIGURE 9.5 The chemical structure and some important dimensions of cyclodextrins.

Source: Almasi et al. (2016).

Waals force and dipole–dipole interaction, may also be involved in the combination of objects (Li et al., 2018).

To investigate the effects of CDs and guest molecular types on the formation of inclusion complexes, Astray et al. (2010) investigated the binding constants of 13 different EOs (maltol, furanol, vanillin, methyl cinnamate, cineole, citral, menthol, geraniol, camphor, nocarone, eugenol, *p*-vinyl guaiacol, and limonene) with cyclodextrins (α-CD and β-CD). The results showed that the binding constant of essential oil to β-CD was higher than that of α-CD. Lower hydration capacity and larger cavity size are the main reasons for the formation of β-CD complex.

Moreover, they found that the driving force for host–guest complex formation was hydrophobic–hydrophilic interactions and the compounds with higher hydrophobicity had better binding to CD's cavity. Barba et al. (2015) prepared the powder of eugenol and carvacrol inclusion complex with α-, β-, and γ-CD via spray-drying method and reported higher retention for eugenol (50.4%, molar basis) and carvacrol (79.6%, molar basis) for β-CD in comparison to other CDs.

Generally, the β-CD is considered as the best type of CDs for the complexation of hydrophobic active compounds, thanks to the solubility, the formation of hydrophobic–hydrophilic interactions, the stability of degradation, and the simplicity of inclusion formation. Barba et al. (2015) evaluated the usefulness of β-CD as a controlled-release tool for eugenol- and carvacrol-incorporated WPI films. The release of active compounds to food simulant medium (ethanol:water, 1:1, v/v) with and without the use of β-CD was compared. They suggested that adding β-CD inclusion complex to the WPI matrix resulted in a decreased release rate of eugenol and carvacrol. Specifically, the release time extension for carvacrol was considerable, attributed to the higher hydrophobicity of carvacrol and its better hiddenness in β-CD's cavity. Li et al. (2018) prepared β-CD–thymol inclusion complex and incorporated into gelatin film's casting solution. The release rate of thymol decreased after complexation and sustained release (235 h) of thymol was achieved by incorporating β-CD–thymol inclusion complexes into the gelatin films. Chen and Liu (2016) prepared antimicrobial cellulose sulfate film by incorporating β-CD and mustard EOs. The films incorporated with β-CD reduced the loss of mustard EOs in the preparation and preservation steps of films, confirming the better protection and slow release of EOs.

9.13 CONCLUDING REMARKS

In the past few years, consumers' pursuit for safety, friendliness, and natural products has promoted the development of mild fresh-keeping technology which not only can improve the quality and safety of the product, but also will not cause the loss of nutrition and sense. Natural plant EO just meets this demands. However, the special odor and volatility of essential oil, to some extent, limits its wide application in food industry. But the defect can be solved by the starch microcapsule. After having been encapsulated in starch microcapsule, the biological activity and physical and chemical stability of EOs are protected to a certain extent. At present, due to the inherent properties of the starch microcapsule, its application in some special foods is limited to some extent, as we described earlier, its application mainly concentrated on cereals and meat products. Therefore, it is very important to choose suitable microencapsulation technology and materials according to the nature of a specific product. In the future, the investigation of starch flavor interactions needs to place more emphasis on the structure–property relationships of amorphous or partly crystalline complexes, since these are the relevant structures in food systems. Finally, we should establish the correspondence between the formula of starch microcapsules and EOs and the microorganism species in order to select the corresponding EOs–starch microcapsules according to different corrupt microbes. Only in this way can EOs–starch microcapsules play their antibacterial function better.

ACKNOWLEDGEMENTS

This work was supported by the National Natural Science Foundation of China (32202192), Special fund for Taishan Scholars Project.

REFERENCES

Al-Maqtari, Q. A., Al-Gheethi, A. A. S., Ghaleb, A. D., Mahdi, A. A., Al-Ansi, W., Noman, A. E., and Yao, W. 2022. Fabrication and characterization of chitosan/gelatin films loaded with microcapsules of *Pulicaria jaubertii* extract. *Food Hydrocoll.* 129: 107624.

Arvisenet, G., Voilley, A., and Cayot, N. 2002. Retention of aroma compounds in starch matrices: Competition between aroma compounds towards amylose and amylopectin. *J. Agric. Food Chem.* 50: 7345–7349.

Astray, G., Mejuto, C. J., Morales, R., Rial, O., and Gandara, J. S. 2010. Factors controlling flavors binding constants to cyclodextrins and their applications in foods. *Food Res. Int.* 43(4): 1212–1218.

Baiocco, D., Preece, J. A., and Zhang, Z. 2021. Microcapsules with a fungal chitosan-gum Arabic-maltodextrin shell to encapsulate health-beneficial peppermint oil. *Food Hydrocoll. Health.* 1: 100016.

Bajac, J., Nikolovski, B., Lončarević, I., Petrović, J., Bajac, B., Đurović, S., and Petrović, L. 2022. Microencapsulation of juniper berry essential oil (*Juniperus communis* L.) by spray drying: Microcapsule characterization and release kinetics of the oil. *Food Hydrocoll.* 125: 107430.

Barba, C., Eguinoa, A., and Mate, J. I. 2015. Preparation and characterization of b-cyclodextrin inclusion complexes as a tool of a controlled antimicrobial release in whey protein edible films. *LWT-Food Sci. Technol.* 64(2): 1362–1369.

Belingheri, C., Giussani, B., Rodriguez-Estrada, M. T., Ferrillo, A., and Vittadini, E. 2015. Oxidative stability of high-oleic sunflower oil in a porous starch carrier. *Food Chem.* 166: 346–351.

Campelo, P. H., do Carmo, E. L., Zacarias, R. D., Yoshida, M. I., Ferraz, V. P., de Barros Fernandes, R. V., and Borges, S. V. 2017. Effect of dextrose equivalent on physical and chemical properties of lime essential oil microparticles. *Ind. Crops Prod.* 102: 105–114.

Chen, G., and Liu. B. 2016. Cellulose sulfatebased film with slowrelease antimicrobial properties prepared by incorporation of mustard essential oil and b-cyclodextrin. *Food Hydrocoll.* 55: 100–107.

Coelho, L. M., Gonçalves, I., Ferreira, P., Pinheiro, A. C., Vicente, A. A., and Martins, J. T. 2022. Exploring the performance of amaranth grain starch and protein microcapsules as β-carotene carrier systems for food applications. *Food Structure.* 33: 100287.

Condepetit, B., Escher, F., and Nuessli, J. 2006. Structural features of starch-flavor complexation in food model systems. *Trends Food Sci. Technol.* 17(5): 227–235.

Cortez-Trejo, M. C., Wall-Medrano, A., Gaytán-Martínez, M., and Mendoza, S. 2021. Microencapsulation of pomegranate seed oil using a succinylated taro starch: Characterization and bioaccessibility study. *Food Biosci.* 41: 100929.

Cruz-Tirado, J. P., Ferreira, R. S. B., Lizárraga, E., Tapia-Blacido, D. R., Silva, N. C. C., Angelats-Silva, L., and Siche, R. 2020. Bioactive Andean sweet potato starch-based foam incorporated with oregano or thyme essential oil. *Food Packag. Shelf Life.* 23: 100457.

Dajic Stevanovic, Z., Sieniawska, E., Glowniak, K., Obradovic, N., and Pajic-Lijakovic, I. 2020. Natural macromolecules as carriers for essential oils: From extraction to biomedical application. *Front. Bioeng. Biotechnol.* 8: 563.

Das, A., and Sit, N. 2021. Modification of taro starch and starch nanoparticles by various physical methods and their characterization. *Starch-Stärke.* 73(5–6): 2000227.

Dima, C., Patrascu, L., Cantaragiu, A., Alexe, P., and Dima. 2016. The kinetics of the swelling process and the release mechanisms of *Coriandrum sativum* L. essential oil from chitosan/alginate/inulin microcapsules. *Food Chem.* 195: 39–48.

Emadzadeh, B., Ghorani, B., Naji-Tabasi, S., Charpashlo, E., and Molaveisi, M. 2021. Fate of β-cyclodextrin-sugar beet pectin microcapsules containing garlic essential oil in an acidic food beverage. *Food Biosci.* 42: 101029.

Glenn, G. M., Klamczynski, A. P., Woods, D. F., Chiou, B., Orts, W. J., and Imam, S. H. 2010. Encapsulation of plant oils in porous starch microspheres. *J. Agric. Food Chem.* 58(7): 4180–4184.

Hasan, M., Khaldun, I., Zatya, I., Rusman, R., and Nasir, M. 2022. Facile fabrication and characterization of an economical active packaging film based on corn starch–chitosan biocomposites incorporated with clove oil. *J. Food Measure. Charac.* 1–11.

Heinemann, C., Escher, F., and Conde-Petit, B. 2003. Structural features of starch inclusion complexes in aqueous potato starch dispersions: The role of amylose and amylopectin. *Carbohydr. Polym.* 51: 159–168.

Hosseinnia, M., Khaledabad, M. A., and Almasi, H. 2017. Optimization of *Ziziphora clinopodiodes* essential oil microencapsulation by whey protein isolate and pectin: A comparative study. *Int. J. Biol. Macromol.* 101: 958–966.

Huang, J., Shen, L. L., Chen, J. P., and W, F. 2016. The self-assembly preparation of thyme oil microcapsule optimization. *Food Sci. Technol.* 37(2): 51–57.

Huang, S. Z., Hu, J. P., and Zhu, L. 2017. The essential oil of Artemisia/hydroxypropyl beta cyclodextrin inclusion compound and preparation of sustained-release antibacterial properties. *Food Sci. Technol.* (10): 98–101.

Huang, Y. P., Ye, J. F., Deng, K. B., and Zheng, B. D. 2018. Research Progress of Resistant Starch for Microencapsulation. *Appl. Food Res.* (40): 218–224.

Jadhav, N. V., and Vavia, P. R. 2017. Supercritical processed starch nanosponge as a carrier for enhancement of dissolution and pharmacological efficacy of fenofibrate. *Int. J. Biol. Macromol.* 99: 713–720.

Jaymand, M. 2021. Sulfur functionality-modified starches: Review of synthesis strategies, properties, and applications. *Int. J. Biol. Macromol.* 197: 111–120.

Ju, J., Xie, Y., Guo, Y., Cheng, Y., Qian, H., and Yao, W. 2018. Application of starch microcapsules containing essential oil in food preservation. *Crit. Rev. Food Sci. Nutr.* 1–12.

Kalateh-Seifari, F., Yousefi, S., Ahari, H., and Hosseini, S. H. 2021. Corn starch-chitosan nanocomposite film containing nettle essential oil nanoemulsions and starch nanocrystals: Optimization and characterization. *Polymers.* 13(13): 2113.

Kujur, A., Kiran, S., Dubey, N. K., and Prakash, B. 2017. Microencapsulation of *Gaultheria procumbens* essential oil using chitosan-cinnamic acid microgel: Improvement of antimicrobial activity, stability and mode of action. *LWT-Food Sci. Technol.* 86: 132–138.

Li, H., Cui, Y., Wang, H., Zhu, Y., and Wang, B. 2017. Preparation and application of polysulfone microcapsules containing tung oil in self-healing and self-lubricating epoxy coating. *Colloids Surf. A.* 518: 181–187.

Li, S. Y., Li, X. K., and Zhang, H. 2015. Preparation of garlic oil microcapsules and its storage stability. *J. Food Sci.* 40(2): 40–43.

Lima, P. S., Lucchese, A. M., Araujo-Filho, H. G., Menezes, P. P., Araujo, A. A., Quintans-Junior, L. J., and Quintans, J. S. 2016. Inclusion of terpenes in cyclodextrins: Preparation, characterization and pharmacological approaches. *Carbohydr. Polym.* 151: 965–987.

Lin, Q., Liang, R., Zhong, F., Ye, A., and Singh, H. 2017. Interactions between octenyl-succinic-anhydride-modified starches and calcium in oil-in-water emulsions. *Food Hydrocoll.* 19–30.

Majeed, H., Liu, F., Hategekimana, J., Sharif, H. R., Qi, J., Ali, B., Bian, Y. Y., and Zhong, F. 2016b. Bactericidal action mechanism of negatively charged food grade clove oil nanoemulsions. *Food Chem.* 197: 75–83.

Mamusa, M., Resta, C., Sofroniou, C., and Baglioni, P. 2021. Encapsulation of volatile compounds in liquid media: Fragrances, flavors, and essential oils in commercial formulations. *Adv. Colloid Interface Sci.* 298: 102544.

Marefati, A., Bertrand, M., Sjöö, M., Dejmek, P., and Rayner, M. 2017. Storage and digestion stability of encapsulated curcumin in emulsions based on starch granule Pickering stabilization. *Food Hydrocoll.* 63: 309–320.

Misra, S., Pandey, P., and Mishra, H. N. 2021. Novel approaches for co-encapsulation of probiotic bacteria with bioactive compounds, their health benefits and functional food product development: A review. *Trends Food Sci. Technol.* 109: 340–351.

Moghimi, R., Ghaderi, L., Rafati, H., Aliahmadi, A., and Mcclements, D. J. 2016. Superior antibacterial activity of nanoemulsion of *Thymus daenensis* essential oil against *E. coli*. *Food Chem.* 194: 410–415.

Moura, W. S., Oliveira, E. E., Haddi, K., Correa, R. F., Piau, T. B., Moura, D. S., and Aguiar, R. W. S. 2021. Cassava starch-based essential oil microparticles preparations:

Functionalities in mosquito control and selectivity against non-target organisms. *Ind. Crops Prod.* 162: 113289.

Munhuweyi, K., Caleb, O. J., van Reenen, A. J., and Opara, U. L. 2018. Physical and antifungal properties of β-cyclodextrin microcapsules and nanofibre films containing cinnamon and oregano essential oils. *LWT-Food Sci. Technol.* 87: 413–422.

Porras-Saavedra, J., Pérez-Pérez, N. C., Villalobos-Castillejos, F., Alamilla-Beltrán, L., and Tovar-Benítez, T. 2021. Influence of *Sechium edule* starch on the physical and chemical properties of multicomponent microcapsules obtained by spray-drying. *Food Biosci.* 43: 101275.

Qiu, C., Chang, R., Yang, J., Ge, S., Xiong, L., Zhao, M., and Sun, Q. 2017. Preparation and characterization of essential oil-loaded starch nanoparticles formed by short glucan chains. *Food Chem.* 221: 1426–1433.

Qiu, L., Ma, H., Luo, Q., Bai, C., Xiong, G., Jin, S. and Liao, T. 2022. Preparation, characterization, and application of modified starch/chitosan/sweet orange oil microcapsules. *Foods*. 11(15): 2306.

Reis, D. R., Ambrosi, A., and Di Luccio, M. 2022. Encapsulated essential oils: A perspective in food preservation. *Future Foods*. 100126.

Sani, I. K., Geshlaghi, S. P., Pirsa, S., and Asdagh, A. 2021. Composite film based on potato starch/apple peel pectin/ZrO_2 nanoparticles/microencapsulated *Zataria multiflora* essential oil; investigation of physicochemical properties and use in quail meat packaging. *Food Hydrocoll.* 117: 106719.

Sharif, H. R., Goff, H. D., Majeed, H., Shamoon, M., Liu, F., Nsor-Atindana, J., and Zhong, F. 2017. Physicochemical properties of β-carotene and eugenol co-encapsulated flax seed oil powders using OSA starches as wall material. *Food Hydrocoll.* 73: 274–283.

Sheng, Z. H., Hu, J. P., and Zhu, L. 2017. The essential oil of artemisia/cyclodextrin inclusion compound and preparation of sustained-release antibacterial properties. *Indian Food Ind.* (10): 98–101.

Shi, C., Zhou, A., Fang, D., Lu, T., Wang, J., Song, Y., and Li, W. 2022. Oregano essential oil/β-cyclodextrin inclusion compound polylactic acid/polycaprolactone electrospun nanofibers for active food packaging. *Chem. Eng. J.* 445: 136746.

Singh, N., and Sheikh, J. 2022. Novel Chitosan-Gelatin microcapsules containing rosemary essential oil for the preparation of bioactive and protective linen. *Ind. Crops Prod.* 178: 114549.

Song, X., Wang, L., Liu, T., Liu, Y., Wu, X., and Liu, L. 2021. Mandarin (*Citrus reticulata* L.) essential oil incorporated into chitosan nanoparticles: Characterization, anti-biofilm properties and application in pork preservation. *Int. J. Biol. Macromol.* 185: 620–628.

Sotelo-Boyás, M., Correa-Pacheco, Z., Bautista-Baños, S., and Gómez, Y. G. 2017. Release study and inhibitory activity of thyme essential oil-loaded chitosan nanoparticles and nanocapsules against foodborne bacteria. *Int. J. Biol. Macromol.* 103: 409–414.

Souza, A. C., Goto, G. E. O., Mainardi, J. A., Coelho, A. C. V., and Tadini, C. C. 2013. Cassava starch composite films incorporated with cinnamon essential oil: Antimicrobial activity, microstructure, mechanical and barrier properties. *LWT-Food Sci. Technol.* 54(2): 346–352.

Sun, Y., Zhang, M., Bhandari, B., and Bai, B. 2021. Fennel essential oil loaded porous starch-based microencapsulation as an efficient delivery system for the quality improvement of ground pork. *Int. J. Biol. Macromol.* 172: 464–474.

Teng, X., Zhang, M., Mujumdar, A. S., and Wang, H. 2022. Garlic essential oil microcapsules prepared using gallic acid grafted chitosan: Effect on nitrite control of prepared vegetable dishes during storage. *Food Chem.* 388: 132945.

Tian, Y. Q., Li, Y. X., Zhang, W., and Wang, Y. P. 2016. Oregano essential oil microcapsule was buried process and antibacterial effect. *Packaging Engineering*. 17: 22–31.

Tian, Y. Q., Lu, X. Y., He, X. J., Yi, K., and Huang, C. J. 2005. Microcapsule preparation technology and its application research. *Adv. Mater. Sci., Energy Technol. Environ. Eng.: Proc. Int. Conf.* 11(1): 44–47.

Wang, N., Wang, J. Q., Wang, Y, F., and Ding, H. 2016. Preparation of sodium alginate/porous starch oreoil microcapsules. *Food Sci. Technol. Int.* 9: 037–044.

Wu, H., Zhao, F., Li, Q., Huang, J., and Ju, J. 2022. Antifungal mechanism of essential oil against foodborne fungi and its application in the preservation of baked food. *Crit. Rev. Food Sci. Nutr.* 1–13.

Wulff, G., Steinert, A., and Höller, O. 1998. Modification of amylose and investigation of its inclusion behavior. *Carbohydr. Res.* 307: 19–31.

Xue, J., Michael Davidson, P., and Zhong, Q. 2015. Antimicrobial activity of thyme oil co-nanoemulsified with sodium caseinate and lecithin. *Int. J. Food Microbiol.* 210: 1–8.

Zhao, B., Li, L., Lv, X., Du, J., Gu, Z., Li, Z., and Hong, Y. 2022. Progress and prospects of modified starch-based carriers in anticancer drug delivery. *J. Controlled Release.* 349: 662–678.

Zhong, Y., Xu, J., Liu, X., Ding, L., Svensson, B., Herburger, K., and Blennow, A. 2022. Recent advances in enzyme biotechnology on modifying gelatinized and granular starch. *Trends Food Sci. Technol.* 123. 343–354.

Zhou, D., Pan, Y., Ye, J., Jia, J., Ma, J., and Ge, F. 2017. Preparation of walnut oil microcapsules employing soybean protein isolate and maltodextrin with enhanced oxidation stability of walnut oil. *LWT-Food Sci. Technol.* 83: 292–297.

Zhou, X., Liu, L., and Zeng, X. 2021. Research progress on the utilisation of embedding technology and suitable delivery systems for improving the bioavailability of nattokinase: A review. *Food Struct.* 30: 100219.

Zhou, Y., Wu, X., Chen, J., and He, J. 2021. Effects of cinnamon essential oil on the physical, mechanical, structural and thermal properties of cassava starch-based edible films. *Int. J. Biol. Macromol.* 184: 574–583.

Zhu, H. M., Tian, J. W., Zhang, Y. J., Xu, F., and Chu, Z. 2017. Flavor quality analysis of vanilla essential oil microcapsules prepared from jackfruit seed starch. *Food Sci. Technol.* 22: 050–061.

Zhu, H. M., Zhang, Y., Tian, J., and Chu, Z. 2018. Effect of a new shell material—Jackfruit seed starch on novel flavor microcapsules containing vanilla oil. *Ind. Crops Prod.* 112: 47–52.

10 Preparation Strategy and Application of Edible Coating Containing Essential Oils

10.1 INTRODUCTION

Compared with other types of packaging, edible coatings are becoming more and more popular because of their more environment-friendly properties and active ingredients carrying ability.

The edible coating can reduce the influence of essential oils on the flavor of the product and also can prolong the action time of EOs through the slow-release effect, which effectively promote the application of EOs in food. Understanding the different combinations of edible coatings and EOs as well as their antimicrobial effects on different microorganisms will be more powerful and targeted to promote the application of EOs in real food systems.

10.2 THE CONCEPT AND DEVELOPMENT OF EDIBLE COATINGS

Since the 12th and 13th centuries, wax has been used to delay the dehydration of citrus fruits, although this was not the first use of edible coatings. At the beginning of the 16th century, this technology was applied to meat products to prevent shrinkage and maintain color. At the end of the last century, it was suggested that meat and other foods could be coated with gelatin film to extend their shelf life. As coating technologies have matured and improved, edible polysaccharide coatings, including alginate, carrageenan, pectin, and starch derivatives, have been used to improve the quality of stored meat products (Khaled and Berardi, 2021).

Over the past few years, consumers' increasing demand for safe, convenient, and green food, and their awareness of the negative impact of nonbiodegradable packaging materials have increased people's interest in edible coatings (Acevedo-Fani et al., 2017). Therefore, a lot of work has been done on formulation, application, and characterization of edible coatings in scientific literature and patent literature. Edible coating can be defined as thin layers of material covering the surface of food and can be eaten as part of the whole product (Aayush et al., 2022). Therefore, the composition of edible coatings must conform to the provisions applicable to food. The structure of the edible coating is shown in Figure 10.1.

DOI: 10.1201/9781003329268-10

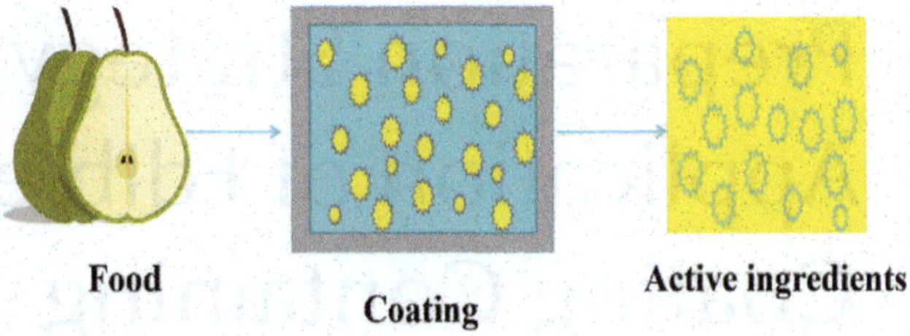

FIGURE 10.1 The structure of the edible coating.

10.3 REQUIREMENTS FOR THE USE OF EDIBLE COATING MATERIALS

When coating is an indispensable part of food and is eaten, it is called "edible coating." We should be more cautious about whether edible coatings are considered a food ingredient or a food additive. Nowadays, there is no specific classification of edible coatings in the European Community.

Food is described in the code of food as all unprocessed, partially treated, or treated substances used for human nutrition and feeding. This involves all ingredients used in formulations, preparation, process of beverages, chewing gum and food, but refuses to use substances such as drugs, cosmetics, and tobacco. If we consider the definition of the latter, edible coatings can be classified as food. However, in most cases, edible coatings do not provide meaningful nutritional value. Therefore, they should be considered more like an additive than an ingredient.

Edible coatings can also be used to improve the nutritional quality of food, thus becoming a qualified food ingredient. As food ingredients, edible coatings should be as tasteless as possible to avoid affecting the quality of the original food. When edible coatings have significant or special flavors, their sensory characteristics must be consistent with those of food. The edible coating must meet the following requirements. First, it has good sensory quality. Second, it has high barrier and mechanical efficiency. Third, it has sufficient biochemical, physical, chemical, and microbial stability. Fourth, it is nontoxic, nonpolluting, and degradable. Fifth, the process is simple and the raw material cost is low.

10.4 MATERIALS USED FOR ESSENTIAL OILS–EDIBLE COATINGS

Coatings can be formulated from different materials and have desired properties. The basic materials for the production of EO–edible coatings are mainly divided into three categories: polysaccharides, proteins, and lipids (Table 10.1) (Ju et al., 2018). EO–edible coatings made of different polysaccharides, proteins, and lipids are conducive to reduce the oxidative degradation of foods, inhibit the growth of spoilage microorganisms, and prolong the shelf life of food. Therefore, the development of EO–edible coatings has attracted considerable attention in recent years. Figure 10.2 shows potential food to be considered for coatings applications.

TABLE 10.1
The Basic Materials for the Production of EO–Edible Coatings and Applications in Food

Coating types	Essential oils	Functions/Results	Product application	Reference
Polysaccharides				
Sweet potato starch	Cumin essential oil	Overall, the antifungal edible coating improved the storage quality of the pears and maintained consumable and better sensory quality compared to uncoated samples. The results suggest that modified sweet potato starch enriched with 0.2–0.4% (v/v) cumin essential oil could be useful as a preservation method to address issues of postharvest losses, and agricultural sustainability in the future	Pear	Oyom et al. (2022)
Alginate and galbanum gum–based edible coatings	*Ziziphora persica* essential oil	Edible coating has obvious antibacterial and antioxidant activity, which significantly improve the shelf life of chicken	Chicken fillet	Hamedi et al. (2017)
Gum Arabic	*Zataria multiflora* Boiss essential oil	The results showed that 6% GA combined with thyme at the concentrations of 0.3% and 0.5% decreased color change and PPO activity of fresh pistachio in comparison with untreated control fruits	Pistachio	Hashemi et al. (2021)
Alginate-based coating	Cinnamon, palmarosa or lemongrass essential oil	The coating improves the shelf life of fruits and vegetables and meat products from the viewpoint of physical chemistry and microbiology	Fresh-cut melon Poached and deli turkey products	Khazaei et al. (2016)
Jujube gum	Nettle oil	The coating improves the shelf life of Beluga sturgeon fillets from sensory, microbiological, and physicochemical perspectives	Beluga sturgeon fillets	Gharibzahedi et al. (2017)
Flaxseed gum	Lemongrass essential oil	Twelve days of storage in the coating effectively inhibit the growth of yeasts and molds, and reduce the color change of the samples	Pomegranate arils	Yousuf and Srivastava (2017)
Basil seed gum	Thymol on oil; viride essential oil	The coating significantly reduced the increase of PV and TBA value during frying and obtained better taste and appearance	Shrimp; fresh-cut apricots	Hashemi et al. (2017); Khazaei et al. (2016)

(*Continued*)

TABLE 10.1
(Continued)

Coating types	Essential oils	Functions/Results	Product application	Reference
Plantago major seed mucilage	*Anethum graveolens* essential oil	Compared with the control group, the experimental group prolonged the shelf life of beef for 9 days	Beef	Behbahani et al. (2017)
Chitosan coating	Mentha essential oil; oregano essential oil; lemon essential oil	The coating significantly inhibited the fungal infection of grape during storage and maintained the original quality better. The coating effectively inhibited the growth of *Listeria monocytogenes* and *Yersinia enterocolitica* during storage of barbecued chicken. The coating inhibits the incidence of gray mold disease during storage of the strawberry	Grape; ready-to-eat barbecued chicken; strawberry	Perdones et al. (2016); Guerra et al. (2017)
Hydroxypropyl methylcellulose	Oregano and bergamot essential oils	The "Formosa" plum in the experimental group had better appearance and quality	"Formosa" plum	Choi et al. (2016)
Starch-based coating	Nigella sativa oil	The coating treatment significantly reduces the softening rate of pomegranate, reducing the loss of vitamins and anthocyanins	Pomegranate arils	Singh et al. (2022)
A sweet potato starch-based coating	Thyme essential oil	Edible coatings effectively extend the shelf life of shrimp during chilling	Shrimp meat	Alotaibi and Tahergorabi (2018)
Chitosan	Cinnamon essential oil	The coatings used in this work were efficient in the conservation of minimally processed pineapple, when compared to the control sample, it also reduced the growth of molds and yeasts, and the loss of weight and firmness. The treatment T4 showed the best results, lower weight loss (14.60%), lower decrease of L* (59.95%), proving to be efficient to retard microbial growth and prolong the shelf life of pineapple	Pineapple	Basaglia et al. (2021)
Proteins				
Whey protein	*Origanum virens* essential oils	The coating effectively inhibited the growth and reproduction of microorganisms, inhibited the evaporation of water, and prolonged the shelf life of sausage for about 20 days	Sausages	Catarino et al. (2017)
Zein	*Zataria multiflora* Boiss. essential oil	The *Lester bacteria* in beef were significantly inhibited when the coating was used	Beef	Moradi et al. (2016)
Whey protein	Clove essential oil	The results depicted that all formulated coatings have potential to maintain quality characteristics and increase the shelf life of tomatoes	Tomato	Kumar and Saini (2021)

Myofibrillar protein	Oregano, cinnamon, and rosemary essential oils	Meanwhile, composite films could significantly reduce the plasma-induced lipid oxidation from 0.16 to 0.06 mg MDA/kg, and decrease the existence of 10 volatile compounds (including alcohols, aldehydes, ketones, and esters) in beef patties after plasma treatment	Beef	Qian et al. (2022)
Whey protein	*Artemisia dracunculus* essential oil	The coated samples also retarded the increase in the contents of TVB-N, pH, TBARS, and FFA during storage. The score less than critical score of 3 was made at day 8 and 12 for fillet coated with control and coated samples except of fillets coated with chitosan, respectively	Fish fillet	Farsanipour et al. (2020)
Composite coatings				
Fish bone gelatin and chitosan	Gallic acid and clove essential oil	Overall, the combination of gelatine, chitosan, gallic acid, and clove oil (GE–CH–GA-CO) had the best performance on salmon fillet preservation and prolonged the shelf life for at least 5 days	Salmon fillet	Xiong et al. (2021)
Whey protein and xanthan gum	Clove essential oil	The results depicted that all formulated coatings have potential to maintain quality characteristics and increase the shelf life of tomatoes	Tomato	Kumar et al. (2021)
Antioxidant pectin and pullulan	*Vitis vinifera* grape seed extract	Raw and roasted peanuts (*Arachis hypogaea*) coated with PEC/PUL/GSE film had a 75% reduction in the peroxide values than uncoated peanuts after 30 days under ambient conditions	Peanut	Priyadarshi et al. (2022)
Gelatin and chitosan containing	*Origanum vulgare* L. essential oil	The composite coating has obvious inhibitory effect on *Staphylococcus aureus* and *Listeria monocytogenes*	Sweet potato	Hosseini et al. (2015)
Lard and starch edible coatings	Black pepper essential oil	The results showed that antimicrobial and antioxidant coating can effectively reduce lipid oxidation and inhibit the growth of microorganisms in semimembranosus in dry-cured ham	Dry-cured ham	Bai et al. (2019)
Sodium bicarbonate and rhamnolipid	Innamon essential oil	The results show that a certain concentration of sodium bicarbonate can effectively reduce the use of laurin oil, and laurin essential oil has some synergies	Cherry and tomato	Xu et al. (2016)
Shrimp chitosan and the essential oil from *Mentha piperita* L. (MPEO) or *M. villosa* Huds (MVEO)	*Mentha piperita* L. or *M. villosa* Huds essential oil	The coating can effectively inhibit the pollution and growth of plant pathogenic fungi	Grape	Guerra et al. (2017)

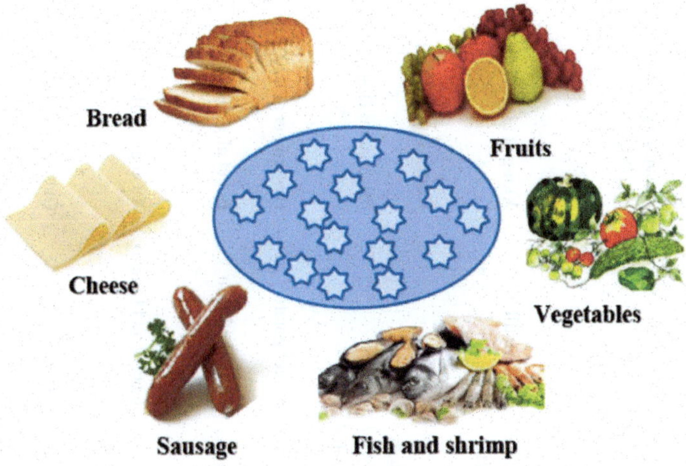

FIGURE 10.2 Potential food to be considered for EO–edible coatings applications.

10.4.1 Polysaccharide-Based Coatings

In recent years, there are many literatures about polysaccharide edible coatings applied in products. Due to the hydrophilic nature, polysaccharide-based coatings may provide only a minimal moisture barrier (Dehghani et al., 2018). Although polysaccharide coatings may not provide a good water vapor barrier, these coatings can be used as "sacrificial agents." Delay the loss of moisture in food by adding extra moisture to the surface of the food. At the same time, after the combination of polysaccharide coating and functional EOs, the conjugate has strong antibacterial and antioxidant effect, which is better able to maintain the original quality and prolong the shelf life of food. The polysaccharide coating mainly includes natural gum, chitosan, starch, alginate, carrageenan, and gellan gum. Hamedi et al. (2017) had researched the effect of alginate coating combined with Ziziphora EOs on the quality of the chicken slices during refrigerated storage and found that the coating could restrain the growth and reproduction of *E. coli*, *P. aeruginosa*, and monocytic lester significantly ($P < 0.05$) during storage and also could reduce the formation of peroxide. Perdones et al. (2016) combined pure chitosan coating with lemon EO for strawberry preservation, and found that chitosan coating can promote the formation of ester in a very short time; besides, the conjugate can transfer the terpene volatiles to the fruit, which made the original fruit flavor more obvious. Recently, some researchers have investigated the effect of sweet potato starch coating containing cumin essential oil on the storage quality of jujube and the effect of chitosan coating containing cinnamon essential oil on the shelf life of pineapple. In their research results, it is confirmed that the polysaccharide-based edible coating containing essential oil can effectively prolong the shelf life of food (Kong et al., 2022; Paidari et al., 2021; Popyrina et al., 2022).

Natural rubber obtained from various plant species has been proved to be effective natural paint. These types of natural rubber are mainly derived from plantain seed, perilla seed, psyllium husk, aloe vera, and locust bean (Kehinde et al., 2020; Choi et al., 2020; Behbahani et al., 2017; Hashemi et al., 2017). The plantago asiatica seed gum and chitosan coating were used in the preservation of fresh-cut apples, respectively. The results showed that the seed coating of plantago asiatica could keep the color of fresh-cut apples or prevent enzymatic browning compared with chitosan coating (Banasaz et al., 2013). Similarly, aloe gel antibacterial coating containing lemon grass essential oil can be used for jujube fruit preservation (Mohammadi et al., 2021). In addition, aloe gel containing cinnamon essential oil and thyme essential oil can prolong the shelf life of nectarines and plums (Khaliq et al., 2019; Saxena et al., 2020).

Chitosan obtained by deacetylation of chitin, also called deacetylation chitin, is a safe, nontoxic, and biodegradable natural polysaccharide. It has been proved to be an effective natural antibacterial substance and its antibacterial activity is considered to be one of the most interesting characteristics of chitosan. Many studies reported the application of chitosan in food. Guerra et al. (2017) studied the effect of chitosan coated with peppermint EO on the quality of fresh grapes. The results showed that the coating can significantly ($P < 0.05$) inhibit the fungal infection in grape during storage. Azevedo et al. (2014) also investigated the effects of edible coating formed by chitosan and carvacrol on postharvest quality of strawberries. The results showed that the coating could effectively ($P < 0.05$) inhibit the decay rate of strawberry during storage. In addition, a number of researchers have used chitosan coatings for the preservation of mangoes, grapes, and guavas (Oh et al., 2017; Frazão et al., 2017; de Aquino et al., 2015).

At present, many native and modified starches from plants have been used in the formulation of edible coatings. Related research used the edible coating of natural or modified corn and cassava starch for the protection of carotene in pumpkin during drying and the result showed that the coating has a significant ($P < 0.05$) inhibit effect on the degradation of carotenoids in pumpkin (Tavassoli-Kafrani et al., 2016). Alotaibi et al. (2017) studied the effects of sweet potato starch combined with thyme flavor coating on the quality of shrimp during freezing. The result showed that the coated shrimp still had higher sensory acceptability and lower colony total value at the end of storage.

Pectin is the soluble component of plant fiber derived from plant cell wall. These polysaccharide derived from plant has poor water holding capacity, so they seem to have a good match with low moisture foods. Gharibzahedi and Mohammadnabi (2017) used the mixture coating of jujube gum and nettle EO to extend the shelf life of sturgeon fillets. Guerreiro et al. (2015) studied the effect of pectin and citral composite coating on the quality of fresh raspberries. Minh et al. (2020) used pectin and other materials to develop a multilayer coating system to extend the shelf life of freshly cut cantaloupe.

10.4.2 Protein-Based Coatings

It has been found that proteins derived from corn, wheat, soybeans, peanuts, milk, or gelatin are suitable for the coating of many foods. Most EO–edible coatings based on protein show good hydrophilic, but in most cases they exhibit poor water vapor diffusivity.

Gluten, collagen, zein, casein, and whey proteins all can be used for the preparation of EO–edible coatings. Moradi et al. (2016) studied the inhibitory effects of *Zataria multiflora* Boiss EOs combined with edible coating based on corn protein on the *L. monocytogenes* and *E. coli* in beef. The results showed that this edible coating could inhibit the growth and reproduction of *L. monocytogenes* and *E. coli* significantly ($P < 0.05$) during the storage of beef and also showed that *L. monocytogenes* was more sensitive than *E. coli* to the coating. Catarino et al. (2017) developed a new edible coating of EO by combining whey protein with oregano EO, which was successfully applied in the processing of meat products. Prior to this, Correa-Betanzo et al. (2011) treated cactus with edible coating based on sodium caseinate and then evaluated its effects on the chemical composition of cactus. It was found that the coating can help to retain the chemical composition of the plant, but the authors proposed the composite coating be used to prevent the rapid deterioration of cactus structure. Recently, related research found that edible coating containing cinnamon EOs can significantly ($P < 0.05$) reduce the oxidation of ham and inhibit the growth of microorganisms during storage (Dai et al., 2017). In addition, a related scholar had investigated the effect of the combination of peanut protein, chitosan, and cinnamon EOs on the shelf life of fish. The results showed that this method can significantly ($P < 0.05$) inhibit fish spoilage (Li et al., 2014).

10.4.3 Lipid-Based Coatings

The lipid coating is usually made up of wax, acyl glycerol, or fatty acids. This kind of hydrophobic edible coating not only can help to reduce the effects of water, light, oxygen, and other external factors on food quality during storage, but also can reduce the water evaporation rate of food itself. In the early days, researchers have explored the inhibitory effect of lipid coating containing oregano EO on *L. monocytogenes*, and the result showed that the EO encapsulated by liposomes had stronger antimicrobial activity than the unsealed EO (Liolios et al., 2009). And some other researchers also found that the lipid coating containing lemon EO and rosemary EO has good preservation effect on spinach (Alikhani-Koupaei, 2014). However, according to investigation, there is relatively little research on the lipid coating containing EOs, and the research on the composite coatings containing EOs has gained more attention.

10.4.4 Composite Coatings

Composite coating mainly refers to the combination of water colloids (proteins or polysaccharides) and lipids (Yousuf and Srivastava, 2017). The coating prepared by single material can exhibit good correlation, but it may not provide multiple functions. For example, polysaccharides and proteins are good film-forming materials which can provide excellent mechanical and structural properties, but their moisture proof effect is poor. Fortunately, the use of hydrophobic lipid components can compensate the drawback. That is to say the composite coating combines the advantages of single coating and avoids its disadvantages at the same time. Therefore, the development of composite coating has gained more attention at present. Azarakhsh et al. (2014) evaluated the effects of composite coatings (glycerol, sodium alginate, and lemon EO) on respiration rate, physicochemical properties, and microbial and sensory quality of fresh pineapple, and the result showed that the coating could significantly ($P < 0.05$) prolong the shelf life of fresh pineapple. Chiumarelli and Hubinger (2014) investigated the effect of the composite edible coating composed of cassava starch, glycerol, Brazil palm wax, and stearic acid on the quality of apple slices. The results showed that the coatings exhibited the best preservation performance when their mass ratios were 3%, 1.5%, 0.2%, and 0.8%, respectively. Fan et al. (2009) prepared a composite coating containing fish gelatin, chitosan, and oregano EO by solvent casting method and composite coating had proved to have a good inhibitory effect on *S. aureus* and *L. monocytogenes*. In addition, the related researchers explored the effect of the thyme EO combined with composite packaging on the shelf life of fresh-cut lettuce (Deng et al., 2016). Other researchers have investigated the compound coating of whey protein concentrate or hydroxypropyl methylcellulose as hydrophilic phase, beeswax, or carnauba wax as a lipid phase. The results showed that the composite coating of whey protein and beeswax could delay the enzymatic browning reaction of apple slices (Perez-Gago et al., 2005).

10.5 DEFINITION AND PRODUCTION METHOD OF COATING

10.5.1 Definition of Essential Oils–Edible Coating

EOs–edible coating is a layer of paint of the mixture of EOs and biological polymers which are able to carry oil (protein, natural gum, modified starch and lipids, etc.). It can not only prevent the exchange of oxygen, water, and carbon dioxide from external and other substances with food, but also can delay the deterioration of food, so as to play a role in preservation. When eaten, they can be eaten together with food, but also can be washed away. Figure 10.3 introduces the production method of the EOs–edible coating and the application process in the food.

10.5.2 Production Method of Coating

The preparation methods of EOs–edible coating mainly include spraying method, dipping method, spreading method, and thin film hydration method. The manufacturing methods and details of EOs–edible coating are given in Table 10.2.

10.5.2.1 Dipping Method

Among the whole methods, only dipping technique can form high thickness coatings. Dipping method is mainly used for fruit, vegetables, and meat products. The density, viscosity, and surface tension of the coating solution have an important influence on the properties of the coating. In dipping method, the food is directly immersed in the corresponding coating solution and removed after a certain period of time. After natural drying, a thin coating layer is formed on the surface of the product. Another method is named as foam application method. This method is usually prepared in emulsion. In this method, the foam will break down by repeated rolling action, so the uniform distribution of coating solution will be over

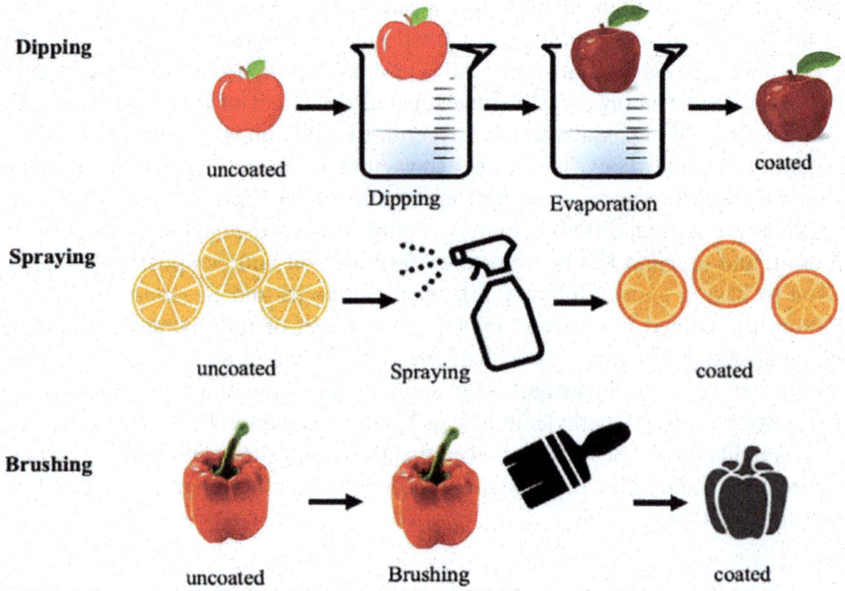

FIGURE 10.3 The production method of the EOs–edible coating and the application process in the food.

the surface of the product. Hamzah et al. (2013) had prepared the coating by dipping method and then studied its effect on the texture and color of papaya. The sodium hydroxide and sodium alginate composite coatings were also prepared by this method, and the effect of the coating on the shelf life of fresh carrots was studied (Mastromatteo et al., 2012). The results showed that this coating could prolong the shelf life of carrots by 7 days. However, it is difficult to form a good attachment to fresh-cut fruits. Therefore, there are some limitations in the formation of edible coatings on micro-machined fruits. At present, multilayer technique is developed to overcome this defect. Here, two or more layers of material are combined with each other by physical or chemical means. For example, Sipahi et al. (2013) investigated the effect of the composite coatings of multilayer antimicrobial algal base and cinnamon EO on the shelf life of fresh watermelon. The results showed that this composite coating can keep the quality and sensory of fresh watermelon and also can prolong its shelf life.

10.5.2.2 Spraying Method

Spraying method is suitable for low-viscosity solution, which can be easily sprayed under high pressure (60–80 psi). The droplet diameter of the solution formed by this method can reach about 20 μm. For example, Song et al. (2015) prepared a perilla seed EO–edible coating successfully by spraying method. The droplet formed by this method has homogeneous and beautiful appearance. However, the polymer coatings formed by spray systems may be affected by other factors, such as drying time, drying temperature, and drying method.

TABLE 10.2
The Manufacturing Methods and Details of EOs–Edible Coating

Method	Preparation conditions	Influencing factors	Product properties	Advantages and disadvantages	Reference
Dipping method	Medium viscosity coating solution	Solution density, viscosity, surface tension, etc.	The coating is thick and uniform	*Advantages:* Easy to operate, do not require high technical means and equipment. *Disadvantages:* Does not apply to the appearance of the higher requirements of the product	Dhanapal et al. (2012)
Spraying method	Low-viscosity coating solution	Drying time, temperature, experimental methods, etc.	The coating is thin and uniform	*Advantages:* It has good product performance and satisfactory appearance. *Disadvantages:* Not suitable for high-viscosity solution, the equipment and technical requirements are relatively high	Ratha et al. (2021)
Spreading method	High-viscosity coating solution	Affected by human factors	Coating thickness between the aforementioned two methods, a little uniform	*Advantages:* Flexible and convenient; suitable for high-viscosity solution. *Disadvantages:* Affected by human factors (nonuniform coating)	Ju et al. (2018)

10.5.2.3 Spreading Method

Spreading method that the coating solution spread directly on the product is a kind of method which is affected by human factors seriously. Ju et al. (2018) used cinnamon and clove EO coating for the preservation of baking by this method. The result showed that this method can significantly ($P < 0.05$) prolong the shelf life of baked foods and also had no significant influence on the sensory quality of baked foods. González-Forte et al. (2014) also brushed two different coatings on pet biscuits using this method. The alginate solution was first brushed onto the surface of the biscuit and then the $CaCl_2$ solution was sprayed to form the gel. Finally, the suspension obtained from gelatinized corn starch was sprinkled on the cookie.

10.5.2.4 Thin Film Hydration Method

Film hydration method is mainly to dissolve phospholipid, cholesterol, and other membrane forming substances in organic solvents, and then remove organic solvents with rotary evaporator, add appropriate buffer, and fully hydrate and disperse finally. In order to reduce the particle size of liposomes and increase the uniformity of liposome size, the products produced by this method need to pass the filter membrane with certain pore size under certain pressure. This method is the most primitive and the most basic method. It is not only easy to operate but also has high encapsulation efficiency. In the past few years, some researchers used this method to explore the effect of liposome embedded lemon EO (EO–liposomes) on spinach quality. The results showed that the EO–liposomes prepared by this method could maintain spinach quality better (Alikhani-Koupaei, 2014). However, due to the limitation of film-forming area, the output of the products produced by this method is small and the continuity is poor, so it is not suitable for large-scale industrial production.

10.6 MECHANISMS OF FOOD PROTECTION

The decline of food quality during storage is not only related to the external environment, but is also related to its tissue decomposition and changes in physiological and chemical structure. For example, one of the core tasks of fresh-cut fruit industry (mainly apples and pears) is how to control pulp browning caused by polyphenol oxidase activity after peeling and cutting (Chumyam et al., 2019; Liu et al., 2019; Li et al., 2017). At present, browning inhibitors are mainly used to cure or inhibit the development of brown discoloration. Different kinds of browning inhibitors can provide different functional mechanisms, including the change of pH and the inhibition of PPO activity. Anti-browning agents acting on enzymes or substrates and/or enzymatic catalytic products can inhibit the formation of browning pigments. Chemical food additives such as citric acid, malic acid, phosphoric acid, ascorbic acid, and calcium ascorbate have been used to control browning in fresh and fresh-cut fruits or vegetables (Liu et al., 2018; Plazzotta et al., 2017; Li, Long et al., 2017). Coating material can be hydrophilic or hydrophobic in nature, but in order to maintain its edibility, water or ethanol should be used as solvent in the process. Other active components can also be added to the matrix to enhance its functional properties. These additives include those additives that can improve or enhance the basic functions of coating materials, such as emulsifiers (SPAN, TWEEN, sodium stearyl lactate-calcium, sodium stearyl lactate, triglyceride, propylene glycol fatty acid ester, sucrose ester, soybean phospholipid, lauric acid monoglyceride, etc.), Plasticizers (glycerol, sorbitol, propylene glycol, sucrose, fatty acids, polyethylene glycol, glycerol monoesters, etc.) cross-linkers (such as transglutaminase for proteins and citric acid or tannic acid for polysaccharides), antimicrobial agents (benzoic acid and its salts, sorbic acid and its salts, calcium propionate, sodium dehydroacetate, natamycin, etc.), and antioxidants (butyl hydroxy anisole, dibutyl hydroxy toluene, propyl gallate, *tert*-butyl hydroquinone, etc.). Different food additives can give different functions to the coating. For example, emulsifiers can be used to stabilize composite coatings and improve their adhesion (such as tween, fatty acid salts, and lecithin). Antioxidants can prevent food oxidation, odor development, and nutritional loss, while antimicrobials can prevent bacterial spoilage and microbial proliferation in food.

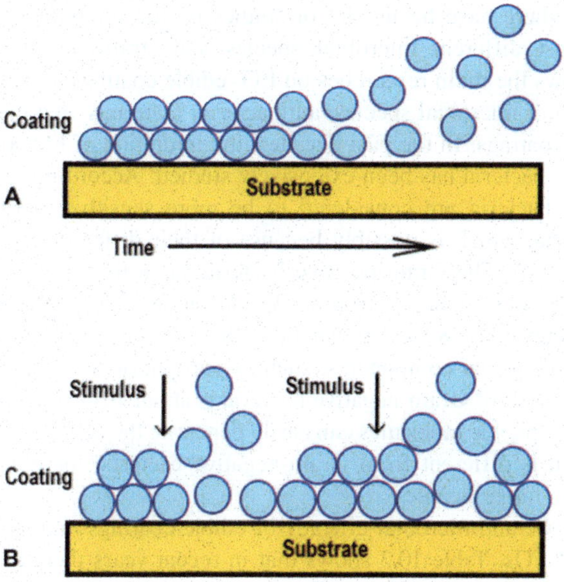

FIGURE 10.4 Two common different diffusion mechanisms of coatings containing antimicrobials.

Source: Atefyekta et al. (2016).

When the coating matrix contains antimicrobial agents, the antibacterial substances must be diffused into the food or packaging environment in order to achieve the purpose of inhibiting the growth of microorganisms. The two common release mechanisms of antimicrobial coatings are simple release (Figure 10.4a) and stimulated release (Figure 10.4b), respectively. In the case of simple release, antimicrobials are usually consumed faster. Stimulated release antimicrobials are released only when certain conditions are met (Dominguez et al., 2019).

10.7 ANTIBACTERIAL ACTIVITY OF ESSENTIAL OILS–EDIBLE COATING

In recent years, the demand for natural, friendly, and safe derivatives to replace synthetic food additives is becoming more and more intense. Hence, EOs are widely recognized as an antiseptic. Although it had been determined that the antimicrobial activity of EOs mainly depended on multiple targets in the cell but not a specific mechanism, their antimicrobial effect mainly based on their molecular hydrophobicity in the final analysis (Salvia-Trujillo et al., 2015). Phenolic compounds in EOs have strong interaction with lipids in cell membranes, thus increasing the permeability of cell membrane, disrupting or destroying the cellular structure, and causing the leakage of intracellular material (Di Pasqua et al., 2007). Formulation of EO–edible coatings tested against different microbial species and strains.

At present, there have been a lot of studies on the inhibitory effects of edible coatings against different microbial species and strains from different foods. Table 10.3 shows the main researches on EO–edible coatings and also summarized the mechanism of microbial species, antibacterial materials, and mechanism action of EO–edible coatings. In the past decades, the inhibition of EOs on food spoilage and pathogenic bacteria has been extensively studied. According to related reports, Gram-positive bacteria are considered to be more sensitive to EOs than Gram-negative bacteria, which is possibly because of their different cell wall structures (Seow et al., 2014). The Gram-positive bacteria has a thick cell wall, the average thickness of which is about 20~80 mm. Besides, it has a rich content of peptidoglycan, which occupies 50–80% dry weight of the cell wall. In addition, it contains a large number of special components, such as phosphoric acid. However, the thickness of the cell wall of Gram-negative bacteria is about 10–15 nm (Wu et al., 2014). Therefore, hydrophobic molecules can easily penetrate the cell wall of Gram-positive bacteria, which is different from Gram-negative bacteria, the outer membrane of which is almost impermeable.

In general, the antimicrobial efficacy of edible coatings mainly depends on the constituents of EOs. Table 10.3 shows that in recent years there have been a large number of researches on different Gram-positive bacteria, such as *S. aureus*, *S. hemolyc*, and *L. monocytogenes* and different Gram-negative bacteria, such as *E. coli*, *S. enteritidis*, and *S. typhimurium*. In addition, some studies have also solved the effect of edible coating of EOs on yeast cells. Yousuf, Qadri et al. (2017) pointed out that the lemongrass EO–edible coating could effectively inhibit the growth of yeast. Artiga-Artigas et al. (2017) studied the effect of the oregano EO–edible coating on the shelf life of the cheese. And the result showed that the coating could effectively inhibit the growth of the yeast in the storage of cheese. Frazão et al. (2017) also studied the effects of *Myrcia ovata Cambessedes* EO–edible coatings on the quality of mangaba fruits storage. The result showed that this method could effectively control the growth and reproduction of yeast and mold of mangaba fruits during storage. According to relevant studies, many EO–edible coatings have been found to have stronger antimicrobial activity and longer duration than free EOs. This may be because the formulation and surface charge of edible coatings affect the mechanism of EOs to the cell membrane. Even in many cases, EO–edible coating can interact with multiple molecular sites on microbial cell membrane. According to relevant literature, the use of large molecules may help to improve the antimicrobial activity of EOs. For example, both the combination of sodium caseinate and thyme EO (Xue et al., 2015) or the combination of sodium alginate and citronella EO can play a synergistic inhibitory (Salvia-Trujillo et al., 2015). Interestingly, however, it is reported that the interaction of some macromolecules as emulsifiers and EO can reduce the antimicrobial activity of EO to a certain extent. For example, the coating of whey protein and maltose dextrin combined with lilac EO shows lower antimicrobial activity than free EO. The reason for this phenomenon may be that some polymer matrices are made into EO–edible coatings after the interaction with EOs, which can produce electrostatic interactions with corresponding cell membranes. Or because the proportion of the water phase contained in the coating is large, which

TABLE 10.3
Formulation of EO–Edible Coatings Tested against Different Microbial Species and Strains

Classification	Microorganism	Coating composition	Product	Result	Antibacterial mechanism	Reference
Gram-positive bacteria	*Listeria monocytogenes*	Whey protein, *origanum virens* essential oils	Sausages	Coating inhibited the growth of *Listeria monocytogenes* and delayed the lipid oxidation of sausage during storage	EO coating prolongs the antibacterial activity of EO by sustained release; coating improves the solubility of EO	Catarino et al. (2017)
		Sodium alginate-carboxy methyl cellulose and *Ziziphora clinopodioides* essential oil	Silver carp fillet	This method significantly inhibits the growth and reproduction of *Listeria monocytogenes* during storage	The combination of EO and coating increases the ability of EO to enter the cell and increases the antibacterial activity of EO	Rezaei and Shahbazi (2018)
		Sodium alginate, galbanum gum, and *Ziziphora persica* essential oil	Chicken fillet	EO coating significantly inhibited the growth and reproduction of microorganisms in chicken fillets during late storage	Coating prolongs the antimicrobial activity of the EO	Hamedi et al. (2017)
		Turmeric residue, gelatin hydrogelor cassava starch, gelatin hydrogels, and purified curcumin	Sausages	The coating has a great antibacterial potential for sausages	The combination of the coating with EO enhances the antibacterial activity of EO	Tosati et al. (2018)
	Staphylococcus aureus	Chitosan-cassava starch, *Lippia gracilis* essential oils	Guavas	Good to keep the quality of guavas during storage	Coating improves EO dispersibility	de Aquino et al. (2015)
		Carboxymethyl cellulose and garlic essential oil	Strawberries	The coating for food preservation has greater potential applications	The coating improves the solubility of the EO and enhances the ability of the EO to enter the cell	Dong and Wang (2017)

(Continued)

TABLE 10.3
(Continued)

Classification	Microorganism	Coating composition	Product	Result	Antibacterial mechanism	Reference
		Sodium alginate, mandarin fiber, Tween 80 and oregano essential oil	Low-fat cut cheese	Edible coating can better inhibit the growth of microorganisms, so that the appearance and nutrition of cheese has been better maintained	Coatings improve the dispersion of EO and prolong its effective antimicrobial time	Artiga-Artigas et al. (2017)
	Bacillus subtilis	Chitosan-cassava starch, *Lippia gracilis* essential oils	Guavas	Good to keep the quality of guavas during storage	The combination of coating and EO increases the antibacterial activity of EO	de Aquino et al. (2015)
		Cassava starch, chitosan, thymol, and carvacrol	Strawberries	Edible coating effectively inhibits the growth and reproduction of microorganisms, ensures the microbiological safety of perishable foods such as fruits	The interaction of cassava starch with the cell wall of microorganisms increases the activity of the coating	Frazão et al. (2017)
	Lactobacillus	Carboxyl methyl cellulose and rosemary essential oil	Smoked eel	The method effectively reduces the number of lactic acid bacteria and inhibits the formation of oxidation products	The combination of extract and coating played a synergistic inhibitory effect	Azarakhsh et al. (2014)
		Sodium alginate, galbanum gum, and *Ziziphora persica* essential oil	Chicken fillet	The edible coating significantly inhibited the growth and reproduction of microorganisms after 12 days of storage	The edible coating improved the ability of EO to enter cells and prolonged the antibacterial time of EO	Hamedi et al. (2017)
Gram-bacteria	*Escherichia coli*	Whey protein, *Origanum virens* essential oils	Sausages	Significantly inhibited the growth of foodborne pathogens in the sausage	The coating prolongs the antibacterial activity of EO by sustained release	Catarino et al. (2017)
		Chitosan-cassava starch, *Lippia gracilis* essential oils	Guavas	Good to keep the quality of guavas during storage	The combination of coating and EO increases the solubility of EO	de Aquino et al. (2015)

	Alginate-based eugenol and citral	Arbutus unedo	The edible coating better maintains the sensory attributes and taste quality of *Arbutus unedo* during storage, and effectively inhibits the growth of spoilage microorganisms	Edible coatings improve the dispersibility of EO	Guerreiro et al. (2015)
	Sodium alginate and lemongrass essential oil	Fresh-cut Fuji apples	This method significantly reduces the rate of ethylene production, delaying the browning of the fruit	EO solubility and adhesion were improved	Salvia-Trujillo et al. (2015)
Salmonella	Whey protein, *Origanum virens* essential oils	Sausages	Significantly inhibited the growth of foodborne pathogens in the sausage	The combination of EO with the coating prolongs the antibacterial activity of the edible coating by sustained release and improves its dispersibility	Catarino et al. (2017)
	Chitosan-cassava starch, *Lippia gracilis* essential oils	Guavas	Good to keep the quality of guavas during storage	The combination of EO and chitosan-cassava starch significantly improved the antibacterial activity of the coating	de Aquino et al. (2015)
	Chitosan and lemongrass oil	Grape berries	Coatings can better preserve the microbial safety of grape berries	Coating improves the solubility of essential oils	Oh et al. (2017)
	Alginate-based eugenol and citral	Arbutus unedo	Coating effectively extends shelf life of arbutus unedo	EO coatings prolong antibacterial activity through sustained release	Guerreiro et al. (2015)
Pseudomonas	Chitosan-cassava starch, *Lippia gracilis* essential oils	Guavas	This approach has the effect of inhibiting the growth and reproduction of *Pseudomonas* during storage	The combination of EO and chitosan-cassava starch significantly improved the antibacterial activity of the coating	de Aquino et al. (2015)

(*Continued*)

TABLE 10.3
(Continued)

Classification	Microorganism	Coating composition	Product	Result	Antibacterial mechanism	Reference
		Carboxyl methyl cellulose and rosemary essential oil	Smoked eel	EO–edible coatings inhibit the growth of *Pseudomonas* and the formation of oxidation products	The combination of extract and coating played a synergistic inhibitory effect	Azarakhsh et al. (2014)
		Tragacanth gum, *Satureja khuzistanica* essential oil	Button mushroom	Coating significantly inhibited the growth and reproduction of microorganisms	The combination of EO and coating significantly increases the antibacterial activity of the coating	Nasiri M et al. (2017)
Fungi	Yeast	Chitosan and bergamot essential oil	Grapes	Edible coatings preferably maintain the quality of the grapes during storage	The combination of coating and EO improves the antibacterial effect of EO	Sánchez-González et al. (2011)
		Sodium alginate and citral and eugeno	Fresh raspberries	This method can effectively inhibit the growth of yeast, extending the shelf life of raspberries	EO combined with coating has played a joint antibacterial effect	Guerreiro et al. (2015)
		Alginate-based and lemongrass essential oil	Fresh-cut pineapples	The coating may extend the shelf life of fresh-cut pineapples	The coating improves the dispersibility and solubility of EO	Azarakhsh et al. (2014)
		Chitosan and lemongrass oil	Grape berries	The coating can effectively inhibit the growth of yeast, and make the product maintain good color	The combination of the coating with EO has synergistic bacteriostasis and prolongs the antibacterial time of EO	Oh et al. (2017)
		Basil seed gum, *Origanum vulgare* spp. and *viride* essential oil	Fresh-cut apricots	Yeast growth is inhibited, extending the shelf life of apricot	Coating improves EO dispersibility	Hashemi et al. (2017)
		Tragacanth gum, *Satureja khuzistanica* essential oil	Button mushroom	The coating effectively extends the shelf life of mushrooms to 16 days	The combination of EO and coating significantly increases the antibacterial activity of the coating	Nasiri M et al. (2017)

		Flaxseed gum, lemongrass essential oil	Pomegranate arils	Coating effectively inhibits the increase in the number of microbial populations, better maintain the intrinsic color of the samples	The coating increases the ability of the EO to enter the cell and improves the interaction between the EO and the cell	Yousuf and Srivastava (2017)
		Cassava-starch, chitosan, and *Myrcia ovata Cambessedes* essential oils	Mangaba	This method is particularly suitable for controlling the growth of foodborne bacteria during storage of mangoes	Edible coatings increase the solubility and prolong the antibacterial time of EO	Frazão et al. (2017)
		Tragacanth gum and *Zataria multiflora* Boiss. essential oil	Button mushrooms	The method significantly inhibits the growth and propagation of yeast and prolongs the shelf life of mushrooms	The combination of EO and coating significantly improved the antibacterial activity of the coating and delayed the release of EO	Nasiri M et al. (2017)
	Botrytis cinerea	Shrimp chitosan and *Mentha piperita* L. essential oil	Grapes	The method effectively reduces the rate of decay of grapes during storage	Coating combined with EO increases the antibacterial activity of the coating	Guerra et al. (2017)
		Chitosan and lemon essential oil	Strawberry	The antifungal activity of strawberry was enhanced, and the respiration rate of strawberry during storage was reduced	EO combined with chitosan coating has a synergistic antimicrobial effect	Perdones et al. (2016)
Total bacterial count		Hydroxypropyl methylcellulose; oregano and bergamot essential oil	Fresh "Formosa" plum	The coating significantly inhibited the increase in the total bacterial count of apples during storage	The interaction between coatings and EO improves the dispersibility of EO	Choi et al. (2016)
		Basil seed gum, *Origanum vulgare* spp. and *viride* essential oil	Fresh-cut apricots	EO coating significantly inhibited the increase of total bacterial count, better maintained the quality of apricots	The addition of the essential oil reduces the water vapor transmission rate of the coating and enhances the antimicrobial activity of the coating by associating with the coating	Hashemi et al. (2017)

(*Continued*)

TABLE 10.3
(Continued)

Classification	Microorganism	Coating composition	Product	Result	Antibacterial mechanism	Reference
		Sodium alginate, pectin, citral essential oils and eugenol essential oils	Strawberries	The coating better preserves the physicochemical properties of the strawberry during storage and prolongs the shelf life of the strawberry	The coating increases the antimicrobial activity of the EO	Guerreiro et al. (2015)
		Sweet potato starch and thyme essential oil	Shrimp	This method effectively prolongs the shelf life of frozen shrimp	The coating increases the solubility and prolongs the antibacterial activity of EO	Alotaibi et al. (2017)
		Jujube gum and nettle essential oil	*Beluga sturgeon*	The coating significantly inhibited the growth of microorganism in fish and lipid oxidation during refrigerated storage	The effect of the coating on EO solubility improves their interaction with the cell membrane	Gharibzahedi et al. (2017)

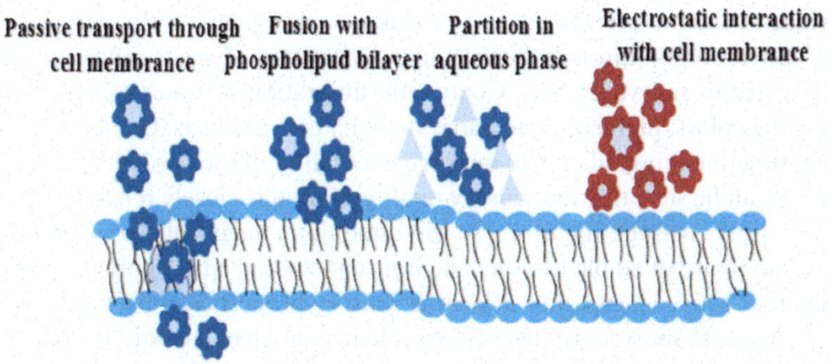

FIGURE 10.5 Schematics of the different routes promoted by the EOs–edible coating for the interaction of EOs with the microbial cell membranes.

greatly reduces the interaction between the EO and the cell membrane. The different interpretations proposed earlier are illustrated in Figure 10.5.

The interaction between EO–edible coating and the membrane of the microorganism may be promoted mainly through the following ways. First of all, the interaction of EO–edible coating with cell membrane can be improved by increasing the contact area between EO and cell membrane (Donsì et al., 2012). In addition, EO can also destroy the cell membrane by altering the integrity of the phospholipid bimolecular layer or by inhibiting the active transporter protein (Moghimi et al., 2016). Second, the edible coating with emulsification can fuse with the phospholipid bilayer of cell membrane, which may promote the targeted release of EO on cell site. In addition, the specific interaction between the emulsifier and the cell membrane has been reported to be able to increase the antibacterial activity of EO (Salvia-Trujillo et al., 2014, 2015). Third, the edible coating with positive charge can interact with negatively charged microorganism cell walls, which can increase the concentration of EO. However, this hypothesis is still controversial. For example, cationic surfactant such as lauric acid also has strong antimicrobial activity (Xue et al., 2015; Chang et al., 2015). And it is also reported that the charge of the emulsion does not affect the antimicrobial activity of the EO (Majeed et al., 2016). Finally, the edible coating may delay the release of the EO components and prolong the action time of it. In accordance with this hypothesis, related research had found that the EO–edible coating could significantly ($P < 0.05$) prolong the action time compared with free EO (Sánchez-González et al., 2011; Nair et al., 2020; Maringgal et al., 2020). The mechanisms of the different routes described in Figure 10.5 may coexist and are difficult to appear alone. Therefore, further explorations are needed to improve the understanding of the basic mechanism of EO–edible coating, so as to improve the specific antibacterial effect of antibacterial agents on target microorganisms.

10.8 FOOD SENSORY PROPERTIES RELATED TO EDIBLE COATINGS

In theory, edible coating as a protective coating of food can maintain or improve the quality of food. For example, edible coating can form an effective barrier on the surface of food to prevent the loss of water, so as to achieve the purpose of controlling dehydration—through the selective penetration of gas to delay the ripening process of fruits and vegetables. Control the migration of water-soluble solutes to maintain the colors, pigments, and nutrients of natural products. Combine with other spices or functional ingredients to improve the quality of the product.

Thus, it can be seen that the sensory quality of food is closely related to the application of edible coatings and films. At the same time, many active compounds used in the manufacture of edible films and coatings, including edible polymers, plasticizers, and other active agents, may affect the sensory properties of packaged or coated products, because most active agents have their own characteristic flavors and colors, and may produce other unique flavors when combined with other compounds in food. Since the functional edible film and coating are part of the packaging or food, all ingredients in the edible film and coating should not interfere with the sensory properties of the food.

Generally speaking, tasteless edible films and coatings are ideal. Fortunately, most people focus their attention on the combination of edible films and paints with active compounds, and the concentrations of these active compounds are usually very low, so their taste effects are negligible. When high concentrations of natural active agents are added to edible films and coatings, strong flavors may be produced. This phenomenon becomes more pronounced especially when essential oils or flavors are added to the edible film. Unfortunately, there are few studies on the effects of edible films on the sensory quality of food. Therefore, the following content will focus on the analysis of the effects of edible coatings or films on the sensory properties of food.

10.8.1 Appearance of the Product

The appearance changes of food mainly include surface dehydration, whitening, discoloration (enzymatic browning), and glossiness decrease. The use of edible coating can significantly affect or change the appearance of the product. For example, by reducing moisture loss, edible coatings can control product surface dehydration and discoloration. By selecting the right material, the coating can enhance the surface gloss of the food. The oxygen-blocking function of edible paint can also prevent the enzymatic browning of some fresh-cut fruits and vegetables. Therefore, the functional coating has the potential to maintain or enhance the surface appearance of the product during storage.

The delay of surface color of fresh carrots and fresh strawberries has been confirmed by edible coating experiments. The white discoloration on the surface of small carrots is the result of dehydration on the surface of lignin, and lignin affects its storage quality and shelf life. The edible coating acts as a moisture proof or surface moisturizing agent to control the change of the surface color of small carrots and enhance the orange intensity. The same study proved that baby carrots coated with xanthan gum had brighter colors and higher orange intensity levels than uncoated ones, so the coated products had the best color properties (Mei et al., 2002).

Enzymatic browning is the main factor leading to the quality deterioration of some fruits and vegetables, especially fresh-cut fruits and vegetables. The use of edible coating can reduce the transfer of O_2 to the surface of the product, and anti-browning agent can be added to the formulation of the coating to prevent enzymatic browning of fruits and vegetables (Ballesteros et al., 2022; Dwivany et al., 2020; Duguma, 2022). With the increasing demand for fresh and least processed fruits and vegetables, the use of edible coatings to prevent enzymatic browning has become more attractive and important.

The color, glossiness, and transparency of edible coatings and films vary significantly depending on the chemical composition and structure of the polymers used, thus affecting the surface appearance of packaged or coated foods. For example, shellac-based coatings are commonly used in fresh fruits (especially sugar cane) to increase their gloss and marketability. Shellac-based paint is also often used in the fruit industry to keep fresh. However, although shellac-based coatings can also give fruit luster, the lack of durability of shellac-based coatings limits their use when fruits must be stored and transported for a long time. If the fruits with shellac-based coating experience too many changes in temperature and humidity, it will cause them to "sweat," then the shellac-based coating may become white, thus affecting the color of the fruit. This phenomenon becomes more serious, especially when the fruit is transferred from the freezer to a humid environment. The color stability of the coating of whey protein isolates is also a problem because milk proteins undergo Millard reaction during storage, resulting in yellowing of the coating. The rate of brown pigment formation in whey powders has been shown to increase as storage temperature and water activity increase from 25 to 45°C and 0.33 to 0.6, respectively. Trezza and Krochta (2000) compared the yellowing rates of commercial coatings to those of whey protein coatings during storage at 23, 40, and 55°C at 75% RH, and found that WPI coatings had lower yellowing rates than whey protein concentrate (WPC) and the same rates as shellac; hydroxypropyl methylcellulose (HPMC) coatings had the lowest yellowing rates, and zein coatings became less yellow during storage. It was concluded that WPI coatings could be used as an alternative coating to shellac or HPMC when low color change is desired, and WPC coatings have potential applications when color development is desired.

10.8.2 Texture of the Product

Hardness and brittleness are important texture indexes of fresh agricultural products. In the process of postharvest storage, the taste of the product decreased due to the loss of moisture and the loss of postharvest brittleness. However, edible coating can reduce the water loss of fruits and vegetables and delay the ripening process of fresh fruits and vegetables. In addition, edible coatings can improve the mechanical integrity or handling properties of the product. Many studies have fully demonstrated that the coating has the potential to maintain the hardness of fresh fruits and vegetables compared with uncoated products (Nicolau-Lapena et al., 2021; Du et al., 2022; Agriopoulou et al., 2020).

The use of paint can improve the integrity of food, especially frozen food. Chitosan-based coating can reduce the loss of raspberries by 24% after thawing. This may be because the edible coating on the surface of the fruit helps to retain moisture in the fruit and prevent moisture from moving from high temperature to the environment during freezing. In the case of freeze-thawed raspberries, the chitosan coating containing calcium increased the hardness by about 25% compared with uncoated fruits. The addition of calcium to the paint formula provides additional benefits in maintaining the integrity and texture of frozen-thawed berries (Han et al., 2004).

10.8.3 Taste, Smell, and Other Sensory Properties of the Product

Due to the selective permeability of O_2 and CO_2, edible coatings can delay the production of ethylene and the ripening process of fruits and vegetables, thus preventing the smell of fresh agricultural products during storage. On the contrary, improperly designed coatings may produce anaerobic conditions, resulting in peculiar smell and odor of the product. Plasticizer is the main component of edible film and coating, which can affect the flavor and taste of film and coating. Common edible plasticizers are glycerin, sorbitol, and polyethylene glycol. Polyethylene glycol is tasteless. The taste of glycerol and sorbitol is sweet, but the sweetness of glycerol in the protein film is negligible, and the sweetness of sorbitol is obvious. Therefore, the possible taste of plasticizer needs to be considered when adding plasticizer.

In addition, some coating materials may also produce an undesirable smell due to the interaction between the components used to make the film and the coating. The use of shellac coating can achieve a "de-astringent" effect on citrus fruits and some apple varieties. However, chitosan-based coatings often produce a special flavor during application. This is mainly because acid (acetic acid or lactic acid) is usually needed to assist the dissolution of chitosan in aqueous solution when preparing chitosan-based edible layer. The pH value of the solution is usually in the range of 3.9–4.2. However, in this pH range, chitosan will show a strong bitter and astringent taste, which is the main factor restricting the entry of chitosan-coated food into the market. The convergence of acids in chemical reactions is not clear. A common hypothesis is that acids precipitate proteins in saliva or cause enough conformational changes that lose their sense of lubrication when they form complexes with salivary proteins or mucopolysaccharides. The convergence related to chitosan may be due to the increase of mineralized groups caused by the dissolution of chitosan in acidic solution. The pK_a of the amino group of glucosamine residue is about 6.3; hence chitosan is polycationic at acidic pH value. Chitosan can selectively bind desired materials, such as proteins. When binding with salivary proteins, the affinity of chitosan in acidic solutions might be increased.

10.9 CONCLUDING REMARKS

Essential oils encapsulated in edible coating have better physical and chemical stability, which shows more fresh-keeping advantage for food. However, several challenges still remain for the full exploitation of EO–edible coating within a mild strategy for food preservation. As mentioned earlier, some researchers have found that the interaction of some individual macromolecules used as emulsifiers with EO can reduce the antimicrobial activity of EO to some extent. Therefore, finding a proper combination of food and coating materials is an important factor to be considered. In addition, an important role of the EO–edible coating is the controlled release of EOs. In this respect, it will be a great value to combine it with food packaging materials in the packaging industry. But how to solve the activity of EOs during the high temperature production of packaging materials will be an important problem. And we also should establish a relationship between the prescription of the EO–edible coating and the species that it inhibits, so as to select the optimum EO–edible coating based on different spoilage microbes. In this way, it can play a better role in its antimicrobial activities. In addition, EOs usually have larger pungent odors. If they are applied directly to the food surface, they will have a certain effect on the quality of food finally. Therefore, it is also the focus of our future research on how to cover the odor of EO under the premise of guaranteeing the antibacterial activity of EO and the slow-release effect of the coating.

ACKNOWLEDGEMENTS

This work was supported by the National Natural Science Foundation of China (32202192), Special fund for Taishan Scholars Project.

REFERENCES

Aayush, K., McClements, D. J., Sharma, S., Sharma, R., Singh, G. P., Sharma, K., and Oberoi, K. 2022. Innovations in the development and application of edible coatings for fresh and minimally processed Apple. *Food Control*. 109188.

Acevedo-Fani, A., Soliva-Fortuny, R., and Martín-Belloso, O. 2017. Nanoemulsions as edible coatings. *Curr. Opin. Food Sci.* S2214799316301369.

Agriopoulou, S., Stamatelopoulou, E., Sachadyn-Król, M., and Varzakas, T. 2020. Lactic acid bacteria as antibacterial agents to extend the shelf life of fresh and minimally processed fruits and vegetables: Quality and safety aspects. *Microorganisms*. 8(6): 952.

Alikhani-Koupaei, M. 2014. Liposome-entrapped essential oils on in vitro and in vivo antioxidant activity in leafy vegetables. *Qual. Assur. Saf. Crops Foods*. 7: 369–373.

Alotaibi, S., and Tahergorabi, R. 2018. Development of a sweet potato starch-based coating and its effect on quality attributes of shrimp during refrigerated storage. *LWT-Food Sci. Technol.* 88: 203–209.

Artiga-Artigas, M., Acevedo-Fani, A., and Martín-Belloso, O. 2017. Improving the shelf life of low-fat cut cheese using nanoemulsion-based edible coatings containing oregano essential oil and mandarin fiber. *Food Control*. 76: 1–12.

Atefyekta, S., Ercan, B., Karlsson, J., Taylor, E., Chung, S., Webster, T. J., and Andersson, M. 2016. Antimicrobial performance of mesoporous titania thin films: Role of pore size, hydrophobicity, and antibiotic release. *Int. J. Nanomed.* 11: 977–990.

Azarakhsh, N., Osman, A., Ghazali, H. M., Tan, C. P., and Adzahan, N. M. 2014. Lemongrass essential oil incorporated into alginate-based edible coating for shelf-life extension and quality retention of fresh-cut pineapple. *Postharvest Biol. Technol.* 88: 1–7.

Azevedo, A. N., Buarque, P. R., Cruz, E. M. O., Blank, A. F., Alves, P. B., Nunes, M. L., and de Aquino Santana, L. C. L. 2014. Response surface methodology for optimisation of edible chitosan coating formulations incorporating essential oil against several foodborne pathogenic bacteria. *Food Control*. 43: 1–9.

Ballesteros, L. F., Teixeira, J. A., and Cerqueira, M. A. 2022. Active carboxymethyl cellulose-based edible coatings for the extension of fresh goldenberries shelf-life. *Horticulturae*. 8(10): 936.

Banasaz, S., Hojatoleslami, M., Razavi, S. H., Hosseini, E., and Shariaty, M, A. 2013. The Effect of psyllium seed gum as an edible coating and in comparison to chitosan on the textural properties and color changes of Red Delicious apple. *Int. J. Food Sci. Nutr.* 2: 651–657.

Basaglia, R. R., Pizato, S., Santiago, N. G., de Almeida, M. M. M., Pinedo, R. A., and Cortez-Vega, W. R. 2021. Effect of edible chitosan and cinnamon essential oil coatings on the shelf life of minimally processed pineapple (*Smooth cayenne*). *Food Bioscience*. 41: 100966.

Behbahani, B. A., Shahidi, F., Yazdi, F. T., Mortazavi, S. A., and Mohebbi, M. 2017. Use of Plantago major seed mucilage as a novel edible coating incorporated with *Anethum graveolens* essential oil on shelf-life extension of beef in refrigerated storage. *Int. J. Biol. Macromol.* 94: 515–526.

Bi, Y., Zhou, G., Pan, D., Wang, Y., Dang, Y., Liu, J., and Cao, J. 2019. The effect of coating incorporated with black pepper essential oil on the lipid deterioration and aroma quality of Jinhua ham. *J. Food Measure. Charac.* 13(4): 2740–2750.

Catarino, M. D., Alves-Silva, J. M., Fernandes, R. P., Gonçalves, M. J., Salgueiro, L. R., Henriques, M. F., and Cardoso, S. M. 2017. Development and performance of whey protein active coatings with *Origanum virens* essential oils in the quality and shelf-life improvement of processed meat products. *Food Control*. 80: 273–280.

Chang, Y., McLandsborough, L., and McClements, D. J. 2015. Fabrication, stability and efficacy of dual-component antimicrobial nanoemulsions: Essential oil (thyme oil) and cationic surfactant (*Lauric arginate*). *Food Chem.* 172: 298–304.

Chauhan, O. P., Raju, P. S., Singh, A., and Bawa, A. S. 2011. Shellac and aloe-gel-based surface coatings for maintaining keeping quality of apple slices. *Food Chem.* 126: 961–966.

Chiumarelli, M., and Hubinger, M. D. 2014. Evaluation of edible films and coatings formulated with cassava starch, glycerol, carnauba wax and stearic acid. *Food Hydrocoll.* 38: 20–27.

Choi, H. J., Song, B. R., Kim, J. E., Bae, S. J., Choi, Y. J., Lee, S. J., and Hwang, D. Y. 2020. Therapeutic effects of cold-pressed perilla oil mainly consisting of linolenic acid, oleic acid and linoleic acid on UV-induced photoaging in NHDF cells and SKH-1 hairless mice. *Molecules*. 25(4): 989.

Choi, W. S., Singh, S., and Lee, Y. S. 2016. Characterization of edible film containing essential oils in hydroxypropyl methylcellulose and its effect on quality attributes of Formosa plum (*Prunus salicina* L.). *LWT-Food Sci. Technol.* 70: 213–222.

Chumyam, A., Faiyue, B., and Saengnil, K. 2019. Reduction of enzymatic browning of fresh-cut guava fruit by exogenous hydrogen peroxide-activated peroxiredoxin/thioredoxin system. *Scientia Horticulturae*. 255: 260–268.

Correa-Betanzo, J., Jacob, J. K., Perez-Perez, C., and Paliyath, G. 2011. Effect of a sodium caseinate edible coating on berry cactus fruit (*Myrtillocactus geometrizans*) phytochemicals. *Food Res. Int.* 44: 1897–1904.

Dai, Z. Q., Luo, J., Xiao, S. L., Xu, X. L., and Zhang, J. H. 2017. Effect of compound essential oil coating on antibacterial and anti-oxidation of dried cured ham. *Meat Res.* 31(8): 1–5.

de Aquino, A. B., Blank, A. F., and de Aquino Santana, L. C. L. 2015. Impact of edible chitosan—cassava starch coatings enriched with *Lippia gracilis* Schauer genotype mixtures on the shelf life of guavas (*Psidium guajava* L.) during storage at room temperature. *Food Chem.* 171: 108–116.

Dehghani, S., Hosseini, S. V., and Regenstein, J. M. 2018. Edible films and coatings in seafood preservation: A review. *Food Chem.* 240: 505–513.

Deng, W. J., Jiang, W. L., Chen, A. J., and Lan, W. J. 2016. The antibacterial effect of coating packaging on the shelf life of fresh cut lettuce physicochemical and microbial quality of thyme oil. *Indian Food Ind.* 07: 247–253.

Dhanapal, A., Sasikala, P., Rajamani, L., Kavitha, V., Yazhini, G., and Banu, M. S. 2012. Edible films from polysaccharides. *Handb. Food Sci. Technol. 1: Food Alteration Food Qual.* 3: 9–18.

Di Pasqua, R., Betts, G., Hoskins, N., Edwards, M., Ercolini, D., and Mauriello, G. 2007. Membrane toxicity of antimicrobial compounds from essential oils. *J. Agric. Food Chem.* 55(12): 4863–4870.

Dominguez, E. T., Nguyen, P. H., Hunt, H. K., and Mustapha, A. 2019. Antimicrobial Coatings for Food Contact Surfaces: Legal Framework, Mechanical Properties, and Potential Applications. *Compr. Rev. Food Sci. Food Saf.* 18(6): 1825–1858.

Dong, F., and Wang, X. 2017. Effects of carboxymethyl cellulose incorporated with garlic essential oil composite coatings for improving quality of strawberries. *Int. J. Biol. Macromol.* 104: 821–826.

Donsì, F., Annunziata, M., Vincensi, M., and Ferrari, G. 2012. Design of nanoemulsion-based delivery systems of natural antimicrobials: Effect of the emulsifier. *J. Biotechnol.* 159: 342–350.

Du, Y., Yang, F., Yu, H., Yao, W., and Xie, Y. 2022. Controllable fabrication of edible coatings to improve the match between barrier and fruits respiration through layer-by-layer assembly. *Food Bioprocess Technol.* 15(8): 1778–1793.

Duguma, H. T. 2022. Potential applications and limitations of edible coatings for maintaining tomato quality and shelf life. *Int. J. Food Sci. Technol.* 57(3): 1353–1366.

Dwivany, F. M., Aprilyandi, A. N., Suendo, V., and Sukriandi, N. 2020. Carrageenan edible coating application prolongs Cavendish banana shelf life. *Int. J. Food Sci. Technol.* 2020: 8861610.

Fan, W., Sun, J., Chen, Y., Qiu, J., Zhang, Y., and Chi, Y. 2009. Effects of chitosan coating on quality and shelf life of silver carp during frozen storage. *Food Chem.* 115: 66–70.

Farsanipour, A., Khodanazary, A., and Hosseini, S. M. 2020. Effect of chitosan-whey protein isolated coatings incorporated with tarragon *Artemisia dracunculus* essential oil on the quality of *Scomberoides commersonnianus* fillets at refrigerated condition. *Int. J. Biol. Macromol.* 155: 766–771.

Frazão, G. G. S., Blank, A. F., and de Aquino Santana, L. C. L. 2017. Optimisation of edible chitosan coatings formulations incorporating Myrcia ovata *Cambessedes* essential oil with antimicrobial potential against foodborne bacteria and natural microflora of mangaba fruits. *LWT-Food Sci. Technol.* 79: 1–10.

Gharibzahedi, S. M. T., and Mohammadnabi, S. 2017. Effect of novel bioactive edible coatings based on jujube gum and nettle oil-loaded nanoemulsions on the shelf-life of Beluga sturgeon fillets. *Int. J. Biol. Macromol.* 95: 769–777.

González-Forte, L., Bruno, E., and Martino, M. 2014. Application of coating on dog biscuits for extended survival of probiotic bacteria. *Anim. Feed Sci. Technol.* 95: 76–84.

Guerra, I. C. D., de Oliveira, P. D. L., Santos, M. M. F., Lúcio, A. S. S. C., Tavares, J. F., and de Souza, E. L. 2017. The effects of composite coatings containing chitosan and *Mentha (piperita* L. or x villosa Huds) essential oil on postharvest mold occurrence and quality of table grape cv. Isabella. *Innovative Food Sci. Emerging Technol.* 34: 112–121.

Guerreiro, A. C., Gago, C. M., Faleiro, M. L., Miguel, M. G., and Antunes, M. D. 2015. Raspberry fresh fruit quality as affected by pectin-and alginate-based edible coatings enriched with essential oils. *Sci. Hortic.* 194: 138–146.

Hamedi, H., Kargozari, M., Shotorbani, P. M., Mogadam, N. B., and Fahimdanesh, M. 2017. A novel bioactive edible coating based on sodium alginate and galbanum gum incorporated with essential oil of *Ziziphora persica*: The antioxidant and antimicrobial activity, and application in food model. *Food Hydrocoll.* 72: 35–46.

Hamzah, H. M., Osman, A., Tan, C. P., and Mohamad Ghazali, F. 2013. Carrageenan as an alternative coating for papaya (*Carica papaya* L. cv. Eksotika). *Postharvest Biol. Technol.* 75: 142–146.

Han, C., Zhao, Y., Leonard, S. W., and Traber, M. G. 2004. Edible coatings to improve storability and enhance nutritional value of fresh and frozen strawberries (Fragaria × ananassa) and raspberries (Rubus ideaus). *Postharvest Biol. Technol.* 33(1): 67–78.

Hashemi, M., Dastjerdi, A. M., Mirdehghan, S. H., Shakerardekani, A., and Golding, J. B. 2021. Incorporation of *Zataria multiflora* Boiss essential oil into gum Arabic edible coating to maintain the quality properties of fresh in-hull pistachio (*Pistacia vera* L.). *Food Packag. Shelf Life*. 30: 100724.

Hashemi, S. M. B., Khaneghah, A. M., and Ghahfarrokhi, M. G., Eş, I. 2017. Basil-seed gum containing *Origanum vulgare* subsp. Viride essential oil as edible coating for fresh cut apricots. *Postharvest Biol. Technol.* 125: 26–34.

Hosseini, S. F., Rezaei, M., and Zandi, M., Farahmandghavi, F. 2015. Bio-based composite edible films containing *Origanum vulgare* L. essential oil. *Ind. Crops Prod.* 67: 403–413.

Ju, J., Xie, Y., Guo, Y., Cheng, Y., Qian, H., and Yao, W. 2019. Application of edible coating with essential oil in food preservation. *Crit. Rev. Food Sci. Nutr.* 59(15): 2467–2480.

Ju, J., Xu, X., Xie, Y., Guo, Y., Cheng, Y., Qian, H., and Yao, W. 2018. Inhibitory effects of cinnamon and clove essential oils on mold growth on baked foods. *Food Chem.* 240: 850–855.

Kehinde, B. A., Majid, I., Hussain, S., and Nanda, V. 2020. 1Department of Biosystems and Agricultural Engineering, University of Kentucky, Lexington, Kentucky, United States; 2Department of Food Technology, Islamic University of Science and Technology, Awantipora, Jammu and Kashmir, India; 3Division of Fishery Biology, Faculty of Fisheries, Sher-e-Kashmir University of Agricultural Sciences and Technology, Srinagar. *Funct. Preserv. Prop. Phytochem.* 377.

Khaled, K., and Berardi, U. 2021. Current and future coating technologies for architectural glazing applications. *Energy and Buildings*. 244: 111022.

Khaliq, G., Abbas, H. T., Ali, I., and Waseem, M. 2019. Aloe vera gel enriched with garlic essential oil effectively controls anthracnose disease and maintains postharvest quality of banana fruit during storage. *Hortic., Environ. Biotechnol.* 60(5): 659–669.

Khazaei, N., Esmaiili, M., and Emam-Djomeh, Z. 2016. Effect of active edible coatings made by basil seed gum and thymol on oil uptake and oxidation in shrimp during deep-fat frying. *Carbohydr. Polym.* 137: 249–254.

Kong, I., Degraeve, P., and Pui, L. P. 2022. Polysaccharide-based edible films incorporated with essential oil nanoemulsions: Physico-chemical, mechanical properties and its application in food preservation—A review. *Foods*. 11(4): 555.

Kumar, A., and Saini, C. S. 2021. Edible composite bi-layer coating based on whey protein isolate, xanthan gum and clove oil for prolonging shelf life of tomatoes. *Measurement: Food*. 2: 100005.

Li, Peng., Feng, Y. L., and Yang, W. Q. 2014. Preparation of edible film containing cinnamon oil chitosan-peanut protein isolate. *Food Industry*. 35(12): 140–143.

Li, X., Long, Q., Gao, F., Han, C., Jin, P., and Zheng, Y. 2017b. Effect of cutting styles on quality and antioxidant activity in fresh-cut pitaya fruit. *Postharvest Biol. Technol.* 124: 1–7.

Li, Z., Zhang, Y., and Ge, H. 2017. The membrane may be an important factor in browning of fresh-cut pear. *Food Chem.* 230: 265–270.

Liolios, C., Gortzi, O., Lalas, S., Tsaknis, J., and Chinou, I. 2009. Liposomal incorporation of carvacrol and thymol isolated from the essential oil of *Origanum dictamnus* L. and in vitro antimicrobial activity. *Food Chem.* 112: 77–83.

Liu, X., Lu, Y., Yang, Q., Yang, H., Li, Y., Zhou, B., and Qiao, L. 2018. Cod peptides inhibit browning in fresh-cut potato slices: A potential anti-browning agent of random peptides for regulating food properties. *Postharvest Biol. Technol.* 146: 36–42.

Liu, X., Yang, Q., Lu, Y., Li, Y., Li, T., Zhou, B., and Qiao, L. 2019. Effect of purslane (Portulaca oleracea L.) extract on anti-browning of fresh-cut potato slices during storage. *Food Chem.* 283: 445–453.

Majeed, H., Liu, F., Hategekimana, J., Sharif, H. R., Qi, J., Ali, B., Bian, Y. Y., and Zhong, F. 2016. Bactericidal action mechanism of negatively charged food grade clove oil nanoemulsions. *Food Chem.* 197: 75–83.

Maringgal, B., Hashim, N., Tawakkal, I. S. M. A., and Mohamed, M. T. M. 2020. Recent advance in edible coating and its effect on fresh/fresh-cut fruits quality. *Trends Food Sci. Technol.* 96: 253–267.

Mastromatteo, M., Conte, A., and Del Nobile, M. A. 2012. Packaging strategies to prolong the shelf life of fresh carrots (Daucus carota L.). *Innovative Food Sci. Emerging Technol.* 13: 215–220.

Mei, Y., Zhao, Y., and Farr, H. 2002. Enhancement of nutritional and sensory qualities of fresh baby carrots by edible coatings. *J. Food Sci.* 67(5): 1964–1968.

Minh, N. P. 2020. Influence of modified atmospheric packaging and storage temperature on the physico-chemical, microbial and organoleptic properties of cantaloupe (*Cucumis melo*) fruit. *Research on Crops*. 21(3): 506–2511.

Moghimi, R., Ghaderi, L., Rafati, H., Aliahmadi, A., and Mcclements, D. J. 2016. Superior antibacterial activity of nanoemulsion of *Thymus daenensis* essential oil against *E. coli*. *Food Chem*. 194: 410–415.

Mohammadi, L., Tanaka, F., and Tanaka, F. 2021. Preservation of strawberry fruit with an Aloe vera gel and basil (*Ocimum basilicum*) essential oil coating at ambient temperature. *J. Food Process. Preserv.* 45(10): e15836.

Moradi, M., Tajik, H., Rohani, S. M. R., and Mahmoudian, A. 2016. Antioxidant and antimicrobial effects of zein edible film impregnated with *Zataria multiflora* Boiss. essential oil and monolaurin. *LWT-Food Sci. Technol.* 72: 37–43.

Nair, M. S., Tomar, M., Punia, S., Kukula-Koch, W., and Kumar, M. 2020. Enhancing the functionality of chitosan-and alginate-based active edible coatings/films for the preservation of fruits and vegetables: A review. *Int. J. Biol. Macromol.* 164: 304–320.

Nasiri M, Barzegar M, and Sahari M. A. 2017. Application of Tragacanth gum impregnated with *Satureja khuzistanica* essential oil as a natural coating for enhancement of postharvest quality and shelf life of button mushroom (*Agaricus bisporus*). *Int. J. Biol. Macromol.* 08(1): 98–103.

Nicolau-Lapena, I., Colas-Meda, P., Alegre, I., Aguilo-Aguayo, I., Muranyi, P., and Vinas, I. 2021. Aloe vera gel: An update on its use as a functional edible coating to preserve fruits and vegetables. *Prog. Org. Coat.* 151: 106007.

Oh, Y. A., Oh, Y. J., Song, A. Y., Won, J. S., Song, K. B., and Min, S. C. 2017. Comparison of effectiveness of edible coatings using emulsions containing lemongrass oil of different size droplets on grape berry safety and preservation. *LWT-Food Sci. Technol.* 75: 742–750.

Oyom, W., Yu, L., Dai, X., Li, Y. C., Zhang, Z., Bi, Y., and Tahergorabi, R. 2022. Starch-based composite coatings modulate cell wall modification and softening in Zaosu pears. *Prog. Org. Coat.* 171: 107014.

Paidari, S., Zamindar, N., Tahergorabi, R., Kargar, M., Ezzati, S., and Musavi, S. H. 2021. Edible coating and films as promising packaging: A mini review. *J. Food Measure. Charact.* 15(5): 4205–4214.

Perdones, Á., Escriche, I., Chiralt, A., and Vargas, M. 2016. Effect of chitosan—lemon essential oil coatings on volatile profile of strawberries during storage. *Food Chem.* 197: 979–986.

Perez-Gago, M. B., Serra, M., Alonso, M., Mateos, M., and del Rio, M. A. 2005. Effect of whey protein- and hydroxypropyl methylcellulose-based edible composite coatings on color change of fresh-cut apples. *Postharvest Biol. Technol.* 36: 77–85.

Plazzotta, S., Manzocco, L., and Nicoli, M. C. 2017. Fruit and vegetable waste management and the challenge of fresh-cut salad. *Trends Food Sci. Technol.* 63: 51–59.

Popyrina, T. N., Demina, T. S., and Akopova, T. A. 2022. Polysaccharide-based films: From packaging materials to functional food. *J. Food Sci. Technol.* 1–12.

Priyadarshi, R., Riahi, Z., and Rhim, J. W. 2022. Antioxidant pectin/pullulan edible coating incorporated with *Vitis vinifera* grape seed extract for extending the shelf life of peanuts. *Postharvest Biol. Technol.* 183: 111740.

Qian, J., Zhao, Y., Yan, L., Luo, J., Yan, W., and Zhang, J. 2022. Improving the lipid oxidation of beef patties by plasma-modified essential oil/protein edible composite films. *LWT-Food Sci. Technol.* 154: 112662.

Ratha, I., Datta, P., Balla, V. K., Nandi, S. K., and Kundu, B. 2021. Effect of doping in hydroxyapatite as coating material on biomedical implants by plasma spraying method: A review. *Ceramics Int.* 47(4): 4426–4445.

Salvia-Trujillo, L., Rojas-Grau, M. A., Soliva-Fortuny, R., and Martin-Belloso, O. 2014. Formulation of antimicrobial edible nanoemulsions with pseudo-ternary phase experimental design. *Food Bioprocess Technol.* 7: 3022–3032.

Salvia-Trujillo, L., Rojas-Graü, M. A., Soliva-Fortuny, R., and Martín-Belloso, O. 2015. Physicochemical characterization and antimicrobial activity of food-grade emulsions and nanoemulsions incorporating essential oils. *Food Hydrocoll.* 43: 547–556.

Sánchez-González, L., Pastor, C., Vargas, M., Chiralt, A., González-Martínez, C., and Cháfer, M. 2011. Effect of hydroxypropyl methylcellulose and chitosan coatings with and without bergamot essential oil on quality and safety of cold-stored grapes. *Postharvest Biol. Technol.* 60(1): 57–66.

Sánchez-González, L., Vargas, M., González-Martínez, C., Chiralt, A., and Chafer, M. 2011. Use of essential oils in bioactive edible coatings: A review. *Food Eng. Rev.* 3(1): 1–16.

Saxena, A., Sharma, L., and Maity, T. 2020. Enrichment of edible coatings and films with plant extracts or essential oils for the preservation of fruits and vegetables. *In Biopolymer-based formulations* (pp. 859–880). New York: Elsevier.

Seow, Y. X., Yeo, C. R., Chung, H. L., and Yuk, H. G. 2014. Plant essential oils as active antimicrobial agents. *Crit. Rev. Food Sci. Nutr.* 54: 625–644.

Singh, J., Pareek, S., Maurya, V. K., Sagar, N. A., Kumar, Y., Badgujar, P. C., and Fawole, O. A. 2022. Application of aloe vera gel coating enriched with cinnamon and rosehip oils to maintain quality and extend shelf life of pomegranate arils. *Foods.* 11(16): 2497.

Sipahi, R. E., Castell-Perez, M. E., Moreira, R. G., Gomes, C., and Castillo, A. 2013. Improved multilayered antimicrobial alginate-based edible coating extends the shelf life of fresh-cut watermelon (*Citrullus lanatus*). *LWT-Food Sci. Technol.* 51(1): 9–15.

Song, N. B., Lee, J. H., and Song, K. B. 2015. Preparation of perilla seed meal protein composite films containing various essential oils and their application in sausage packaging. *J. Korean Soc. Appl. Biol. Chem.* 58(1): 83–90.

Tavassoli-Kafrani, E., Shekarchizadeh, H., and Masoudpour-Behabadi, M. 2016. Development of edible films and coatings from alginates and carrageenans. *Carbohydr. Polym.* 137: 360–374.

Tosati, J. V., de Oliveira, E. F., Oliveira, J. V., Nitin, N., and Monteiro, A. R. 2018. Light-activated antimicrobial activity of turmeric residue edible coatings against cross-contamination of *Listeria innocua* on sausages. *Food Control.* 84: 177–185.

Trezza, T. A., and Krochta, J. M. 2000. Color stability of edible coatings during prolonged storage. *Afr. J. Food Sci.* 65(7): 1166–1169.

Wu, J. E., Lin, J., and Zhong, Q. 2014. Physical and antimicrobial characteristics of thyme oil emulsified with soluble soybean polysaccharide. *Food Hydrocolloids.* 39: 144–150.

Xiong, Y., Kamboj, M., Ajlouni, S., and Fang, Z. 2021. Incorporation of salmon bone gelatine with chitosan, gallic acid and clove oil as edible coating for the cold storage of fresh salmon fillet. *Food Control.* 125: 107994.

Xue, J., Michael Davidson, P., and Zhong, Q. 2015. Antimicrobial activity of thyme oil co-nanoemulsified with sodium caseinate and lecithin. *Int. J. Food Microbiol.* 210: 1–8.

Yousuf, B., Qadri, O. S., and Srivastava, A. K. 2017. Recent developments in shelf-life extension of fresh-cut fruits and vegetables by application of different edible coatings: A review. *LWT-Food Sci. Technol.* 45(7): 88–93.

Yousuf, B., and Srivastava, A. K. 2017. Flaxseed gum in combination with lemongrass essential oil as an effective edible coating for ready-to-eat pomegranate arils. *Int. J. Biol. Macromol.* 104: 1030–1038.

11 Preparation Strategy and Application of Nanoemulsion Containing Essential Oils

11.1 INTRODUCTION

In order to effectively solve the problem of high hydrophobicity of essential oils, while minimizing the impact of antimicrobials on the sensory characteristics of food, EOs need to be encapsulated in a transport system compatible with food applications (Ju et al., 2018). Emulsifier-based delivery systems can be prepared from food-grade raw materials and easily dispersed in areas of food where microorganisms are likely to grow and reproduce. In addition, nanoemulsions offer advantages such as minimal impact on food sensory quality and increased bioactivity of transported components due to their subcellular size and high diffusivity. The wettability of surfactants and emulsifiers also contributes to the antimicrobial and antibiofilm activity of nanoemulsions. Using nanomaterials as carriers of EOs is a promising way to improve food quality, safety, and functionality (Garcia et al., 2022). Therefore, the application of nanotechnology in the food industry has been one of the fastest growing areas of food research in recent years.

This chapter will discuss the effective contribution of nanoemulsion containing essential oils (EOs-NE) to food preservation, describe the preparation method and formation mechanism of EOs-NE, examine the antimicrobial mechanism of EOs-NE, and explore the effect of nano-packaging on the antimicrobial activity of EOs. In addition, applications of EOs-NE in food preservation developed in recent years will be discussed, as well as their potential for commercial applications, limitations, and future development directions.

11.2 DEVELOPMENT OF NANOEMULSION

The concept of nanotechnology was first proposed by Richard Feynman in 1959 (Hulla et al., 2015). Later in 1974, Nario Taniguchi coined the term nanotechnology and used it to treat particles less than 1 mm, which provided technical support for the subsequent preparation of nanoemulsions with various biological activities. By the early 1980s, with the emergence of some excellent properties of nanoemulsion,

DOI: 10.1201/9781003329268-11

nanotechnology has attracted wide attention. With the maturity of nanotechnology, the research of drug delivery system of nanoemulsion has attracted great interest of international medical scholars (Moghassemi et al., 2022). At present, nanoemulsification technology is widely used in many fields, such as daily chemical industry, food preservation, textile technology, and material science. Nanoemulsion, also known as microemulsion (ME), is a thermodynamically stable colloidal dispersion system formed by water, oil, surfactant, and cosurfactant in appropriate proportion. Its particle size is generally between 10 and 100 nm, and the whole system has transparent appearance and low viscosity.

11.3 PHYSICOCHEMICAL PROPERTIES AND ADVANTAGES OF NANOEMULSION

The choice of preparation methods has an important influence on the physical and chemical properties of nanoemulsions. At the same time, oil, water, surfactant, and cosurfactant are also important factors to determine the properties of nanoemulsions. Nanoemulsions need to be homogeneous, stable, nonstratified, good appearance, good color, and no bad odor. Figure 11.1 provides a schematic representation of add order of different components on nanoemulsions formation.

Oil-in-water (O/W) nanoemulsions consist of oil droplets with a mean droplet size of 20–200 nm dispersed in an aqueous medium and stabilized by an emulsifier layer. Food-grade surfactants (i.e., polysorbates, sugar esters, and lecithins) or biopolymers (i.e., natural gums, vegetable or animal proteins, and modified starches) are frequently used in food applications not only as emulsifying agents but also

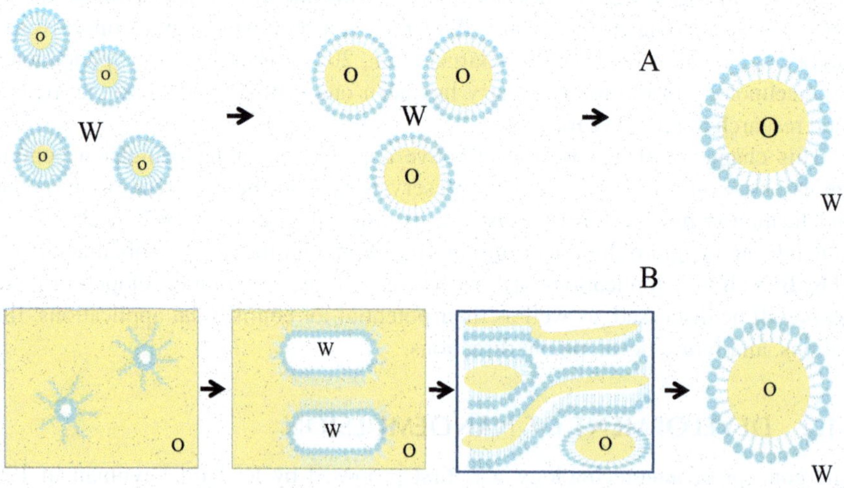

FIGURE 11.1 Schematic representation of add order of different components on nanoemulsions formation. (a) Add oil into the mixture of water and surfactant. (b) Add water into the mixture of oil and surfactant.

to impart desired features such as specific interfacial behaviors (i.e., electrostatic forces, steric repulsion, and rheology), loading capabilities, and the ability to respond to environmental stressors. A schematic diagram of an O/W nanoemulsion is shown in Figure 11.2.

The main characteristics of nanoemulsions are as follows:

(1) Nanoemulsions are translucent with low viscosity, whereas typical emulsions have a higher viscosity.
(2) Nanoemulsions are thermodynamically unstable nonequilibrium systems. Although some nanoemulsions can remain stable for several years, there is still a tendency for them to spontaneously separate into two phases.
(3) In terms of composition, the content of surfactant required is approximately 5–30%, which is much higher than that of typical emulsifiers. In addition, a cosurfactant is necessary to form nanoemulsions with oil and water phases.
(4) Nanoemulsions have low interfacial tension and strong transport capabilities.
(5) W/O nanoemulsions can improve the stability of easily hydrolyzed drugs and prolong the release time of water-soluble active substances (Patel et al., 2023).
(6) Nanoemulsions have sustained release and targeting effects. The absorption of many peptide drugs in the gastrointestinal tract can be improved by nanoemulsification (Salvia-Trujillo et al., 2015).

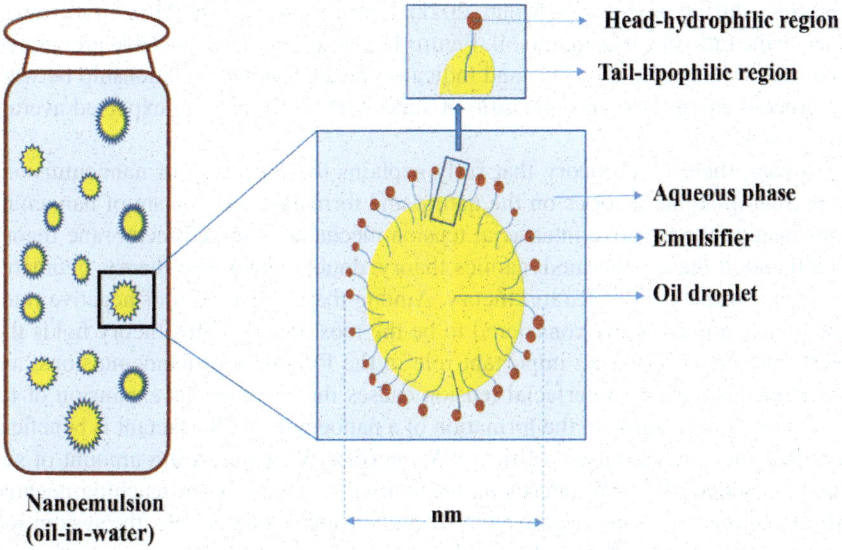

FIGURE 11.2 The schematic diagram of O/W nanoemulsion.

Source: Karthik and Anandharamakrishnan (2018).

Advantages of nanoemulsions include the following:

(1) Nanoemulsions have good physical stability.
(2) Nanoemulsions can fuse two different components through the hydrocarbon chain of the internal oil phase and surfactant, which greatly improves the solubility of insoluble substances in water.
(3) The particle size of nanoemulsions is small and uniform, which promotes the absorption of effective substances and improves their bioavailability.
(4) As drug carriers, nanoemulsions can achieve slow release of drugs and other substances. As novel carriers of bioactive substances, nanoemulsions have broad application prospects.

11.4 PREPARATION METHODS AND FORMATION MECHANISM

Nanoemulsion preparation methods can be divided into two types, top-down and bottom-up. The top-down method mainly uses mechanical pressure to reduce the size of the nanoemulsion. In this method, the emulsification process is divided into two stages. First, coarse droplets are decomposed into smaller droplets. The emulsifier is then absorbed into the newly formed interface to prevent recrystallization and promote dynamic stability. Therefore, the concentration and properties of emulsifiers play an important role in this process. In contrast, the bottom-up method is driven by attraction and repulsion in a physicochemical process formed by the spontaneous combination of surfactants around EO molecules. The bottom-up method can effectively produce fine droplets at low cost while preventing degradation of encapsulated molecules during processing (Moud, 2022). However, this method has strict restrictions on the ratio of surfactant to oil. Figure 11.3 describes some possible preparation routes for nanoemulsion of EOs and indicates the approximate relationship between the process energy demand, the ratio of surfactant to oil, and the expected average droplet size.

To date, there is no theory that fully explains the formation of nanoemulsions. At present, possible theories on the nature and formation mechanism of nanoemulsions include the negative interfacial tension mechanism, mixed membrane theory, solubilization theory, thermodynamics theory, double membrane theory, geometric arrangement theory, and R ratio theory. Among these, the theory of negative interfacial tension is currently considered to be the most mature. This theory holds that interfacial tension plays an important role in the formation of nanoemulsions, and instantaneous negative interfacial tension causes the spontaneous expansion of the system interface leading to the formation of a nanoemulsion. Surfactant is beneficial to reduce the surface tension of the O/W interface. When a certain amount of surfactant is added, the O/W interfacial tension will decrease, even reaching negative values. However, because negative interfacial tension is not possible, the system will spontaneously expand the interface and increase adsorption through more surfactants and cosurfactants. This reduces the volume concentration until interfacial tension is restored to zero or to a small positive value. Once the components of the nanoemulsion system coalesce, the interfacial area decreases and negative interfacial tension is produced, which prevents further coalescence of the nanoemulsion and increases

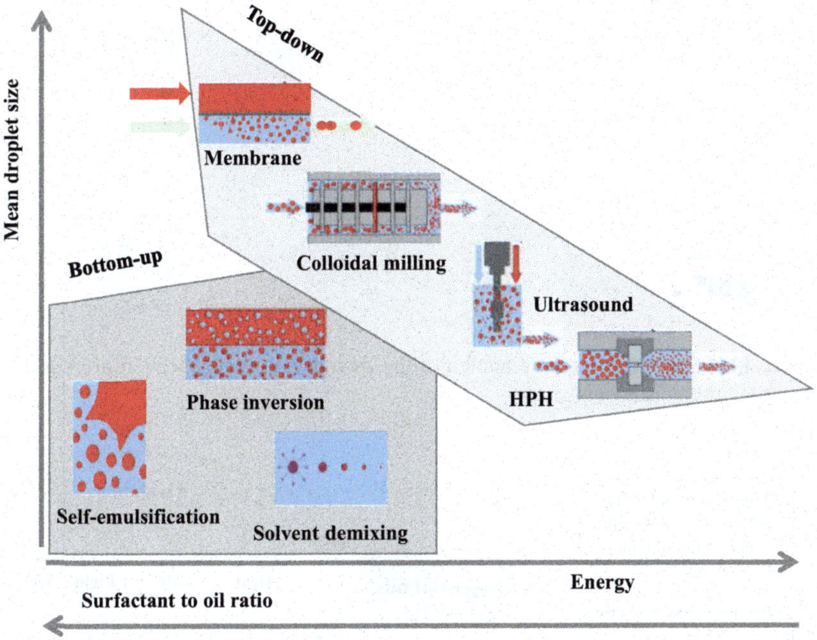

FIGURE 11.3 Some possible routes preparing nanoemulsion of essential oils and indicating the approximate relationship between the process energy, the demand of surfactant relative to the oil phase, and the expected average droplet size.

Source: Donsì and Ferrari (2016).

its stability. However, because negative interfacial tension is difficult to measure, this theory explaining the phenomenon of automatic emulsification of nanoemulsions lacks strong evidence. A schematic representation of the spontaneous emulsification process used to form antimicrobial nanoemulsions is shown in Figure 11.4.

11.5 NANOCARRIERS AS ESSENTIAL OIL TRANSPORTERS

As food preservation technologies continue to develop, applications of nanotechnology in the field of food preservation are becoming more and more important, especially in improving the antiseptic potential of EOs and prolonging their release time. Encapsulation materials such as starch, chitosan, cyclodextrin, cellulose, alginate, and albumin are used in different transport systems to protect EOs from internal and external factors such as temperature, relative humidity, water activity, pH, enzymatic degradation, and the storage environment (Mamusa et al., 2021; Maqsoudlou et al., 2020; Prakash et al., 2018). In recent years, a variety of nano-preparation methods (i.e., polymer nanoparticle, solid lipid nanoparticle, and nanostructured lipid carrier preparation) have been proposed, but each method has its own limitations, such as high production costs, poor physical and chemical stability, poor biocompatibility, and lack of food-grade coating materials. Therefore,

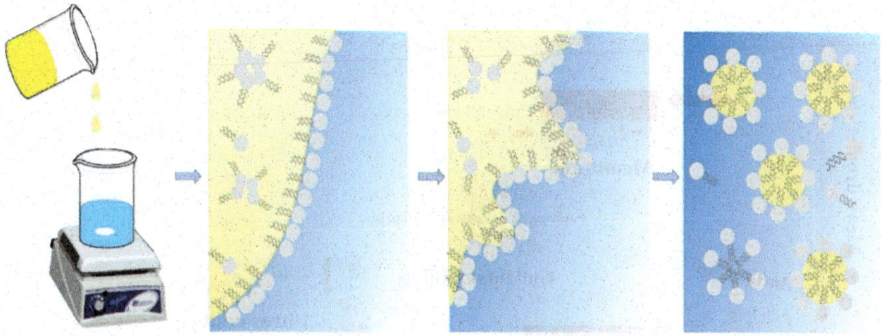

FIGURE 11.4 Schematic representation of the spontaneous emulsification process used to form antimicrobial nanoemulsions.

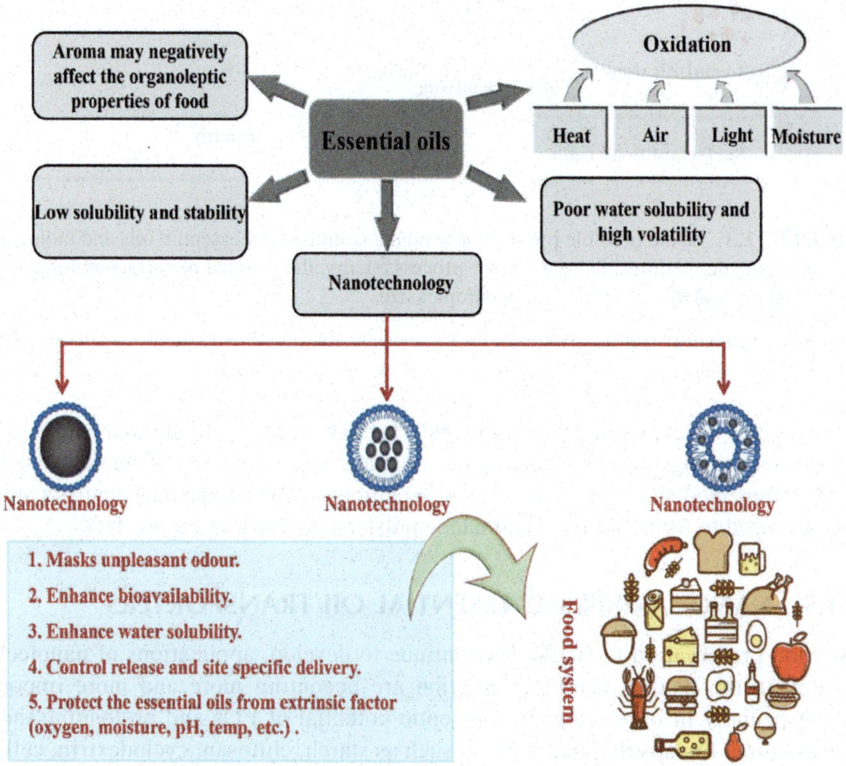

FIGURE 11.5 The preparation of different nano-formulations and their expected significance in food system.

the polarity, solubility, volatility, compatibility with food substrates, and safety of EOs must be considered when preparing nanocarriers. Figure 11.5 systematically illustrates the preparation of different nano-formulations and their expected performance in food systems.

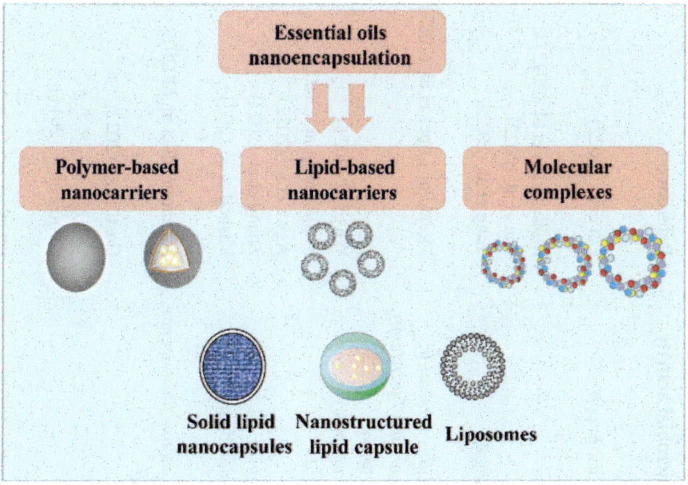

FIGURE 11.6 Schematic illustration of different approaches for the encapsulation of EOs using different nanomaterials.

For EO formulations, three main types of nano-carriers are employed: (i) polymer-based nanoparticles, (ii) lipid-based nanoparticles, and (iii) molecular complexes (Figure 11.6), and describe in detail the different formulations used to combine EOs and nanoparticles in order to use them in food preservation and biomedical applications (Table 11.1).

11.6 EFFECT OF EMBEDDING ON ANTIMICROBIAL ACTIVITY OF ESSENTIAL OIL

EOs have many biological activities that promote their applications in food preservation (Figure 11.7). However, EOs are volatile, oxidize and decompose easily, and have low water solubilities and strong odors, which restricts their wide application. Embedding technology solves many of the problems faced by the application of EOs by improving their stability and reducing their chemical degradation. At the same time, embedding technologies can effectively reduce the impact of plant EOs on food sensory and texture qualities, which is conducive to their application in food preservation. Nanoemulsion is one important embedding technique that can improve the water solubility and chemical stability of EOs. Compared with traditional emulsifying techniques, nanoemulsion results in smaller particle sizes and transparent appearances. Therefore, it is a more acceptable food preservation method for consumers.

Previous studies investigated the antimicrobial effects of clove, oregano, rosemary, and thyme Eos on *Staphylococcus aureus*, *Escherichia coli*, and *Listeria* spp. and showed that the low water solubility and oxidability of EOs reduced their antimicrobial activity (Laranjo et al., 2022; He et al., 2022; Carvalho et al., 2018). Other researchers have shown that the preparation of nanoparticles by embedding EOs in appropriate materials can improve their water solubility and stability, thereby

TABLE 11.1
Different Combinations of EOs and Nanoparticles Used for Antimicrobial Purpose

S. No.	Source of essential oils/EOs constituents	Type of NP	Combination type	Antimicrobial activity	Reference
1	*Ocimum basilicum* L.	Chitosan	Lotion and ionic gel encapsulation	*S. aureus*, *E. coli*	Cai et al. (2022)
2	Lemongrass oil	Metal NPs	Dispersion	Lichens	Mulwandari et al. (2022)
3	Cinnamaldehyde essential oil	Chitosan	Encapsulation	–	Niu et al. (2022)
4	Tea tree essential oil	TiO_2 nanoparticles	Encapsulation	Antifungal	Yang et al. (2022)
5	Garlic essential oil	Chitosan	Encapsulation	*Aspergillus versicolor*, *A. niger*, and *Fusarium oxysporum*	Mondéjar-López et al. (2022)
6	Oregano essential oil	ZnO	Encapsulation	*Listeria monocytogenes*	Wu et al. (2021)
7	Hemp essential oil	Protein complex nanoparticles	Complexation	–	Majidiyan et al. (2022)
8	Frankincense (FRK) essential oil	Whey protein nanoparticles	Embed	–	Agwa et al. (2022)
9	Oregano essential oil	Corn gluten, pectin	Encapsulation	*S. aureus*, *E. coli*	Zhang et al. (2022)
10	*Zanthoxylum schinifolium* essential oil	Octenyl amber-modified starch	Encapsulation	Insect repellent	Rashed et al. (2021)
11	Cinnamon oil	Chitosan	Encapsulation	*Pseudomonas fluorescence*, *E. coli*	Mohammadi et al. (2020)
12	Clove essential oil	Chitosan, ZnO hybrid nanoparticles	Encapsulation	–	Gasti et al. (2022)
13	Cinnamaldehyde, Eugenol	Poly(DL-lactide-*co*-glycolide) (PLGA)	Nanoencapsulation	–	Gomes et al. (2011)
14	Cinnamon oil	β-Cyclodextrin/gum Arabic	Mix	Insecticidal	Elbehery et al. (2022)
15	*Mosla chinensis* essential oils	Chitosan/zein	Pouring	–	Li et al. (2022)

16	*Mentha spicata* L. essential oil	MgO nanoparticles	Nanobiosystem	–	Eghbalian et al. (2021)
17	*Cinnamomum zeylanicum* essential oil	β-Cyclodextrin/chitosan nanoparticles	Biofilm	Fungi	Matshetshe et al. (2018)
18	Bergamot essential oil	TiO nanoparticles, Ag nanoparticles	Mix	Fungi	Chi et al. (2019)
19	Orange peel essential oil	Silver nanoparticles	Reduction	Fungi	Veisi et al. (2019)
20	*Aegle marmelos* essential oil	Chitosan	Encapsulation	*K. pneumoniae*	Rajivgandhi et al. (2021)
21	*Schinus molle* L. essential oil	Chitosan	Encapsulation	*Aflatoxin*	López-Meneses et al. (2018)
22	*Artemisia argyi*	Hydroxyapatite	Microcapsule	*S. aureus, E. coli*	Hu et al. (2013)
23	*Oreganum* spp.	Silver	Edible film	*S. aureus, Listeria monocytogenes*	Solórzano-Santos and Miranda-Novales (2012)
24	*Anethum graveolens*, *Salvia officinalis*	Magnetic	Nanobiocoated wound dressings	*C. albicans*	Anghel et al. (2013)
25	*Melaleuca alternifolia*	Polymeric	Nanocapsule	*Trychophyton rubrum*	Flores et al. (2013)
26	*Lippia sidoides*	Alginate/cashew gum	Encapsulation	*Fungicide* and bactericide	Paula et al. (2012)
27	*Cocos nucifera*	Polymeric	Nanoencapsulation	*C. albicans, C. glabrata*	Santos et al. (2014)
28	Thymol	Zein NP	Encapsulation	Gram-positive bacteria	Lou et al. (2013)
29	*Cananga odorata*, *Pogostemon cablin*, *Vanilla planifolia*	Magnetic	Nanostructure	*S. aureus*, *Klebsiella pneumoniae*	Bilcu et al. (2014)
30	*Cinnamomum cassia*, lemon, basil, thyme, *geranium*, clove	Silver	Nanocomposite	*Candida krusei*, *C. albican*, *C. glabrata*	Szweda et al. (2015)
31	Peppermint oil and cinnamaldehyde hybrid	Silica	Nanocomposite	*E. coli*, *P. aeruginosa*, *Enterobacter cloacae* complex, MDR *S. aureus* biofilms	Duncan et al. (2015)

(*Continued*)

TABLE 11.1
(Continued)

S. No.	Source of essential oils/EOs constituents	Type of NP	Combination type	Antimicrobial activity	Reference
32	Terpenoides	Nonpolar functionalized NP	Nanocomposite	Microbial biofilm	Mogosanu et al. (2015)
33	*Cymbopogon citratus*, *C. martini*, *Eucalyptus globules*, *Azadirachta indica*, *Ocimum sanctum*	Silver	Nano-functionalized antimicrobial oil	Bacterial and fungal pathogens on animals skin	Bansod et al. (2015)
34	*Copaifera* spp.	Solid lipid	Nanoencapsulation	*Candida krusei*, *C. parapsilosis*, *Trichophyton rubrum*, *Microsporum canis*	Svetlichny et al. (2015)
35	*Lippia sidoides*	Chitosan	Nanoencapsulation	Food microbes	Paula et al. (2010)
36	*Origanum vulgare*	Silver	–	Gram-positive and Gram-negative bacteria, including MDR	Scandorieiro et al. (2016)
37	Lemongrass	Cellulose acetate	Nanoencapsulation	*S. aureus*	Liakos et al. (2016)
38	Cinnamon, garlic, clove	Zinc oxide, silver–copper alloy NP	Polylactide (PLA)-based films	*L. monocytogenes*, *S. typhimurium*	Ahmed et al. (2016)
39	Thyme	Chitosan	Nanoencapsulation	*Enterobacter* sp., *S. aureus*	Ghaderi-Ghahfarokhi et al. (2016)
40	Cinnamon, clove	Mesoporous silica NP	Nanoencapsulation	*S. aureus,* methicillin-resistant *S. aureus*, *E. coli*, *P. aeruginosa*, *C. albicans*	Lillie et al. (2016)

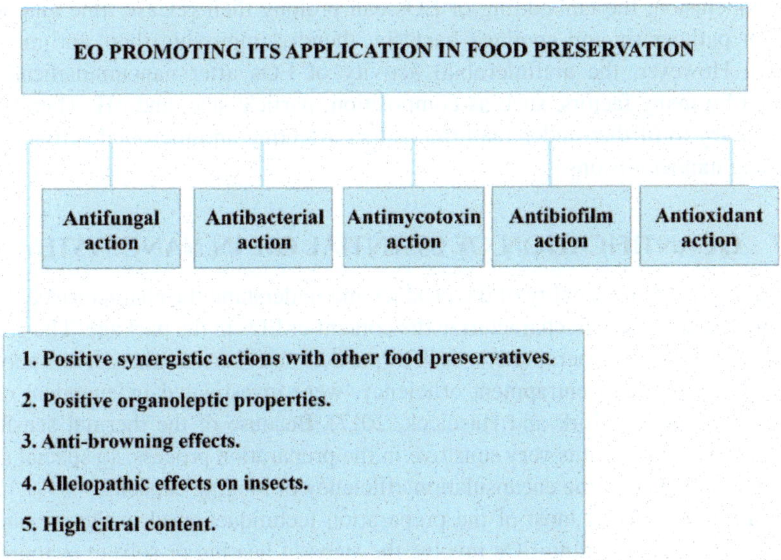

FIGURE 11.7 Characteristics of essential oil to promote application.

enhancing their antimicrobial activity. For example, Salvia et al. (2015) found that nanoemulsified citronella and clove EOs had strong bacteriostatic effects on *E. coli*. de Meneses et al. (2019) also found that nanoemulsion of clove EO enhanced its antimicrobial activity. Najmeh Feizi Langaroudi and Motakef Kazemi (2019) used chitosan to encapsulate parsley EO by nanoemulsion and assessed its slow-release ability and biological activity. The results showed that pH had an important impact on the release of the EO, and gradually raising the pH to neutral increased the release of the EO. Although the antioxidant activity of parsley EO was stronger than that of nanoemulsified parsley EO, the antimicrobial activity of the encapsulated nanoemulsion against *S. aureus*, *E. coli*, *Salmonella typhi*, and *Bacillus subtilis* was stronger than that of the free EO, indicating that encapsulation improves the antimicrobial activity of parsley EO. Similar results were found in a study on the antimicrobial activity of D-limonene and oregano EO after nanoemulsification (Christaki et al., 2022; Ros-Chumillas et al., 2017).

It has been reported that nanoemulsion of citric aldehyde and cinnamaldehyde into different particle sizes had no effect on their antimicrobial activity against spoilage yeast or *E. coli*, but their activity was closely related to the composition of the emulsion system (Donsi et al., 2011). Another study showed that the antimicrobial effect of clove EOs-NE with larger particle sizes was stronger than those with small particle sizes, suggesting that the size of a nanoemulsion influences its antimicrobial effect (Pilong et al., 2022). In addition, nanoemulsification of EOs can prolong their release times, thereby prolonging their action time against pathogenic and spoilage bacteria. However, reports on the antimicrobial effects of EOs-NE are also contradictory. For example, Köllner et al. (2017) did not observe any improvement in the antimicrobial activity of octanoic acid nanodroplets.

In conclusion, the embedding of EOs can prolong their release time and action time on pathogenic and spoilage bacteria, thereby improving their antimicrobial activity. However, the antimicrobial activity of EOs after nanoemulsification is affected by many factors, such as composition, particle size, and pH. Therefore, it is necessary to further understand the factors affecting antimicrobial activity when assessing nanoemulsions.

11.7 QUANTIFICATION OF ESSENTIAL OIL IN NANOSYSTEM

In order to ensure the quality of the product and understand the characteristics of the system, it is necessary to characterize the content of EOs in the package. The general physical and chemical parameters, such as particle size, surface charge, morphology, thermal stability, and entrapment efficiency, were investigated independent of the system used (Sengel-Turk and Hascicek, 2017). Because of the thermal sensitivity and volatility of EOs, it is very sensitive in the preparation process, so special attention should be paid to the encapsulation efficiency or loading capacity.

It can be noted that most of the preparation techniques used in the preparation of nanosystems containing EOs involve the steps of heating or solvent evaporation, which may lead to volatilization and/or degradation of the components, thereby affecting the final composition of the formulation. At this point, in order to understand the changing behavior of EOs during the preparation of nanostructures, quantitative analysis is very important, because it may interfere with the effectiveness and safety of the product. Techniques used in the analysis of EOs are well described in the literature and good reviews have been published. de Matos et al. (2019) discussed classical techniques and trends on sample preparation, analysis, and quantification of volatile components of EOs. In addition, in recent decades, special attention has been paid to reducing the environmental impact of packaging materials and the development of green chemistry. Based on this, Mandrioli et al. (2022) discussed the basic principles of green analytical chemistry, analytical methods, and sustainable alternatives to pre-analytical sample preparation. In a recent report, López-Lorente et al. (2022) suggested 12 principles to guide the development of greener analytical methods.

In this chapter, we investigated the relevant studies on the quantification of EOs in the final formula. It was found that more than half of the selected papers did not mention any quantitative method to evaluate the content of EOs in the final formula. In addition, no more than ten publications reported only the quantification of unencapsulated EOs, which may be unreasonable, as the calculation of association efficiency is usually based on the assumption that there is no loss of EOs content. Finally, the quantitative studies of essential oils are classified according to the applied analytical methods, which mainly include indirect quantitative method, spectrophotometry, and chromatography (Table 11.2).

11.7.1 Spectrophotometry

Spectrophotometry is based on the absorption capacity of certain molecules in the ultraviolet wavelength range. Absorbance is a dimensionless unit that represents the

TABLE 11.2
Overview of the Analytical Methods Employed in the Assessment of Encapsulation Efficiency of EO in Nanosystems

Method	Sample preparation	Drawbacks
Indirect quantification	No sample preparation required	Nonspecific Use of heat and solvents
Spectrophotometry	Need to dilute sample Analyte needs to be in solution	Lack of specificity Light scatter effects
GC	Need to displace water content of formulation	Difficult to analyze nonvolatile fraction of EOs
HPLC	The sample needs to be filtered and sonicated to avoid air bubbles	More expansive Demands use of solvents

light emitted by a light source in a spectrophotometer. Because it is absorbed or scattered in the optical path, it will not reach the detector of the device. Generally speaking, according to Beer's law, the absorbance of a substance can be related to its concentration. However, in some cases, such as high concentration samples and the existence of nonoptical absorption molecules, Beer's law may not apply. In addition, quantification of complex samples containing more than one analyte ideally requires different maximum absorption wavelengths and avoids spectral overlap between analytes.

A considerable number of studies have described the use of spectrophotometry to determine the entrapment efficiency of EOs or the drug loading of nanosystems. In many cases, the preparation steps of the sample mainly include centrifugation, solubilization, reflux, and colorimetric reaction to release the essential oil wrapped in the nanostructure, followed by spectrophotometric measurement using the calibration curve. This can be justified by the Tyndall effect, in which colloidal dispersions scatter the light beams and may interfere with the spectrophotometric reading of the analyte's absorbance. In these studies, de Matos et al. (2019) also determined the entrapment efficiency of oregano essential oil in nanoemulsions, and the results showed that the entrapment efficiency was 5.45–24.72%. In addition, Castangia et al. (2015) measured the content of *Santolina insularis* EO in liposomes and found that the oil content in the system was as high as 70%. Nantrajan et al. (2015) evaluated the combination of citronella essential oil in nanoemulsion and found that the entrapment efficiency of vanilla essential oil was 86.9%.

Although spectrophotometry is a cheap and easily available method, which has been widely used in the quantification of essential oils, it does not seem to be a suitable method. This technique lacks specificity and is not conducive to the analysis of complex matrices such as essential oils. In addition, the light scattering of colloidal suspensions such as polymers and nanoparticles in the dispersion system requires us to pay special attention in the process of sample preparation.

11.7.2 Chromatographic Method

In analytical chemistry, chromatography is widely used as a separation technology. It is often combined with other analytical techniques in an attempt to identify and quantify chemical mixtures. The occurrence of separation is due to the distribution equilibrium of compounds between the mobile phase and the stationary phase, in which each different compound shows different interactions with different phases, resulting in different migration between the compounds.

11.7.2.1 Gas Chromatography

Gas chromatography is a commonly used detection technique, in which gaseous or volatile samples are carried by a gaseous mobile phase through a chromatographic column coated with solid or liquid stationary phase. EOs are mainly composed of nonpolar volatile components, which makes capillary gas chromatography the most widely used analytical method. The fused quartz capillary column on the stationary phase is usually used for the separation of compounds. Identification is achieved by coupling the gas chromatograph with the detector selected according to the needs of the analysis. Flame ionization detector (FID) provides information about retention time and peak intensity, while mass spectrometer provides mass spectrogram.

In addition, in many cases, GC analysis requires sample preparation steps, as the injection of nonvolatile components may damage the chromatographic column. For example, nanostructured systems containing EOs almost always need to extract aqueous and other nonvolatile components from samples before being injected into the chromatographic system. Traditional extraction techniques, such as distillation and liquid-phase extraction, have shortcomings in the preparation of gas chromatographic injection because they may use a large number of organic solvents and/or lead to thermal degradation of analytes. On the other hand, newer technologies, such as headspace (HS), whether coupled with solid-phase microextraction or not, are effective techniques that can be coupled to the chromatographic system and allow the separation of volatile components while avoiding the formation of degradation products, because these preparation techniques do not require high-temperature steps.

Surprisingly, some publications have reported cases of direct quantification of EOs using GC in nanosystems. For example, González-Rivera et al. (2016) used GC and FID to analyze the stability of EOs blends under different storage conditions, and no significant changes in EOs markers were found. In addition, Farzaneh et al. (2016) usedGC-MS detector to evaluate the entrapment efficiency of *Alisma plantago-aquatica* essential oil in nanosystem, and finally determined that the entrapment efficiency was 84%. In a previous study, Dias et al. (2012) used solid-phase microextraction to extract volatile components from caryophyllene nanoemulsion and analyzed it with GC/FID system, and established an analytical method for the determination of β-caryophyllene.

11.7.2.2 High-Performance Liquid Chromatography

Unlike gas chromatography, high-performance liquid chromatography (HPLC) uses a liquid mobile phase in which the analyte is dissolved. Under the action of high pressure, the mobile phase is forced to pass through the packed column. The system can realize high-resolution sorting. However, HPLC is usually more expensive than GC and produces chemical residues (Ma et al., 2008). HPLC is a general technology, which can be used to analyze the content of EOs in nanosystems. Mostafa et al. (2015) used HPLC coupled with ultraviolet (UV) detector to evaluate the entrapment amount of fennel EOs in nanoemulsion by measuring markers. Also using a HPLC/UV system, de Matos et al. (2019) quantified terpinen-4-ol, the major component of Tea Tree EOs in the NE and related it to the final content of EOs in nanosystem, found to be 70 mg/ml, which was equivalent to approximately 58% of the marker content initially added to the system. In addition, Das et al. (2021) quantitatively determined the encapsulated essential oil using HPLC coupled with photodiode array (PDA) detector, and obtained that the content of *Pimpinella anisum* essential oil encapsulated in Chitosan nanoparticles was 90.51%. In both cases, a spectrophotometric detector is used (de Matos et al., 2019).

11.7.3 Indirect Quantization

Indirect quantification is performed in two studies by assessing the mass balance of EOs on the formulation with the help of the Clevenger apparatus. First described by Clevenger in 1928, this allows the extraction of EOs content from the formulation avoiding the use of solvents. Upon heating of the material, the EO is carried by water steam toward a condenser and falls into separator with a graduated tube. Ghayempour and Mortazavi (2013) assessed the encapsulation efficiency of EO in nanosystems by distillation of formulations in Clevenger apparatus. de Matos et al. (2019) used the Clevenger apparatus to extract the EO from formulation after previous extraction with acetonitrile and subsequentially weighing the EO extracted, obtaining an encapsulation efficiency of 96% in NE prepared by SNE and 97.7% on NC prepared by NPP. However, the hydrodistillation by Clevenger apparatus involves heating the sample, which may lead to chemical conversions. Also, the Clevenger apparatus does not allow the identification of single components unless associated with an additional analytical method.

11.8 ANTIMICROBIAL MECHANISMS

Nanoencapsulated EOs with bioactive components that exhibit strong antimicrobial activity could offer safer alternatives to synthetic antimicrobials. Relevant studies have shown that EOs are usually composed of complex mixtures of dozens to hundreds of compounds, and the antimicrobial activity of EOs is determined by their main components or the synergistic action of their various components (Hassan et al., 2020; Hlebová et al., 2021). Furthermore, the antimicrobial mechanisms of the different components may also differ. Therefore, the antimicrobial mechanisms of EOs are generally not single modes of action but involve multiple mechanisms. EOs can also simultaneously interfere with different aspects of the normal growth of microorganisms, which reduces the probability of drug resistance (Ju et al., 2022). In general, the antimicrobial activity of EOs is based on their molecular hydrophobicity. EOs increase membrane permeability through strong interactions between phenolic compounds and lipids in cell membranes, leading to dysfunction or even rupture of the membrane, dysfunction of enzymes and proteins related to normal cell functions, and entry of EO molecules into the cell, where they can interact with organelles and cause leakage of cell contents and cell death (He et al., 2022).

Nanoemulsions usually use macromolecules as emulsifiers, which further improves antimicrobial activity. In other cases, the delivery methods of EO nanoemulsions may promote their interactions with microbial cell membranes through five main mechanisms (Figure 11.8):

1. The emulsified essential oil increased its surface area to some extent, and the interaction with plasma membrane was improved by passive transport of outer membrane when in contact with cell membrane. In addition, the surface of nanoemulsions is usually hydrophilic, which enables the nanoemulsions containing essential oils to pass through the cell membrane smoothly through porins, providing hydrophilic transmembrane channels for Gram-negative bacteria. At the same time, the nanoemulsion carrying essential oil can increase the contact ability between essential oil and microbial cell membrane, thus changing the permeability of cell membrane.
2. The fusion of emulsifier droplets with cell membrane phospholipid layer may promote the targeted release of essential oils at the desired location. The specific interaction between emulsifier and cell membrane has also been reported to increase the antibacterial activity of essential oils.
3. The sustained release of essential oils from nanoemulsion droplets prolonged the activity of essential oils. In addition, emulsifier also plays an important role in the solubility of essential oils through micellization mechanism.

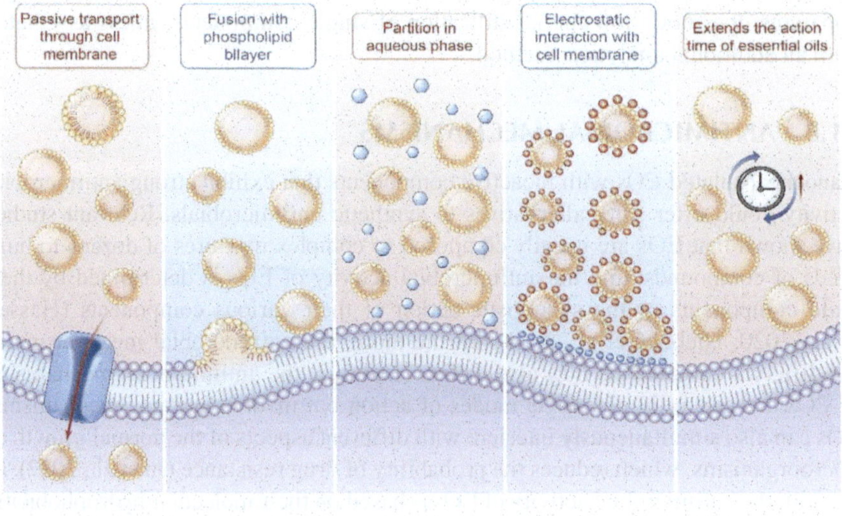

FIGURE 11.8 Schematic diagram of different ways in which nanoemulsion promotes the interaction between essential oils and microbial cell membrane.

4. The electrostatic interaction between positively charged nanoemulsion droplets and negatively charged microbial cell walls increased the concentration of the site essential oils. However, this hypothesis is still controversial because the cationic surfactant used is also a powerful antibacterial drug in the case of enhanced antibacterial activity.
5. The essential oils in nanoemulsion extend the action time of essential oils through sustained release.

11.9 FACTORS AFFECTING ANTIMICROBIAL MECHANISMS

Previous studies have reported that EO-based nanoemulsions have higher antibacterial activities compared with free EOs. For example, Gahruie et al. (2017) showed that encapsulation by nanoemulsion significantly ($P < 0.05$) improved the antimicrobial activity of basil EO. Suresh et al. (2020) found that fennel EO encapsulated by nanoemulsion had better antimicrobial activity than free EO. The same effect was also found in nanoemulsions containing clove and *Salvia miltiorrhiza* EOs (Gago et al., 2019; Jampílek et al., 2019). However, there have been conflicting reports about the antimicrobial effects of EO nanoemulsions. For example, Li et al. (2018) did not observe any improvement in the antimicrobial activity of finger citron EO after encapsulation.

One explanation for these differences is that the antimicrobial activity of EOs is affected by the system of encapsulation by nanoemulsion. This suggests that the overall antimicrobial activity of EO nanoemulsions is the result of multiple factors, although it is difficult to separate the individual mechanisms. In order to further explore these differences and to obtain EO nanoemulsions with significant antimicrobial effects, the main factors affecting the antimicrobial effects of EO nanoemulsions must be considered.

First, the formulation of nanoemulsions not only affects their appearance but also has an important impact on their antimicrobial activity; different formulations can even produce completely different effects. For example, nanoemulsions prepared from modified starches and peppermint EOs-NE prepared from whey protein maltodextrin and eugenol exhibited lower antimicrobial activity than unencapsulated compounds (Popa et al., 2021). However, an EO nanoemulsion prepared with sodium caseinate and thyme EO and a nanoemulsion produced with Tween 80 and sodium alginate combined with lemon grass EO showed higher bioactivity than free EOs (Zhang and Zhong, 2020; Idrees et al., 2015).

Second, the surface charge of a nanoemulsion is an important factor affecting its antimicrobial effect. One study found that positively charged nanoscale droplets generate electrostatic interactions between the cell walls of negatively charged microorganisms, thereby increasing the EO concentration at the site of action (Chang et al., 2015). However, because laurate, which has strong antimicrobial potential, was used as the cationic surfactant in this study, this hypothesis is controversial.

Third, the average droplet size of the nanoemulsion is another important factor affecting its antimicrobial activity. This is mainly because droplet size can affect the interaction between EOs-NE and cell membranes as well as the transport of EOs through cell membranes. Researchers have previously reported that reducing the size of nanoemulsion droplets results in enhanced antibacterial activity (Pandey et al., 2022; Ghosh et al., 2013; Syed et al., 2022). These different results indicate that antimicrobial activity depends not only on average droplet size but also on the structure of the emulsion.

Additional factors affecting the antimicrobial activity of EOs-NE include the surfactant type, preparation method, microbial species, food substrate, and pH. Furthermore, many factors may simultaneously affect the antimicrobial activity of an EOs-NE, making it difficult to assess individual factors. Therefore, further research is needed to better understand the antimicrobial mechanism of EOs-NE and to develop more appropriate formulations for nanoemulsions according to the EO, food type, and target microorganism.

11.10 APPLICATIONS IN FOOD PRESERVATION

The molecular formulas of the active components of EOs commonly used in food preservation are shown in Figure 11.9. These antimicrobial substances are mainly aldehydes, alcohols, and phenols. The molecular diagrams in the figure show that most of these substances have hydroxyls, delocalized electrons, or unsaturated double bonds. Hydroxyl can change the permeability of cell membranes, leading to the leakage of cell contents that are essential for the survival of microorganisms. Delocalized electrons allow protons to be released from hydroxyl groups by decreasing the proton gradient of the cell membrane, resulting in ATP depletion and the death of microbial cells. The presence of unsaturated double bonds also has an important influence on the activity of enzymes. Studies have shown that the existence of unsaturated double bonds can increase the affinity of receptors for lipid-soluble compounds. Similarly,

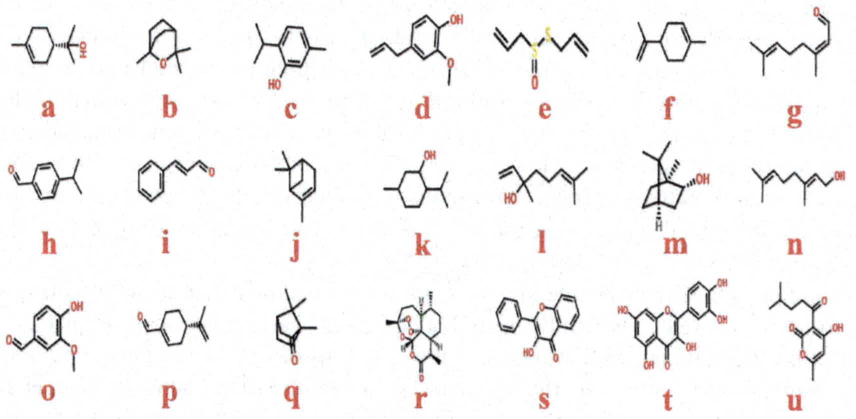

FIGURE 11.9 The molecular formulas of active components of essential oils commonly used for food preservation. (a–u) Terpineol, eucalyptol, thymol, eugenol, garlicin, limonene, citral, cuminal, cinnamaldehyde, α-pinene, menthol, linalool, borneol, geraniol, vanillin, perillaldehyde, camphor, artemisinin, flavonol, quercetin, pogostone.

researchers have demonstrated that the presence of unsaturated double bonds has a significant ($P < 0.05$) effect on enzyme activity, and, as an anti-inflammatory active group, these can enhance the anti-inflammatory activity of compounds (Pontiki and Hadjipavlou-Litina, 2007).

As mentioned earlier, although many active ingredients of EOs have antimicrobial activity, it is challenging to apply EOs directly to food preservation due to their low water solubility, high volatility, and strong odor. In some cases, the active components of Eos may also interact with the components of food (i.e., proteins, lipids, and minerals) leading to changes in food quality. The inhomogeneous distribution of Eos in foods also reduces their antimicrobial activity. In fact, many factors, such as water activity, enzymes, and pH, may cause a decrease in EO activity in complex food systems. Therefore, in the practical application of food preservation, it is often necessary to increase the concentration of EOs to achieve preservation. However, high concentrations of EOs produce unpleasant odors. In contrast, EOs encapsulated by nanoemulsion allow for the slow release of compounds and even exhibit significantly ($P < 0.05$) higher antimicrobial activity, even at low concentrations. For example, Moghimi et al. (2016) showed that the antimicrobial activity of EOs encapsulated by nanoemulsion was ten times higher than that of free EOs. Gundewadi et al. (2018) also found that the antimicrobial activity of basil EO packaged by nanoemulsion was 20% higher than that of the free EO. Therefore, nanoemulsions allow for much lower concentrations of EOs to be used in practical food systems.

Reports have shown that the acceptance of EOs-NE in food is highly dependent on compatibility between the specific EO compound and the type of food. For example, researchers explored the sensory acceptance of six EOs combined with three foods (vegetable soup, tomato juice, and chicken burgers). Results showed that a higher concentration of lemon EO was acceptable in vegetable soup than in tomato juice, which may be due to the higher acidity and citrus flavor already present in the soup (Espina et al., 2014). Therefore, careful selection of EOs for specific food types can help prevent poor sensory qualities. At present, a lot of researches have been carried out about the application of EOs-NE in food preservation. But most of the research focuses on liquid foods, such as fruit juices and milk. The reason for this may be that EOs-NE is more suitable for liquid food systems. However, more attention should be paid to the combination of EOs-NE with active packaging or edible coating in future research. In addition, the combination of EOs-NE and functional food is also the main development trend of future research.

11.11 APPLICATION PROSPECTS AND CHALLENGES

Because of the characteristics of nanomaterials, encapsulation not only maintains the physical and chemical stability and biological activity of EOs but also allows for their slow-release and minimizes their impact on the sensory characteristics of foods. To date, several products based on nanotechnology have been applied in different food science fields, including food processing, quality control, nutritional supplementation, and antimicrobial and intelligent packaging systems. Given its characteristics and the results of recent research, nanotechnology is likely to play a significant role in the development of food technologies. These novel concepts are expected to provide consumers the colors, taste, and nutrition they desire, making nanotechnology an innovative and economic option for food safety developments in the future.

Although nanotechnology has tremendous potential for food applications, there are still many challenges to the development of EO nanoemulsions for food preservation. First, to effectively inhibit food spoilage and prolong the shelf life of foods, the formulation of EO nanoemulsions should correspond with the dominant bacteria found in food. Second, the selection of EOs should be based on the characteristics of food substrates, as EO nanoemulsions can influence the sensory quality of food products. Third, the synergistic antimicrobial actions of different EOs and/or their components should be taken into consideration during the development of nanoemulsion formulations. Owing to these synergistic antimicrobial actions, EOs can have antimicrobial effects at lower concentrations, which can effectively solve the problem of inhibitory concentrations being higher than flavor thresholds. Fourth, controlled or triggered release of EOs from their delivery systems is also desirable to ensure higher responsiveness when needed (e.g., in response to pH changes caused by microbial growth). Finally, with the emergence of drug-resistant bacteria, the antimicrobial mechanisms of EO nanoemulsions need to be critically studied to design new antimicrobial strategies.

11.12 CONCLUDING REMARKS

Increased consumer demand for safe, environment-friendly, and natural products in the past few years has driven the search for alternatives to synthetic preservatives that improve product quality and safety without affecting nutrition or sensory qualities. In this regard, natural plant extracts, in particular EOs, have the potential to provide quality and safety benefits. The application of modern scientific methods, such as metabolic engineering, combinatorial chemistry, and novel extraction methods, has generated new sources of high-yielding raw materials and expanded the potential of EOs as antibiotics. The continuous development of nanotechnology offers great potential for the development of new carrier preparations. Preservatives based on controlled-release EOs have many advantages in food systems, such as maintaining biological activity, enhancing chemical and thermal stability, and reducing impacts on flavor. Although EOs-NE are a promising food preservation technology, the application of nano-products in food systems has not yet been fully approved by the relevant regulatory authorities. Research on the negative effects of nanomaterials on health and the environment is still weak. Therefore, it is necessary to further study the toxicology of EOs-NE in food systems before they can be widely used as plant-based preservatives.

Future research on EOs-NE must be based on the synergistic antimicrobial action of EOs and their nanomaterials, as this addresses the problem of effective EO concentrations being higher than flavor thresholds, reduces the negative impact on the sensory organs of products, and improves antimicrobial efficacy in food matrices.

ACKNOWLEDGEMENTS

This work was supported by the National Natural Science Foundation of China (32202192), Special fund for Taishan Scholars Project.

REFERENCES

Agwa, M. M., Sabra, S., Atwa, N. A., Dahdooh, H. A., Lithy, R. M., and Elmotasem, H. 2022. Potential of frankincense essential oil-loaded whey protein nanoparticles embedded in frankincense resin as a wound healing film based on green technology. *J. Drug Delivery Sci. Technol.* 71: 103291.

Ahmed, J., Hiremath, N., and Jacob, H. 2017. Antimicrobial efficacies of essential oils/nanoparticles incorporated polylactide films against *L. monocytogenes* and *S. typhimurium* on contaminated cheese. *Int. J. Food Prop.* 20(1): 53–67.

Anghel, I., Holban, A. M., Andronescu, E., Grumezescu, A. M., and Chifiriuc, M. C. 2013. Efficient surface functionalization of wound dressings by a phytoactive nanocoating refractory to *Candida albicans* biofilm development. *Biointerphases.* 8(1): 1–8.

Bansod, S. D., Bawaskar, M. S., Gade, A. K., and Rai, M. K. 2015. Development of shampoo, soap and ointment formulated by green synthesised silver nanoparticles functionalised with antimicrobial plants oils in veterinary dermatology: Treatment and prevention strategies. *IET Nanobiotechnology.* 9(4): 165–171.

Bilcu, M., Grumezescu, A. M., Oprea, A. E., Popescu, R. C., Mogoşanu, G. D., Hristu, R., and Chifiriuc, M. C. 2014. Efficiency of vanilla, patchouli and ylang ylang essential oils stabilized by iron oxide@ C14 nanostructures against bacterial adherence and biofilms formed by *Staphylococcus aureus* and *Klebsiella pneumoniae* clinical strains. *Molecules.* 19(11): 17943–17956.

Cai, M., Wang, Y., Wang, R., Li, M., Zhang, W., Yu, J., and Hua, R. 2022. Antibacterial and antibiofilm activities of chitosan nanoparticles loaded with *Ocimum basilicum* L. essential oil. *Int. J. Biol. Macromol.* 202: 122–129.

Carvalho, M., Albano, H., and Teixeira, P. 2018. In vitro antimicrobial activities of various essential oils against pathogenic and spoilage microorganisms. *J. Food Qual. Hazards Control.* 5(2): 41–48.

Castangia, I., Manca, M. L., Caddeo, C., Maxia, A., Murgia, S., Pons, R., and Manconi, M. 2015. Faceted phospholipid vesicles tailored for the delivery of *Santolina insularis* essential oil to the skin. *Colloids Surf., B.* 132: 185–193.

Chang, Y., McLandsborough, L., and McClements, D. J. 2015. Fabrication, stability and efficacy of dual-component antimicrobial nanoemulsions: Essential oil (thyme oil) and cationic surfactant (*Lauric arginate*). *Food Chem.* 172: 298–304.

Chi, H., Song, S., Luo, M., Zhang, C., Li, W., Li, L., and Qin, Y. 2019. Effect of PLA nanocomposite films containing bergamot essential oil, TiO_2 nanoparticles, and Ag nanoparticles on shelf life of mangoes. *Scientia Horticulturae.* 249: 192–198.

Christaki, S., Moschakis, T., Hatzikamari, M., and Mourtzinos, I. 2022. Nanoemulsions of oregano essential oil and green extracts: Characterization and application in whey cheese. *Food Control.* 141: 109190.

de Matos, S. P., Lucca, L. G., and Koester, L. S. 2019. Essential oils in nanostructured systems: Challenges in preparation and analytical methods. *Talanta.* 195: 204–214.

de Meneses, A. C., Sayer, C., Puton, B. M., Cansian, R. L., Araújo, P. H., and de Oliveira, D. 2019. Production of clove oil nanoemulsion with rapid and enhanced antimicrobial activity against gram-positive and gram-negative bacteria. *J. Food Process Eng.* 42(6): e13209.

Dias, D. D. O., Colombo, M., Kelmann, R. G., De Souza, T. P., Bassani, V. L., Teixeira, H. F., . . . Koester, L. S. 2012. Optimization of headspace solid-phase microextraction for analysis of β-caryophyllene in a nanoemulsion dosage form prepared with copaiba (*Copaifera multijuga* Hayne) oil. *Anal. Chim. Acta*. 721: 79–84.

Donsì, F., Annunziata, M., Sessa, M., and Ferrari, G. 2011. Nanoencapsulation of essential oils to enhance their antimicrobial activity in foods. *LWT-Food Sci. Technol.* 44(9): 1908–1914.

Donsì, F., and Ferrari, G. 2016. Essential oil nanoemulsions as antimicrobial agents in food. *J. Biotechnol.* 233: 106–120. Duncan, B., Li, X., Landis, R. F., Kim, S. T., Gupta, A., Wang, L. S., and Rotello, V. M. 2015. Nanoparticle-stabilized capsules for the treatment of bacterial biofilms. *ACS Nano*. 9(8): 7775–7782.

Eghbalian, M., Shavisi, N., Shahbazi, Y., and Dabirian, F. 2021. Active packaging based on sodium caseinate-gelatin nanofiber mats encapsulated with *Mentha spicata* L. essential oil and MgO nanoparticles: Preparation, properties, and food application. *Food Packag. Shelf Life*. 29: 100737.

Elbehery, H. H., and Ibrahim, S. S. 2022. Cinnamon essential oil loaded β-cyclodextrin/gum Arabic nanoparticles affecting life table parameters of potato tuber moth, *Phthorimaea operculella* (Zeller). *Biocatal. Agric. Biotechnol.* 42: 102349.

Espina, L., García-Gonzalo, D., and Pagán, R. 2014. Impact of essential oils on the taste acceptance of tomato juice, vegetable soup, or poultry burgers. *J. Food Sci.* 79(8): S1575–S1583.

Farzaneh, V. 2016. Development of the nutraceutical and pharmaceutical applications of plants selected from Portugal and Iran with presumptive health potentials. *Sapientia*. 119(3): 221–228.

Flores, F. C., De Lima, J. A., Ribeiro, R. F., Alves, S. H., Rolim, C. M. B., Beck, R. C. R., and Da Silva, C. B. 2013. Antifungal activity of nanocapsule suspensions containing tea tree oil on the growth of *Trichophyton rubrum*. *Mycopathologia*. 175(3): 281–286.

Gago, C. M. L., Artiga-Artigas, M., Antunes, M. D. C., Faleiro, M. L., Miguel, M. G., and Martín-Belloso, O. 2019. Effectiveness of nanoemulsions of clove and lemongrass essential oils and their major components against *Escherichia coli* and *Botrytis cinerea*. *J. Food Sci. Technol.* 56(5): 2721–2736.

Gahruie, H. H., Ziaee, E., Eskandari, M. H., and Hosseini, S. M. H. 2017. Characterization of basil seed gum-based edible films incorporated with *Zataria multiflora* essential oil nanoemulsion. *Carbohydr. Polym.* 166: 93–103.

Garcia, C. R., Malik, M. H., Biswas, S., Tam, V. H., Rumbaugh, K. P., Li, W., and Liu, X. 2022. Nanoemulsion delivery systems for enhanced efficacy of antimicrobials and essential oils. *Biomaterials Science*. 10(3): 633–653.

Gasti, T., Dixit, S., Hiremani, V. D., Chougale, R. B., Masti, S. P., Vootla, S. K., and Mudigoudra, B. S. 2022. Chitosan/pullulan based films incorporated with clove essential oil loaded chitosan-ZnO hybrid nanoparticles for active food packaging. *Carbohydr. Polym.* 277: 118866.

Ghaderi-Ghahfarokhi, M., Barzegar, M., Sahari, M. A., and Azizi, M. H. 2016. Nanoencapsulation approach to improve antimicrobial and antioxidant activity of thyme essential oil in beef burgers during refrigerated storage. *Food Bioprocess Technol.* 9(7): 1187–1201.

Ghosh, V., Mukherjee, A., and Chandrasekaran, N. 2013. Ultrasonic emulsification of food-grade nanoemulsion formulation and evaluation of its bactericidal activity. *Ultrasonics sonochemistry*. 20(1): 338–344.

Gomes, C., Moreira, R. G., and Castell-Perez, E. 2011. Poly (DL-lactide-co-glycolide)(PLGA) nanoparticles with entrapped trans-cinnamaldehyde and eugenol for antimicrobial delivery applications. *J. Food Sci.* 76(2): N16–N24.

González-Rivera, J., Duce, C., Falconieri, D., Ferrari, C., Ghezzi, L., Piras, A., and Tine, M. R. 2016. Coaxial microwave assisted hydrodistillation of essential oils from five different herbs (lavender, rosemary, sage, fennel seeds and clove buds): Chemical composition and thermal analysis. *Innovative Food Sci. Emerging Technol.* 33: 308–318.

Gundewadi, G., Rudra, S. G., Sarkar, D. J., and Singh, D. 2018. Nanoemulsion based alginate organic coating for shelf life extension of okra. *Food Packag. Shelf Life*. 18: 1–12.

Hassan, H. A., Genaidy, M. M., Kamel, M. S., and Abdelwahab, S. F. 2020. Synergistic antifungal activity of mixtures of clove, cumin and caraway essential oils and their major active components. *J. Herb. Med.* 24: 100399.

He, Q., Zhang, L., Yang, Z., Ding, T., Ye, X., Liu, D., and Guo, M. 2022. Antibacterial mechanisms of thyme essential oil nanoemulsions against *Escherichia coli* O157: H7 and *Staphylococcus aureus*: Alterations in membrane compositions and characteristics. *Innovative Food Sci. Emerging Technol.* 75: 102902.

Hlebová, M., Hleba, L., Medo, J., Kováčik, A., Čuboň, J., Ivana, C., and Klouček, P. 2021. Antifungal and synergistic activities of some selected essential oils on the growth of significant indoor fungi of the genus *Aspergillus*. *J. Environ. Sci. Health, Part A*. 56(12): 1335–1346.

Hu, Y., Yang, Y., Ning, Y., Wang, C., and Tong, Z. 2013. Facile preparation of artemisia argyi oil-loaded antibacterial microcapsules by hydroxyapatite-stabilized Pickering emulsion templating. *Colloids Surf., B*. 112: 96–102.

Hulla, J. E., Sahu, S. C., and Hayes, A. W. 2015. Nanotechnology: History and future. *Hum. Exp. Toxicol.* 34(12): 1318–1321.

Idrees, M., Dar, T. A., Naeem, M., Aftab, T., Khan, M. M. A., Ali, A., and Varshney, L. 2015. Effects of gamma-irradiated sodium alginate on lemongrass: Field trials monitoring production of essential oil. *Ind. Crops Prod.* 63: 269–275.

Jampílek, J., Kráľová, K., Campos, E. V., and Fraceto, L. F. 2019. Bio-based nanoemulsion formulations applicable in agriculture, medicine, and food industry. In *Nanobiotechnology in bioformulations* (pp. 33–84). Cham: Springer.

Ju, J., Guo, Y., Cheng, Y., and Yaoc, W. 2022. Analysis of the synergistic antifungal mechanism of small molecular combinations of essential oils at the molecular level. *Ind. Crops Prod.* 188: 115612.

Ju, J., Xie, Y., Guo, Y., Cheng, Y., Qian, H., and Yao, W. 2018. Application of edible coating with essential oil in food preservation. *Crit. Rev. Food Sci. Nutr.* 1–14.

Karthik, P., and Anandharamakrishnan, C. 2018. Droplet coalescence as a potential marker for physicochemical fate of nanoemulsions during in-vitro small intestine digestion. *Colloids Surf., A*. 553: 278–287.

Köllner, S., Nardin, I., Markt, R., Griesser, J., Prüfert, F., and Bernkop-Schnürch, A. 2017. Self-emulsifying drug delivery systems: Design of a novel vaginal delivery system for curcumin. *Eur. J. Pharm. Biopharm.* 115: 268–275.

Laranjo, M., Fernández-León, A. M., Agulheiro-Santos, A. C., Potes, M. E., and Elias, M. 2022. Essential oils of aromatic and medicinal plants play a role in food safety. *J. Food Process. Preserv.* 46(8): e14278.

Li, Z. H., Cai, M., Liu, Y. S., and Sun, P. L. 2018. Development of finger citron (*Citrus medica* L. var. sarcodactylis) essential oil loaded nanoemulsion and its antimicrobial activity. *Food Control*. 94: 317–323.

Li, Y., Xie, H., Tang, X., Qi, Y., Li, Y., Wan, N., and Wu, Z. 2022. Application of edible coating pretreatment before drying to prevent loss of plant essential oil: A case study of *Zanthoxylum schinifolium* fruits. *Food Chem.* 389: 132828.

Liakos, I. L., D'autilia, F., Garzoni, A., Bonferoni, C., Scarpellini, A., Brunetti, V., and Athanassiou, A. 2016. All natural cellulose acetate—Lemongrass essential oil antimicrobial nanocapsules. *Int. J. Pharm.* 510(2): 508–515.

Lillie, J. 2016. *Mesoporous silica nanoparticle incorporation of essential oils onto synthetic textiles for tailored antimicrobial activity* (Doctoral dissertation, Manchester Metropolitan University).

López-Lorente, Á. I., Pena-Pereira, F., Pedersen-Bjergaard, S., Zuin, V. G., Ozkan, S. A., and Psillakis, E. 2022. The ten principles of green sample preparation. *TrAC, Trends Anal. Chem.* 116530.

López-Meneses, A. K., Plascencia-Jatomea, M., Lizardi-Mendoza, J., Fernández-Quiroz, D., Rodríguez-Félix, F., Mouriño-Pérez, R. R., and Cortez-Rocha, M. O. 2018. *Schinus molle* L. essential oil-loaded chitosan nanoparticles: Preparation, characterization, antifungal and anti-aflatoxigenic properties. *LWT-Food Sci. Technol.* 96: 597–603.

Luo, Y., Teng, Z., Wang, T. T., and Wang, Q. 2013. Cellular uptake and transport of zein nanoparticles: Effects of sodium caseinate. *J. Agric. Food Chem.* 61(31): 7621–7629.

Ma, Z., Ge, L., Lee, A. S., Yong, J. W. H., Tan, S. N., and Ong, E. S. 2008. Simultaneous analysis of different classes of phytohormones in coconut (*Cocos nucifera* L.) water using high-performance liquid chromatography and liquid chromatography–tandem mass spectrometry after solid-phase extraction. *Analytica Chimica Acta.* 610(2): 274–281.

Majidiyan, N., Hadidi, M., Azadikhah, D., and Moreno, A. 2022. Protein complex nanoparticles reinforced with industrial hemp essential oil: Characterization and application for shelf-life extension of Rainbow trout fillets. *Food Chem: X.* 13: 100202.

Mamusa, M., Resta, C., Sofroniou, C., and Baglioni, P. 2021. Encapsulation of volatile compounds in liquid media: Fragrances, flavors, and essential oils in commercial formulations. *Adv. Colloid Interface Sci.* 298: 102544.

Mandrioli, R., Cirrincione, M., Mladěnka, P., Protti, M., and Mercolini, L. 2022. Green analytical chemistry (GAC) applications in sample preparation for the analysis of anthocyanins in products and by-products from plant sources. *Advances in Sample Preparation.* 3: 100037.

Matshetshe, K. I., Parani, S., Manki, S. M., and Oluwafemi, O. S. 2018. Preparation, characterization and in vitro release study of β-cyclodextrin/chitosan nanoparticles loaded *Cinnamomum zeylanicum* essential oil. *Int. J. Biol. Macromol.* 118: 676–682.

Maqsoudlou, A., Assadpour, E., Mohebodini, H., and Jafari, S. M. 2020. Improving the efficiency of natural antioxidant compounds via different nanocarriers. *Adv. Colloid Interface Sci.* 278: 102122.

Moghassemi, S., Dadashzadeh, A., Azevedo, R. B., and Amorim, C. A. 2022. Nanoemulsion applications in photodynamic therapy. *J. Controlled Release.* 351: 164–173.

Moghimi, R., Ghaderi, L., Rafati, H., Aliahmadi, A., and Mcclements, D. J. 2016. Superior antibacterial activity of nanoemulsion of *Thymus daenensis* essential oil against *E. coli. Food Chem.* 194: 410–415.

Mogosanu, G., M Grumezescu, A., Huang, K. S., E Bejenaru, L., and Bejenaru, C. 2015. Prevention of microbial communities: novel approaches based natural products. *Curr. Pharm. Biotechnol.* 16(2): 94–111.

Mohammadi, A., Hosseini, S. M., and Hashemi, M. 2020. Emerging chitosan nanoparticles loading-system boosted the antibacterial activity of *Cinnamomum zeylanicum* essential oil. *Ind. Crops Prod.* 155: 112824.

Mondéjar-López, M., Rubio-Moraga, A., López-Jimenez, A. J., Martínez, J. C. G., Ahrazem, O., Gómez-Gómez, L., and Niza, E. 2022. Chitosan nanoparticles loaded with garlic essential oil: A new alternative to tebuconazole as seed dressing agent. *Carbohydr. Polym.* 277: 118815.

Mostafa, D. M., Abd El-Alim, S. H., Asfour, M. H., Al-Okbi, S. Y., Mohamed, D. A., and Awad, G. 2015. Transdermal nanoemulsions of *Foeniculum vulgare* Mill. essential oil: Preparation, characterization and evaluation of antidiabetic potential. *J. Drug Delivery Sci. Technol.* 29: 99–106.

Moud, A. A. 2022. Advanced cellulose nanocrystals (CNC) and cellulose nanofibrils (CNF) aerogels: Bottom-up assembly perspective for production of adsorbents. *Int. J. Biol. Macromol.*222: 1–29.

Mulwandari, M. 2022. *Sintesis Nanopartikel Perak Menggunakan Minyak Serai Wangi (Cymbopogon nardus L. Rendle) untuk Menghambat Pertumbuhan Lichen pada Batuan Candi* (Doctoral dissertation, Universitas Islam Indonesia).

Najmeh Feizi Langaroudi, N., and Motakef Kazemi, N. 2019. Preparation and characterization of O/W nanoemulsion with Mint essential oil and Parsley aqueous extract and the presence effect of chitosan. *Nanomed. Res. J.* 4(1): 48–55.

Niu, B., Chen, H., Wu, W., Fang, X., Mu, H., Han, Y., and Gao, H. 2022. Co-encapsulation of chlorogenic acid and cinnamaldehyde essential oil in Pickering emulsion stablized by chitosan nanoparticles. *Food Chem: X*. 14: 100312.

Oki, Y., and Sasaki, H. 2000. Social and environmental impacts of packaging (LCA and assessment of packaging functions). *Packag. Technol. Sci.* 13(2): 45–53.

Pandey, V. K., Islam, R. U., Shams, R., and Dar, A. H. 2022. A comprehensive review on the application of essential oils as bioactive compounds in Nano-emulsion based edible coatings of fruits and vegetables. *Appl. Food Res.* 100042.

Patel, A. S., Balasubramaniam, S. L., Nayak, B., and Camire, M. E. 2023. Lauric acid adsorbed cellulose nanocrystals reduced the in vitro gastrointestinal digestion of oil-water pickering emulsions. *Food Hydrocolloids*. 134: 108120.

Paula, H. C., de Oliveira, E. F., Abreu, F. O., and de Paula, R. C. 2012. Alginate/cashew gum floating bead as a matrix for larvicide release. *Mater. Sci. Eng: C*, 32(6): 1421–1427.

Paula, H. C., Sombra, F. M., de Freitas Cavalcante, R., Abreu, F. O., and de Paula, R. C. 2011. Preparation and characterization of chitosan/cashew gum beads loaded with *Lippia sidoides* essential oil. *Mater. Sci. Eng: C*. 31(2): 173–178.

Pilong, P., Chuesiang, P., Mishra, D. K., and Siripatrawan, U. 2022. Characteristics and antimicrobial activity of microfluidized clove essential oil nanoemulsion optimized using response surface methodology. *J. Food Process. Preserv.* e16886.

Pontiki, E., and Hadjipavlou-Litina, D. 2007. Synthesis and pharmacochemical evaluation of novel aryl-acetic acid inhibitors of lipoxygenase, antioxidants, and anti-inflammatory agents. *Bioorg. Med. Chem.* 15(17): 5819–5827.

Popa, R. M., Fetea, F., and Socaciu, C. 2021. ATR-FTIR-MIR spectrometry and pattern recognition of bioactive volatiles in oily versus microencapsulated food supplements: authenticity, quality, and stability. *Molecules*. 26(16): 4837.

Prakash, B., Kujur, A., Yadav, A., Kumar, A., Singh, P. P., and Dubey, N. K. 2018. Nanoencapsulation: An efficient technology to boost the antimicrobial potential of plant essential oils in food system. *Food Control*. 89: 1–11.

Rajivgandhi, G., Stalin, A., Kanisha, C. C., Ramachandran, G., Manoharan, N., Alharbi, N. S., and Li, W. J. 2021. Physiochemical characterization and anti-carbapenemase activity of chitosan nanoparticles loaded *Aegle marmelos* essential oil against *K. pneumoniae* through DNA fragmentation assay. *Surf. Interfaces*. 23: 100932.

Rashed, M. M., Ghaleb, A. D., Li, J., Al-Hashedi, S. A., and Rehman, A. 2021. Functional-characteristics of *Zanthoxylum schinifolium* (Siebold & Zucc.) essential oil nanoparticles. *Ind. Crops Prod.* 161: 113192.

Ros-Chumillas, M., Garre, A., Maté, J., Palop, A., and Periago, P. M. 2017. Nanoemulsified D-limonene reduces the heat resistance of *Salmonella senftenberg* over 50 times. *Nanomaterials*. 7(3): 65.

Salvia-Trujillo, L., Rojas-Graü, A., Soliva-Fortuny, R., and Martín-Belloso, O. 2015. Physicochemical characterization and antimicrobial activity of food-grade emulsions and nanoemulsions incorporating essential oils. *Food Hydrocolloids*. 43: 547–556.

Santos, S. S. D. 2012. Desenvolvimento de formulações nanotecnológicas para o tratamento da candidíase vulvovaginal. 119: 126–131.

Scandorieiro, S., De Camargo, L. C., Lancheros, C. A., Yamada-Ogatta, S. F., Nakamura, C. V., De Oliveira, A. G., and Kobayashi, R. K. 2016. Synergistic and additive effect of oregano essential oil and biological silver nanoparticles against multidrug-resistant bacterial strains. *Frontiers in microbiology*. 7: 760.

Sengel-Turk, C. T., and Hascicek, C. 2017. Design of lipid-polymer hybrid nanoparticles for therapy of BPH: Part I. Formulation optimization using a design of experiment approach. *J. Drug Delivery Sci. Technol.* 39: 16–27.

Solórzano-Santos, F., and Miranda-Novales, M. G. 2012. Essential oils from aromatic herbs as antimicrobial agents. *Curr. Opin. Biotechnol.* 23(2): 136–141.

Suresh, U., Murugan, K., Panneerselvam, C., Cianfaglione, K., Wang, L., and Maggi, F. 2020. Encapsulation of sea fennel (*Crithmum maritimum*) essential oil in nanoemulsion and SiO2 nanoparticles for treatment of the crop pest *Spodoptera litura* and the dengue vector *Aedes aegypti. Ind. Crops Prod.* 158: 113033.

Svetlichny, G., Külkamp-Guerreiro, I. C., Dalla Lana, D. F., Bianchin, M. D., Pohlmann, A. R., Fuentefria, A. M., and Guterres, S. S. 2017. Assessing the performance of copaiba oil and allantoin nanoparticles on multidrug-resistant *Candida parapsilosis. J. Drug Delivery Sci. Technol.* 40: 59–65.

Syed, U. T., Leonardo, I. C., Mendoza, G., Gaspar, F. B., Gámez, E., Huertas, R. M., and Brazinha, C. 2022. On the role of components of therapeutic hydrophobic deep eutectic solvent-based nanoemulsions sustainably produced by membrane-assisted nanoemulsification for enhanced antimicrobial activity. *Sep. Purif. Technol.* 285: 120319.

Szweda, P., Gucwa, K., Kurzyk, E., Romanowska, E., Dzierżanowska-Fangrat, K., Zielińska Jurek, A., and Milewski, S. 2015. Essential oils, silver nanoparticles and propolis as alternative agents against fluconazole resistant *Candida albicans*, *Candida glabrata* and *Candida krusei* clinical isolates. *Indian J. Microbiol.* 55(2): 175–183.

Teixeira, P. C., Leite, G. M., Domingues, R. J., Silva, J., Gibbs, P. A., and Ferreira, J. P. 2007. Antimicrobial effects of a microemulsion and a nanoemulsion on enteric and other pathogens and biofilms. *Int. J. Food Microbiol.* 118(1): 15–19.

Veisi, H., Dadres, N., Mohammadi, P., and Hemmati, S. 2019. Green synthesis of silver nanoparticles based on oil-water interface method with essential oil of orange peel and its application as nanocatalyst for A3 coupling. *Mater. Sci. Eng: C.*105: 110031.

Wu, M., Zhou, Z., Yang, J., Zhang, M., Cai, F., and Lu, P. 2021. ZnO nanoparticles stabilized oregano essential oil Pickering emulsion for functional cellulose nanofibrils packaging films with antimicrobial and antioxidant activity. *Int. J. Biol. Macromol.* 190: 433–440.

Yang, Z., Li, M., Zhai, X., Zhao, L., Tahir, H. E., Shi, J., and Xiao, J. 2022. Development and characterization of sodium alginate/tea tree essential oil nanoemulsion active film containing TiO2 nanoparticles for banana packaging. *Int. J. Biol. Macromol.* 213: 145–154.

Zhang, Y., and Zhong, Q. 2020. Physical and antimicrobial properties of neutral nanoemulsions self-assembled from alkaline thyme oil and sodium caseinate mixtures. *Int. J. Biol. Macromol.* 148: 1046–1052.

Zhang, L., Zhang, M., Ju, R., Mujumdar, A. S., and Deng, D. 2022. Recent advances in essential oil complex coacervation by efficient physical field technology: A review of enhancing efficient and quality attributes. *Crit. Rev. Food Sci. Nutr.* 1–23.

12 Preparation Strategy and Application of Electrospinning Active Packaging Containing Essential Oils

12.1 INTRODUCTION

Food packaging is essential to the food industry and it can ensure food quality and safety, extend shelf life, improve the commercial value of food, and facilitate storage, transportation, and sales. Most of the food packaging materials currently used are still based on plastics derived from petrochemicals, but used plastic packaging is very challenging to recycle (Ju et al., 2019; Cheng et al., 2022). Therefore, it is increasingly important to switch to biodegradable materials. Although it is impossible to completely replace traditional plastics with environment-friendly substitutes, there is considerable interest in replacing as much traditional packaging as possible with biodegradable alternatives.

Active food packaging is a new approach to prolong the shelf life of food and maintain its safety, quality, and integrity. According to European Regulation (ER) No. 450/2009, active food packaging is defined as being able to interact with food, by absorbing moisture, carbon dioxide, or odors from the food, or by releasing active components such as antibacterial agents, antioxidants, or spices into the food. Although active food packaging has been intensively researched, the existing polymeric materials still have limitations and need to be improved. Such improvements include the addition of useful additives with different functions.

Essential oils are natural products extracted from aromatic plants. Consumer preference for "clean-label" products has stimulated the food industry to replace synthetic chemical food preservatives with natural substitutes. Compared with synthetic preservatives, essential oils are becoming increasingly popular, because they are ideal for development of "green," "safe," and "healthy" food additives (Ju et al., 2020; Mukurumbira et al., 2022). However, there are still many problems to be solved in the application of essential oils to food preservation. Essential oils usually have a strong smell, which may adversely affect the sensory properties of food and are sensitive to light and heat, easily oxidized and decomposed, and have low water solubility. Essential oils are volatile and often of low stability, so they have a short half-life in the food environment, and the minimum inhibitory concentration (MIC)

DOI: 10.1201/9781003329268-12

of an essential oil against pathogenic organisms in a food matrix is usually higher than that in vitro. The odor threshold of essential oils is usually higher than the flavor threshold of the food being preserved and the direct application of essential oils, some of which are mildly toxic to humans, in food processing or preservation, may pose a threat to human health (Ju et al., 2020). Therefore, to solve these problems effectively, new techniques are being developed and of these, electrospun mats containing essential oils appear to be a very promising solution, because of their high porosity, large specific surface area, and controllable fiber diameter, all of which are favorable properties for controlled release of essential oils, from the packaging film to the food surface. Electrospinning technology does not involve heat, so it can be used to encapsulate heat-sensitive active components, which is very important to maintain the antimicrobial activity of essential oils.

At present, although there are a few reports on the application of electrospun fibers or mats containing essential oils to food packaging, most of these reports focus on the preparation methods for electrospun materials. However, a clear understanding of the main factors affecting the quality of electrospun mats and of their loading and release behavior is very important for the application of essential oils in food preservation. Therefore, this chapter first introduces the principle of electrospinning and the main factors affecting the quality of electrospinning. Second, the loading and release behavior of electrospun films containing essential oils are discussed in detail. Finally, the current application status of these films in food preservation is systematically discussed and its future potential is evaluated.

12.2 THE PRINCIPLES OF ELECTROSPINNING

Electrospinning is an effective packaging technology, which can produce ultrathin continuous fibers from polymer and nonpolymer systems, or their composite materials under the action of an electric field, thus generating a single continuous jet with a diameter from 10 nm to several micrometers and a length of several kilometers, under the impetus of static electricity (Khoshnoudi-Nia et al., 2020; Zhang et al., 2022). Although there are several electrospinning systems, literature reports show that electrospun food packaging materials are mainly produced by the traditional solution electrospinning system shown in Figure 12.1.

During the process of electrospinning, the polymer solution is extruded from the rapidly rotating needle tip at a constant speed, by an injection pump, or extruded from a collecting tank at a constant pressure, and then droplets are formed at the needle tip. When the droplets of polymer liquid are exposed to an electric field, they extend to the nearest low potential point, forming a structure called a Taylor cone (Figure 12.2, right diagram) (Hemamalini and Dev, 2017). The charged jet is maintained in the form of droplets by surface tension and viscosity, but a sufficiently strong electric field overcomes the surface tension of the liquid and elongates the droplets into a jet. The Taylor cone is disrupted and a rotating jet of solution is ejected from the point of the cone. Under the influence of the electric field, the jet travels in a trajectory of unstable rotation. While moving toward the collector, the jet is solidified by evaporation of solvent, thus preventing further stretching of the jet and forming fine, dry fibers (Soleimanifar et al., 2020).

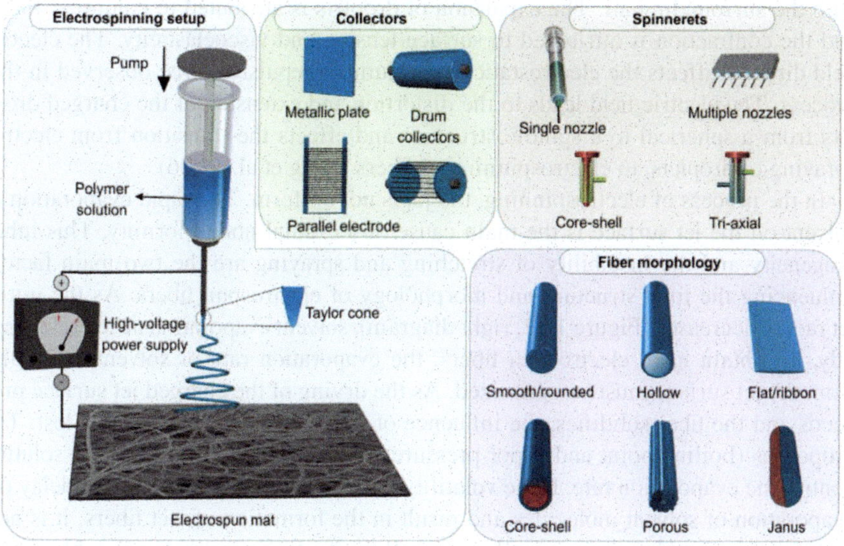

FIGURE 12.1 A schematic diagram of the process of electrospinning and its important components. Typical collectors, spinnerets, and fiber structures are shown.

Source: Fuat and Tamer (2020).

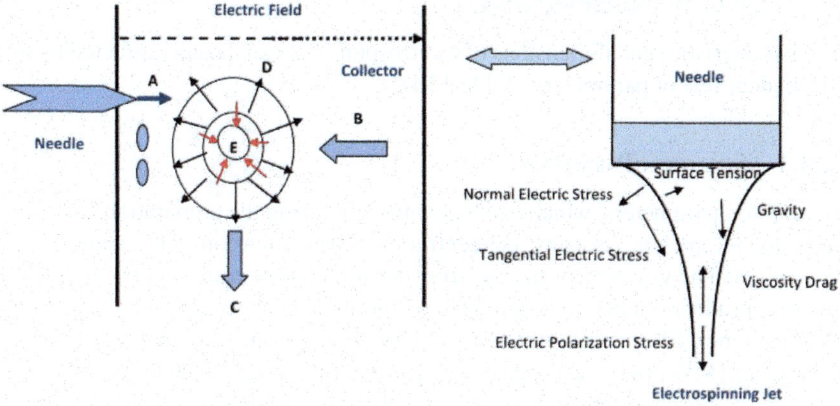

FIGURE 12.2 Force acting on charged droplets during electrospinning. A: electrostatic force, B: drag force, C: gravity, D: cylindrical repulsion force, and E: surface tension and viscoelasticity.

Source: Ding et al. (2006).

Static electricity, drag force, gravity, cylindrical repulsion force, surface tension, and viscoelasticity all act on the charged jet (Figure 12.2) (Dabirian et al., 2007; Luraghi et al., 2021). When the charged droplet stream is projected from the spinning tip to the collector by electrostatic action, the high-speed jet is resisted by interaction with the surrounding air. The expansion of droplets is attributed to columnar force, and the contraction is attributed to surface tension and viscoelasticity. The electric field directly affects the electrostatic and columnar repulsive force observed in this process. The electric field leads to the distortion and extension of the charged droplets from a spherical to a spindle structure and effects the transition from electrospraying of droplets, to electrospinning of fibers (Ding et al., 2006).

In the process of electrospinning, the jet is not uniform. The rapid evaporation of solvent on the jet surface is the main cause of jet radial nonuniformity. This inhomogeneity and the instability of stretching and spraying are the two main factors influencing the final structure and morphology of electrospun fibers. As the initial jet radius decreases (Figure 12.2, right diagram), solvent evaporation accelerates rapidly. To obtain ideal electrospun fibers, the evaporation rate of solvent molecules from the jet surface must be optimized. As the drying of the charged jet surface proceeds and the fiber solidifies, the influence of the electrodynamic force is lost. The properties (boiling point and vapor pressure) of the solvent in the spinning solution control the evaporation rate. If the volatility of the solvent is too low, it will delay the evaporation of solvent molecules and result in the formation of wet fibers. It is normal practice that fibers are not collected until the Taylor cone reaches a stable steady state, in the form of a spinneret tip, which usually takes 1–2 min, and then consistent quality fibers can be collected.

12.3 THE MAIN FACTORS AFFECTING THE QUALITY OF ELECTROSPINNING

The characteristics and dimensions of electrospun fibers are usually affected by three independent sets of parameters (Table 12.1).

12.3.1 Solution Parameters

The solution parameters, which can be controlled during the preparation of the polymer solution, are the viscosity, concentration, surface tension, and conductivity of the solution. The viscosity of the electrospinning solution is an important parameter affecting product quality, and is related to the molecular weight and concentration of the polymer. Electrospun fibers are usually made from solutions of high molecular weight (MW) polymers, because the critical polymer concentration (Ce) must be exceeded for chain entanglement and stable fiber formation to occur. Low MW polymers, or insufficient polymer concentrations produce low-viscosity polymer solutions, which tend to electrospray and form nanodroplets instead of nanofibers, because of insufficient interaction and entanglement between polymer chains during electrospinning (Rostamabadi et al., 2020). As the polymer concentration increases, the viscosity also increases (Figure 12.3) and the material deposited on the collector transitions from particles to beaded fibers (as Ce is exceeded) of increasing fiber/bead ratio. High-viscosity polymer solutions, between 2.0 and 2.5 times Ce, produce bead-free fibers.

The surface tension of the solvent in the polymer solution is also an important factor in the electrospinning process. Formation of bead-free fibers at a given solution

TABLE 12.1
Main Parameters Affecting Electrospinning Quality

Main parameter	Effect	Effect on fiber	Reference
Solution parameters			
Solution properties	It reflects the entanglement of the chain in the polymer solution	The fiber diameter increases with the increase of concentration. At low concentrations, beads form on the fibers	Gupta et al. (2019)
Polymer viscosity	Viscosity is a key parameter affecting fiber morphology. It is related to the concentration and molecular weight of the polymer	At low viscosity, beads form	Deshawar et al. (2020)
Surface tension	It is the main force opposite to voltage in the process of electrohydrodynamics	There is no direct relationship between surface tension and fiber morphology	Silva et al. (2021)
Conductivity	It indicates the charge density of polymer solution and affects the morphology/particle size of fiber	With the increase of conductivity, finer, uniform and bead free fibers are formed, while bead fibers are formed at low conductivity	Liu et al. (2018)
Solution volatility	A process technology for increasing surface area	High-molecular-weight polymers form finer fibers, while low-molecular-weight polymers form bead fibers	Nair and Mathew (2017)
Process parameters			
Applied voltage	The diameter of electrospinning is sometimes changed by providing a surface charge on the electrospinning nozzle	With the increase of voltage, the fiber diameter decreases. The best voltage is needed to obtain the desired morphology	Jain et al. (2020)
Feed speed/flow	The feed rate/flow rate will affect the transfer rate and injection rate of polymer solution. Ideally, the feed rate must match the rate at which the solution is removed from the tip	The decrease of flow rate will reduce the diameter of fiber. Beads can form at high flow rates	Mfaab et al. (2021)

(*Continued*)

TABLE 12.1
(Continued)

Main parameter	Effect	Effect on fiber	Reference
Distance from nozzle to collector	Because the solution jet has enough flight time, adjusting the distance from the needle tip to the collector is the key factor to obtain uniform fiber	For uniform beads, a minimum distance is required. Beads appear when the distance is too high or too low	Alehosseini et al. (2017)
Environmental parameters			
Humidity	It will affect the spinnability of polymer solution	The increase of humidity will lead to the formation of pores on the fiber. At low humidity, bead formation is reduced	Deng et al. (2018)
Temperature	It can induce the formation of different electrospun fiber morphology by changing the characteristics of the solution	As the temperature increases, smaller and finer fibers will be formed	Deng et al. (2017)

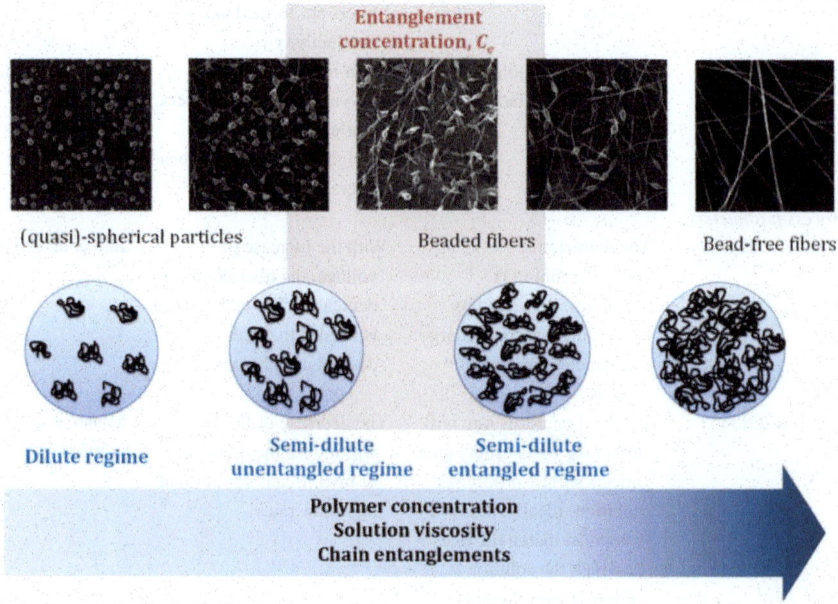

FIGURE 12.3 Polymer solution system and corresponding material morphology.

Source: Heriberto et al. (2019).

viscosity is facilitated by reducing the surface tension of the solution. Decreasing the surface tension of the solution allows bead-free fiber electrospinning to occur at a lower electric field strength.

The polymers used to make nanofibers are mostly electrically conductive and the charged ions present in the polymer solution have an important function in the formation of the polymer jet. The type of polymer and solvent used and the presence of ionizable salts determine the conductivity of the solution and the diameter of electrospun nanofibers decreases as the conductivity of the solution increases. Low solution conductivity results in spray droplet elongation to form uniform fibers, but may not eliminate beads formation (Jain et al., 2020). Various natural polymers, such as gelatin, are polyelectrolytes. The presence of ions in gelatin increases the charge-carrying capacity of the jet, which exposes it to higher electric field tension. This reduces the fiber-forming ability of gelatin compared with synthetic polymers. The conductivity of the solution can be improved by adding different salts.

12.3.2 Process Parameters

Process parameters, such as the voltage applied between the two electrodes, the flow rate of the polymer solution, and the distance between the spinneret and the collector can also significantly affect the electrospinning process (Wang et al., 2019; Angel et al., 2022). The voltage between the electrodes controls the electric field strength between the spinneret and the collection point, thereby modulating the strength of the stretching force. When the electrostatic force in the solution overcomes the surface tension of the solution, the electric field starts the electrospinning process. Depending on the feed rate of the solution, a higher voltage may be required in order to form a stable Taylor cone (Mingjun et al., 2019). The balance between surface tension and electrodynamic force is crucial for forming the initial Taylor cone of the polymer solution at the tip. As the driving voltage increases, the length of a single jet decreases slightly, and the sharp angle of the Taylor cone increases. If the applied voltage is higher, the acceleration speed of the jet will be faster, and the volume of solution drawn from the needle tip will be larger (Liashenko et al., 2020). This may result in a smaller and less stable Taylor cone. As the viscosity increases, more force is required to overcome surface tension and viscoelasticity. Since both the supplied voltage and the generated electric field affect the stretching and acceleration of the jet, they will affect the shape and diameter of the fiber obtained.

The influence of applied voltage and tip collector distance on the diameter of synthetic electrospun fibers has been investigated. The average fiber diameter reached a minimum, and then increased as the applied voltage increased (Figure 12.4A). This trend is attributed to the shortening of the deformation time of forming fibers as the voltage increases. Consistent with this conclusion is the tendency of electrospun fibers to produce relatively large beads at higher voltages (Zhang et al., 2018).

The extension of the polymer jet occurs between the tip of the Taylor cone and the collector. The applied voltage and the distance between the tip and the collector have a direct influence on the jet flight time and the intensity of the electrostatic field. Insufficient drying of electrospun fibers is mainly caused by the insufficient distance between the tip and the collector, allowing too little time for the solvent to

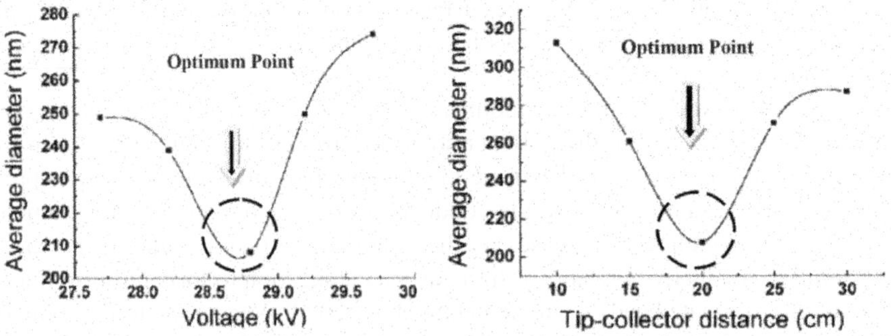

FIGURE 12.4 Relationships between (a) applied voltage and average fiber diameter; (b) collector tip distance and average fiber diameter, in the electrospinning process.

Source: Behrouz and Nick (2015).

evaporate and resulting in thicker fibers. Increasing the distance between the tip and collector initially reduces the average fiber diameter; however, at a fixed voltage, further increasing the distance between the tip and the collector reduces the intensity of the electrostatic field and increases the fiber thickness. Therefore, as the distance increases, the average fiber diameter passes through a minimum (Figure 12.4b).

12.3.3 Environmental Parameters

In addition to intrinsic process factors, as described earlier, environmental factors, i.e., the humidity and temperature, will also affect the characteristics of the fiber. Humidity affects the spinnability of polymer solutions; low humidity (<35%) is ideal for electrospinning, whereas at high humidity (>35%), it is difficult to produce continuous fibers consistently. Relative humidity can also affect the quality of electrospun fibers, especially their surface morphology (Haider et al., 2018; Sílvia et al., 2020). The viscosity of the spinning solution decreases as the temperature increases, which usually leads to the formation of smaller diameter fibers.

12.4 ELECTROSPUN ENTRAPMENT OF BIOACTIVE SUBSTANCES

There are many kinds of bioactive compounds, including hydrophobic small molecular substances, hydrophilic small molecular substances, and macromolecular active substances. However, their application in the food industry is limited because they are very unstable in oxygen, light, temperature, or other extreme conditions. Encapsulation or embedding may be an effective solution to overcome this problem. Among all kinds of packaging technologies, electrospinning technology has great advantages in nano-embedding of bioactive compounds. For example, it can improve the stability and bioavailability of bioactive molecules and improve the viability of probiotics, which is very important for the development of new functional foods (Wen et al., 2016). Another important advantage is that it can mask unpopular smells and tastes to improve product acceptance. This is very important for the

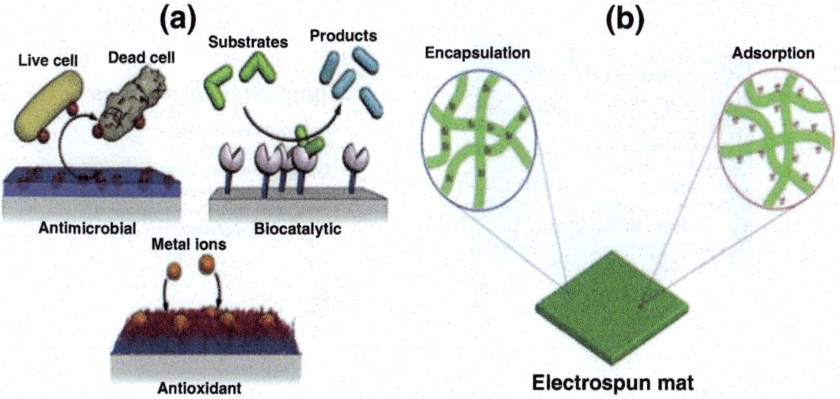

FIGURE 12.5 Schematic diagram of the main types of (a) active food packaging; (b) wrapped in the fiber enzyme and electrospun fiber pad or adsorbed on the surface.

Source: Bastarrachea et al. (2015).

application of plant essential oils in the field of food preservation. In addition, the obtained nanofibers can also be used as the delivery and control system of nutritious food or medicine. In the field of food packaging, electrospinning is a promising technology for the manufacture of active packaging materials or nanostructure layers for food packaging. Its applications range from controlling the growth of microorganisms and inhibiting oxidative degradation to the realization of targeted biocatalysis (Figure 12.5) (Bastarrachea et al., 2015; Tran and Balkus, 2012). In addition to these applications, electrospun fibers can also be used as filtration membranes in food and beverage processing.

12.5 APPLICATIONS OF ELECTROSPINNING TO FOOD PRESERVATION AND PACKAGING

As consumer awareness of food safety and environmental protection increases, demand is increasing for biodegradable functional food packaging materials that can carry bioactive components. Electrospinning technology is well-placed to meet this demand. Electrospun material can carry a high load of bioactive components and its high surface area enhances its mass-transfer characteristics, compared with conventional materials (Peng et al., 2017). In addition, electrospun films are effective with relatively low concentrations of bioactive compounds, so they have a low impact on the sensory quality of the product, and the electrospinning process does not require any heating, which minimizes degradation of heat-sensitive natural products (Pérez-Masiá et al., 2014). Overall, electrospinning to produce nanofibers is effective, relatively simple, and low-cost.

These advantages give electrospinning strong potential for application in various aspects of food science and technology (Figure 12.6), such as active packaging materials, shelf life enhancement of packaged foods, edible packaging films, and development of controlled-release delivery systems.

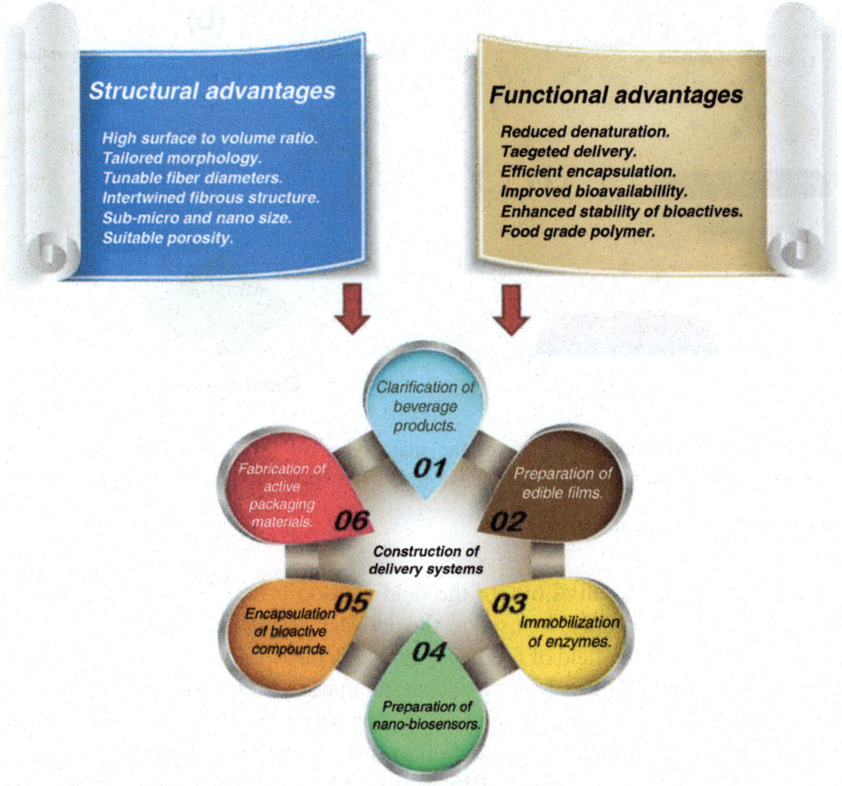

FIGURE 12.6 Advantages and applications of electrospun nanofibers in food science.

12.6 OPTIMIZATION OF ACTIVE-COMPONENT LOADING AND SIMULATION OF RELEASE BEHAVIOR

Incorporation of a bioactive component into electrospun fibers can be achieved by addition to the polymer solution, or adsorption onto the resulting fiber layer after it is formed (Son et al., 2013; Sill and Recum, 2008). The adsorption capacity of electrospun fibers for bioactives can be described by a single-parameter Langmuir model adsorption isotherm:

$$\frac{m}{m_{\max}} = \left(\frac{k_1 C_e}{1 + k_1 C_e}\right)$$

In the formula, m is the amount of bioactive adsorbed per unit weight of fiber matrix, $m_{\max}$ is the maximum amount of bioactive that can be adsorbed to form a monolayer, C_e refers to the residual concentration of the bioactive in solution, when the amount adsorbed is m, and k_L is the Langmuir constant. Alternatively, a two-parameter Freundlich model can be used.

$$m = K_F C_e^{1/n}$$

Here n is an empirical parameter related to the adsorption strength and K_F is the Freundlich constant. Either of these equations can be used to estimate or calculate how much of the bioactive is physically adsorbed and the maximum loading achievable. These models do not provide an analysis of the mechanism of bioactive transport, but can be used to calculate the drug-loading efficiency from the ratio of the bioactive concentration in the fiber to that in the original polymer solution.

To enable control of the bioactive release-rate and release-time from different types of fiber, it is necessary to improve understanding of the release mechanism (Tiwari et al., 2010). In general, the release mechanism of bioactives, or food additives from delivery systems can be classified in three ways: diffusion from the fiber, expansion of the fiber, making it more porous, and degradation of the polymer structure. Some bioactives diffuse from the fiber with no change in its structure or size, whereas others may only be released as the fiber expands by swelling from, e.g., water absorption, or the fiber structure degrades under the action of environmental conditions, or food-enzyme action. There are four main factors affecting the bioactive release rate: (1) fiber physicochemical properties, such as hydrophilicity, hydrophobicity, and biodegradability; (2) fiber morphological properties, such as surface roughness and thickness; (3) location of the bioactive, e.g., surface adsorption, or inside a core-shell fiber; and (4) bioactive solubility, hydrophobicity, stability, and compatibility with the polymer.

The porosity the polymer matrix is an important factor in the release of the bioactive from the fiber. Many models have been proposed to simulate and aid understanding of the diffusivity of polymer matrices. One of the earliest models uses Fick's law of diffusion to relate the release rate of the bioactive from planar surfaces, or spherical particles (Mircioiu et al., 2019). Structural changes such as fiber expansion and stress relaxation may lead to non-Fickian diffusion. The Avrami equation is also a commonly used model to simulate the release rate. The formula can simply assume that when the release mechanism parameter nmol is 1, the corresponding release behavior is first-order kinetic release, and when nmol is 0.54, it is diffusion-limited reaction release. When nmol is between 0.54 and 1, it belongs to the diffusion-limiting kinetics and the first-order release kinetics. The Korsmeyer–Peppas model is generally accepted to be the best model to analyze the release of bioactive components from porous materials. In this model, the kinetic mechanism is determined by diffusion Changshu n: when $n \leq 0.45$, the release mechanism of active components accords with the Fickian diffusion mechanism; when $0.45 < n < 0.89$, it accords with the non-Fickian diffusion mechanism; when $n \geq 0.89$, it accords with the skeleton dissolution mechanism (Jiang et al., 2010). A good understanding of the bioactive release behavior enables optimization of the fiber formulation.

12.7 APPLICATION OF ELECTROSPUN MATS CONTAINING ESSENTIAL OIL IN FOOD PRESERVATION

Essential oils have attracted extensive research interest in recent years, because of their significant antibacterial, antioxidant, anti-inflammatory and antiviral activities, and their natural origin. However, their application to the food industry is still limited, for the following four reasons. (1) The solubility of essential oils in water is poor, making formulation challenging. (2) Some essential oil components are volatile and/or unstable in the presence of oxygen, light, and high temperature, giving them a short half-life. (3) Many essential oils have strong smells which can adversely affect the sensory quality of foods. (4) The practical MIC of an essential oil in the food matrix is usually higher than that of the theoretical value obtained under laboratory conditions. Electrospinning technology may be a potential solution to these problems, because it can not only improve the stability and bioavailability of bioactive molecules, but also mask the undesirable smell of essential oils, so as to improve the consumer acceptability of the product (Lim, 2021; Rostamabadi et al., 2020; Ping et al., 2019). Figure 12.7 shows the manufacturing process of electrospun nanofibers containing essential oil.

There have been many reports on the application of electrospun mats containing essential oil to food preservation. For example, nanofiber rods were prepared by electrospinning after mixing thymol and γ-cyclodextrin with zein (maize protein)

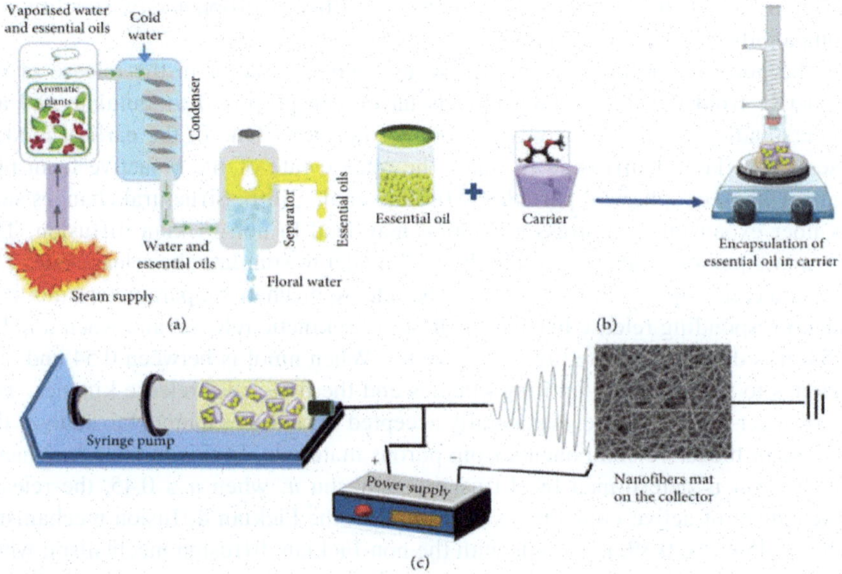

FIGURE 12.7 An overview of the manufacture of electrospun nanofibers containing essential oils. (a) Extraction of essential oil. (b) Incorporation of essential oil into the polymer solution. (c) Electrospinning of composite solution.

Source: Ataei et al. (2020).

solution and effectively inhibited the growth of *Escherichia coli* and *Staphylococcus aureus* in meat products stored at 4°C, suggesting potential application as a food packaging material (Aytac et al., 2017). An electrospun mat prepared by combining rose seed oil with gliadin (a wheat protein) by coaxial electrospinning to form a core-shell fiber extended the shelf life of kumquat and banana. In addition, Yilmaz et al. (2016) reported that curcumin-loaded zein nanofibers can be used as antifungal food packaging materials, which effectively inhibited the decay of apples during storage. Similarly, encapsulation of cinnamon essential oil in a polyvinyl alcohol and β-cyclodextrin matrix effectively extends the shelf life of strawberries.

The addition of cyclodextrin to the electrospinning polymer solution can produce active packaging with higher stability and durability, in addition to improving the solubility of hydrophobic essential oils. For example, cyclodextrin and electrospinning technology were combined to improve the thermal stability and shelf life of geraniol. The γ-cyclodextrin/geraniol fibers lost only 10% of their geraniol content after storage at room temperature for 2 years, indicating that encapsulation with cyclodextrin and electrospun fibers can significantly improve the stability and shelf life of bioactive compounds (Lemma et al., 2015). Mascheroni et al. (2013) developed an edible nanofilm containing pullulan, cyclodextrin, and perillaldehyde, from which the release of perillaldehyde was negligible at 23°C and even at 230°C, but release was triggered easily at a high relative humidity (Aw ≥ 0.9). Thymol and cinnamaldehyde are the two most commonly used essential oil components in food preservation, which may be related to their antibacterial activity and flavor. In addition, the main applications of electrospun mats containing essential oils in food preservation are for fruits, vegetables, and meat products.

12.8 CONCLUDING REMARKS

The growing consumer demand for environment-friendly, safe, and healthy food products has stimulated the improvement of product quality and safety in the food processing industry, mainly by improved control of pathogenic microorganisms and extension of shelf life. In this regard, the study of essential oils is of great significance, because their antibacterial activity can replace that of artificial preservatives. However, their low solubility, low stability, high volatility, and odor limit their application to food preservation. To solve these problems, a new type of antibacterial packaging has been developed, which can be used for food preservation, by using electrospinning to encapsulate essential oils into fiber mats; however, this technology is still in its infancy. To promote the development of this technology and its application in food preservation, this chapter has discussed in detail the principles of electrospinning technology and the main factors affecting its implementation and application.

Although electrospinning containing essential oil has many advantages, there are still problems to be solved. First, the long-term effects of essential oils on the physicochemical properties of polymer fibers are unknown. It is not yet clear whether the presence of essential oil will affect the morphology, stability, or degradation rate of the fibers during long-term storage. Second, low production efficiency, especially for aqueous polymer solutions, is still a major problem, because of their low volatility and slow drying and difficulty of incorporating essential oils into aqueous solutions.

Volatile organic solvents would easily solve this problem, but they are costly and undesirable for food applications. Third, a disadvantage of electrospun mats in food preservation is that loose fibers can easily adhere to the food surface. However, this can be avoided by combining with the film. Future research should include combining electrospinning with intelligent packaging, to develop new products with improved functions. In addition, it is also necessary to develop industrial-scale electrospinning machines, as, to date, only laboratory-scale machines have been developed. Finally, there is a need for governments to develop clear regulatory frameworks for the application of electrospun packaging in the food industry to ensure food safety.

ACKNOWLEDGEMENTS

This work was supported by the National Natural Science Foundation of China (32202192), Special fund for Taishan Scholars Project.

REFERENCES

Alehosseini, A., Ghorani, B., Sarabi-Jamab, M., and Tucker, N. 2017. Principles of electrospraying: A new approach in protection of bioactive compounds in foods. *Crit. Rev. Food Sci. Nutr.* 1–18.

Angel, N., Li, S., Yan, F., and Kong, L. 2022. Recent advances in electrospinning of nanofibers from bio-based carbohydrate polymers and their applications. *Trends Food Sci. Technol.* 120: 308–324.

Ataei, S., Azari, P., Hassan, A., Pingguan-Murphy, B., and Muhamad, F. 2020. Essential oils-loaded electrospun biopolymers: A future perspective for active food packaging. *Adv. Polym. Technol.* (1): 1–21.

Aytac, Z., Ipek, S., Durgun, E., Tekinay, T., and Uyar, T. 2017. Antibacterial electrospun zein nanofibrous web encapsulating thymol/cyclodextrin-inclusion complex for food packaging. *Food Chem.* 233: 117–124.

Bastarrachea, L., Wong, D., Roman, M., Lin, Z., and Goddard, J. 2015. Active packaging coatings. *Coatings*. 5(4): 771–782.

Behrouz, G., and Nick, T. 2015. Fundamentals of electrospinning as a novel delivery vehicle for bioactive compounds in food nanotechnology. *Food Hydrocolloids*. 57: 227–240.

Cheng, H., Xu, H., McClements, D. J., Chen, L., Jiao, A., Tian, Y., and Jin, Z. 2022. Recent advances in intelligent food packaging materials: Principles, preparation and applications. *Food Chem.* 375: 131738.

Dabirian, F., Hosseini, Y., and Ravandi, S. A. H. 2007. Manipulation of the electric field of electrospinning system to produce polyacrylonitrile nanofiber yarn. *J. Text. Inst.* 98(3): 237–241.

Deshawar, D., Gupta, K., and Chokshi, P. 2020. Electrospinning of polymer solutions: An analysis of instability in a thinning jet with solvent evaporation. *Polymer*. 202: 122656.

Deng, L., Kang, X., Liu, Y., Feng, F., and Zhang, H. 2017. Effects of surfactants on the formation of gelatin nanofibres for controlled release of curcumin. *Food Chem.* 231: 70–77.

Deng, L., Zhang, X., Li, Y., Que, F., Kang, X., and Liu, Y. 2018. Characterization of gelatin/zein nanofibers by hybrid electrospinning. *Food Hydrocolloids*. 75: 72–80.

Ding, B., Li, C., Miyauchi1, Y., Kuwaki, O., and Shiratori, S. 2006. Formation of novel 2D polymer nanowebs via electrospinning. *Nanotechnology*. 17(15): 3685–3691.

Forghani, S., Almasi, H., and Moradi, M. 2021. Electrospun nanofibers as food freshness and time-temperature indicators: A new approach in food intelligent packaging. *Innovative Food Sci. Emerging Technol.* 73: 102804.

Fuat, T., and Tamer, U. 2020. Antioxidant, antibacterial and antifungal electrospun nanofibers for food packaging applications. *Food Res. Int.* 130: 108927.

Gupta, D., Jassal, M., and Agrawal, A. K. 2019. Solution properties and electrospinning of poly (galacturonic acid) nanofibers. *Carbohydr. Polym.* 212: 102–111.

Haider, A., Haider, S., and Kang, I. K. 2018. A comprehensive review summarizing the effect of electrospinning parameters and potential applications of nanofibers in biomedical and biotechnology. *Arabian J. Chem.* 11(8): 1165–1188.

Hemamalini, T., and Dev, V. R. G. 2017. Comprehensive review on electrospinning of starch polymer for biomedical applications. *Int. J. Biol. Macromol.* 106: 712–718.

Heriberto, R. T., Graciela, M., and Daniel, G. 2019. Comprehensive review on electrospinning techniques as versatile approaches toward antimicrobial biopolymeric composite fibers. *Recent Res. Dev. Mater. Sci. Eng.* 101: 306–322.

Jain, R., Shetty, S., and Yadav, K. S. 2020. Unfolding the electrospinning potential of biopolymers for preparation of nanofibers. *J. Drug Delivery Sci. Technol.* 57: 101604.

Jiang, Q., Reddy, N., and Yang, Y. 2010. Cytocompatible cross-linking of electrospun zein fibers for the development of water-stable tissue engineering scaffolds. *Acta Biomater.* 6(10): 4042–4051.

Ju, J., Chen, X., Xie, Y., Yu, H., and Yao, W. 2019. Application of essential oil as a sustained release preparation in food packaging. *Trends Food Sci. Technol.* 92: 22–32.

Ju, J., Xie, Y., Yu, H., Guo, Y., Cheng, Y., and Yao, W. 2020. Major components in lilac and *Litsea cubeba* essential oils kill *Penicillium roqueforti* through mitochondrial apoptosis pathway. *Ind. Crops Prod.* 149: 112349.

Ju, J., Xie, Y., Yu, H., Guo, Y., Cheng, Y., and Zhang, R. 2020. Synergistic inhibition effect of citral and eugenol against *Aspergillus niger* and their application in bread preservation. *Food Chem.* 310(25): 125974.1–25974.7.

Khoshnoudi-Nia, S., Sharif, N., and Jafari, S. M. 2020. Loading of phenolic compounds into electrospun nanofibers and electrosprayed nanoparticles. *Trends Food Sci. Technol.* 95: 59–74.

Lemma, S. M., Scampicchio, M., Mahon, P. J., Sbarski, I., Wang, J., and Kingshott, P. 2015. Controlled release of retinyl acetate from β-cyclodextrin functionalized poly (vinyl alcohol) electrospun nanofibers. *J. Agric. Food Chem.* 63(13): 3481–3488.

Liashenko, I., Rosell-Llompart, J., and Cabot, A. 2020. Ultrafast 3D printing with submicrometer features using electrostatic jet deflection. *Nat. Commun.* 11(1): 1–9.

Lim, L. T. 2021. Electrospinning and electrospraying technologies for food and packaging applications. *Polym. Polym. Compos.* 217–259.

Liu, W., Lu, J., Ye, A., Xu, Q., Tian, M., and Kong, Y. 2018. Comparative performances of lactoferrin-loaded liposomes under in vitro adult and infant digestion models. *Food Chem.* 258: 366–373.

Luraghi, A., Peri, F., and Moroni, L. 2021. Electrospinning for drug delivery applications: A review. *J. Controlled Release.* 334: 436–484.

Mascheroni, E., Fuenmayor, C. A., Cosio, M. S., Di Silvestro, G., Piergiovanni, L., and Mannino, S. 2013. Encapsulation of volatiles in nanofibrous polysaccharide membranes for humidity-triggered release. *Carbohydr. Polym.* 98(1): 17–25.

Mfaab, E., Aac, E., Fed, E., and Bcaa, E. 2021. Effect of core-to-shell flowrate ratio on morphology, crystallinity, mechanical properties and wettability of poly (lactic acid) fibers prepared via modified coaxial electrospinning. *Polymer.* 237: 124278.

Mingjun, C., Youchen, Z., Haoyi, L., Xiangnan, L., and Weimin, Y. 2019. An example of industrialization of melt electrospinning: Polymer melt differential electrospinning. *Adv. Ind. Eng. Polym. Res.* 2(3): 110–115.

Mircioiu, C., Voicu, V., Anuta, V., Tudose, A., Celia, C., Paolino, D., and Mircioiu, I. 2019. Mathematical modeling of release kinetics from supramolecular drug delivery systems. *Pharmaceutics.* 11(3): 140.

Mukurumbira, A. R., Shellie, R. A., Keast, R., Palombo, E. A., and Jadhav, S. R. 2022. Encapsulation of essential oils and their application in antimicrobial active packaging. *Food Control*. 108883.

Nair, S. S., and Mathew, A. P. 2017. Porous composite membranes based on cellulose acetate and cellulose nanocrystals via electrospinning and electrospraying. *Carbohydr. Polym.* 175: 149–157.

Peng, W., Min, H. Z., Linhardt, R. J., Kun, F., and Hong, W. 2017. Electrospinning: A novel nano-encapsulation approach for bioactive compounds. *Trends Food Sci. Technol.* 70: 56–68.

Pérez-Masiá, R., Lagaron, J. M., and López-Rubio, A. 2014. Development and optimization of novel encapsulation structures of interest in functional foods through electrospraying. *Food Bioprocess Technol.* 7(11): 3236–3245.

Ping, S., Fangmi, A. I., Peiyu, Q., and Jiefeng, P. 2019. Classification of electrospinning polymer matrix and application of antibacterial packaging. *Food Ferment. Ind.* 45(20): 290–297.

Rostamabadi, H., Assadpour, E., Shahiritabarestani, H., Falsafi, S. R., and Jafari, S. M. 2020. Electrospinning approach for nanoencapsulation of bioactive compounds: Recent advances and innovations. *Trends Food Sci. Technol.* 100–112.

Son, Y. J., Kim, W. J., and Yoo, H. S. 2013. Therapeutic applications of electrospun nanofibers for drug delivery systems. *Arch. Pharmacal Res.* 37(1): 69–78.

Sill, T. J., and Recum, H. A. V. 2008. Electrospinning: Applications in drug delivery and tissue engineering. *Biomaterials*. 29(13): 1989–2006.

Silva, B., Cunha, R., Valério, A., Junior, A., Hotza, D., and González, S. Y. G. 2021. Electrospinning of cellulose using ionic liquids: An overview on processing and applications. *Eur. Polym. J.* 147: 110283.

Sílvia, C. C., Estevinho, B. N., and Rocha, F. 2020. Encapsulation in food industry with emerging electrohydrodynamic techniques: Electrospinning and electrospraying-a review. *Food Chem.* 339: 127850.

Soleimanifar, M., Jafari, S. M., and Assadpour, E. 2020. Encapsulation of olive leaf phenolics within electrosprayed whey protein nanoparticles; production and characterization. *Food Hydrocolloids*. 101: 105572.

Tiwari, S. K., Tzezana, R., Zussman, E., and Venkatraman, S. S. 2010. Optimizing partition-controlled drug release from electrospun core-shell fibers. *Int. J. Pharm.* 392(1–2): 209–217.

Tran, D. N., and Balkus, K. J. 2012. Enzyme immobilization via electrospinning. *Top. Catal.* 55(16): 1057–1069.

Wang, H., Kong, L., and Ziegler, G. R. 2019. Fabrication of starch-nanocellulose composite fibers by electrospinning. *Food Hydrocolloids.* 90: 90–98.

Wen, P., Zhu, D. H., Wu, H., Zong, M. H., Jing, Y. R., and Han, S. Y. 2016. Encapsulation of cinnamon essential oil in electrospun nanofibrous film for active food packaging. *Food Control*. 59: 366–376.

Yilmaz, A., Bozkurt, F., Cicek, P. K., Dertli, E., Durak, M. Z., and Yilmaz, M. T. 2016. A novel antifungal surface-coating application to limit postharvest decay on coated apples: Molecular, thermal and morphological properties of electrospun zein-nanofiber mats loaded with curcumin. *Innovative Food Sci. Emerging Technol.* 37: 74–83.

Zhang, K., Li, Z. J., and Huang, W. M. 2018. Preparation and characterization of tree-like cellulose nanofiber membranes via the electrospinning method. *Carbohydr. Polym.* 183: 62–96.

Zhang, Y., Zhang, X., Silva, S. R. P., Ding, B., Zhang, P., and Shao, G. 2022. Lithium—sulfur batteries meet electrospinning: Recent advances and the key parameters for high gravimetric and volume energy density. *Adv. Sci.* 9(4): 2103879.

13 Preparation Strategy and Application of Active Packaging Containing Essential Oils

13.1 INTRODUCTION

Generally, inert packaging materials are used in traditional food packaging to carry food and isolate environmental factors. With the development of integrated technology of food processing, transportation, and storage, the requirements of food safety, food quality, and shelf life run through food production, distribution, storage, and retail. Modern food packaging should not only meet the consumer demand for long shelf life, fresh quality, and nutrition of food, but also meet the demand for "no addition" or "little addition" of food additives and preservatives (Azman et al., 2022). As a result, it poses new challenges to food safety and quality from processing sites to sales markets and finally to consumers.

At the beginning of the 21st century, the European Union put forward the concept of "active packaging" in food packaging. Active packaging is a kind of packaging method which improves the internal environment of packaging, improves the safety of food, and prolongs the shelf life of food while maintaining the original quality of food. Controllable slow-release packaging is an innovative active packaging technology in recent years, which can protect food by continuously releasing active ingredients from packaging materials to food surface (Lim et al., 2022). Almasi et al. (2021) were the first scholars to introduce controlled release into food packaging applications Sustained-release antimicrobial packaging can release antimicrobial agents slowly to food, so that the concentration of antimicrobial agents in the packaging can be maintained for a long time, so as to achieve the purpose of antimicrobial and antiseptic. It replaces the traditional food preservation method of adding antimicrobial agents to food directly and can maintain the nutrition and flavor of food for a long time, prolong shelf life, and improve safety. Besides, it can effectively solve the problem of antimicrobial and antiseptic of food. Figure 13.1 is the schematic diagram of antimicrobial package for food protection.

Essential oil is an important plant extract, which has aroused great interest in replacing chemical preservatives in the field of food preservation because of its various biological activities. Moreover, the volatility of EO has further expanded its application in antibacterial packaging system. By combining with packaging materials, the effect of EOs' bad flavor on products can be reduced and the action time of EO can be prolonged, which greatly improves the effective utilization rate of EO

DOI: 10.1201/9781003329268-13

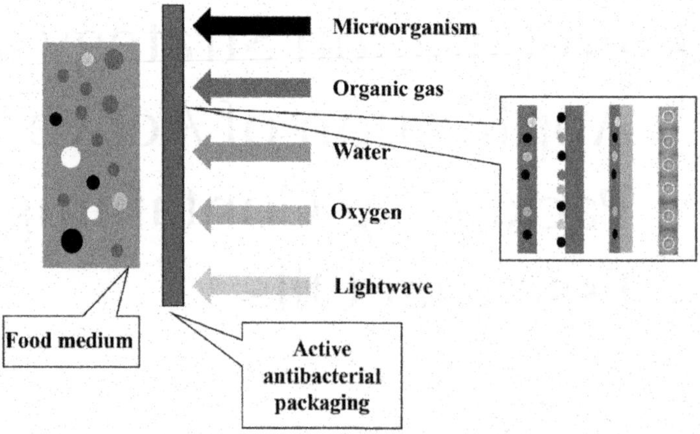

FIGURE 13.1 The schematic diagram of antimicrobial package for food protection.

in food preservation. Understanding the different combination of EO and food packaging, its application in food preservation will be more effective and targeted to promote the application of essential oils-antibacterial packaging (EOs-AP) in real food systems.

13.2 RESEARCH STATUS OF ANTIMICROBIAL PACKAGING THEORY

In theory, Crank's mathematical model for chemical migration in single-layer plastic packaging materials and mathematical model for evaluating the release of antimicrobial agents from active components were used (Alamri et al., 2021). It is based on Crank's prediction model for chemical migration in single-layer plastic packaging materials. Mascheroni et al. (2011) calculated the release of carvacrol from the mixture of wheat gluten and montmorillonite coated with carvacrol on the surface of self-made antimicrobial material-paper. Comparing with the sterilization data of *Botrytis cinerea* of tomato, it was found that the fitting effect was in good agreement. Uz and Altınkaya used Crank's mathematical model of chemical migration in double-layer plastics to verify the release consistency of potassium sorbate when contacted with water with temperature of 4°C and pH value of 7 (Uz et al., 2011). The results show that the fitting effect is in good agreement. Kuorwel et al. (2013) evaluated the migration effect of thymol, carvacrol, and linalool in their coatings and calculate the diffusion coefficient with Crank's prediction model for chemical migration in single-layer plastic materials.

13.3 ESSENTIAL OIL–ACTIVE PACKAGING SYSTEM

Because of its volatile nature, EOs can be used in antibacterial packaging systems. They are often added to the sol used to prepare films and made into antibacterial films by tape casting technology. This can not only avoid the influence of high temperature on EOs, but also reduce the possibility of reactions between EOs and food itself leading to lower activity. At present, the selection of film-forming materials is mostly concentrated in EVOH, PVA, LDPE, and other polymer materials. In the future, degradable and environment-friendly natural polymer materials will become a major force. In EOs-AP system, the EOs molecules exist in the film, and the film is a material with innumerable micropores. After the antibacterial active substances in EO molecules enter the film material, they gradually diffuse to the film surface through micropore, and then contact with food and play a role (Li et al., 2018). It is noteworthy that the solubility of antibacterial agents in food is an important factor affecting the antibacterial effect. If the antibacterial agent is highly soluble in food, the migration curve will follow an unconstrained free diffusion, while the antibacterial agent with little solubility will produce a unilateral system (as shown in Figure 13.2). The unconstrained free diffusion model indicates that the highly soluble antibacterial agents in packaging materials migrate to the food layer, and the concentration of antibacterial agents in packaging decreases with the migration. The concentration of antibacterial agents on food surface decreases with the decrease of the concentration in packaging, and finally falls below the minimum concentration, thus losing its antibacterial activity. What happens in the unilateral system is the migration of antibacterial agents with low solubility or low affinity to the food layer. In this system, the concentration of antibacterial agent on food surface is lower than that of soluble antibacterial agent and migration system, which is highly dependent on the solubility of antibacterial agent in food. Although the total amount of antibacterial agents in packaging is decreasing, the concentration of antibacterial agents on food surface can still be maintained at a constant concentration of antibacterial activity (Khaneghah et al., 2018). In order to obtain the maximum solubility in practice,

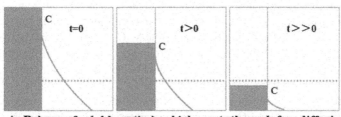

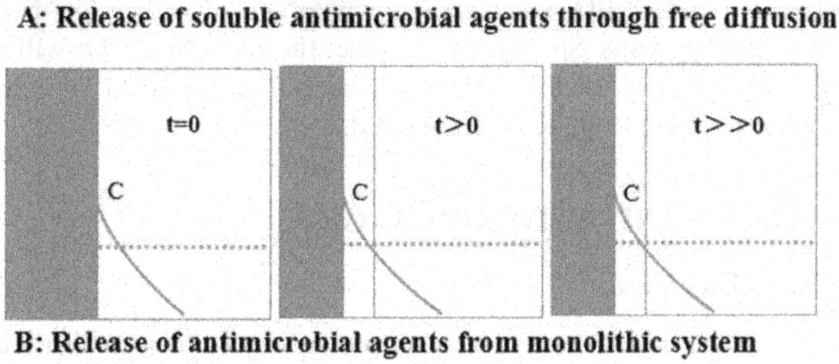

FIGURE 13.2 The migration curve of antimicrobial agents.

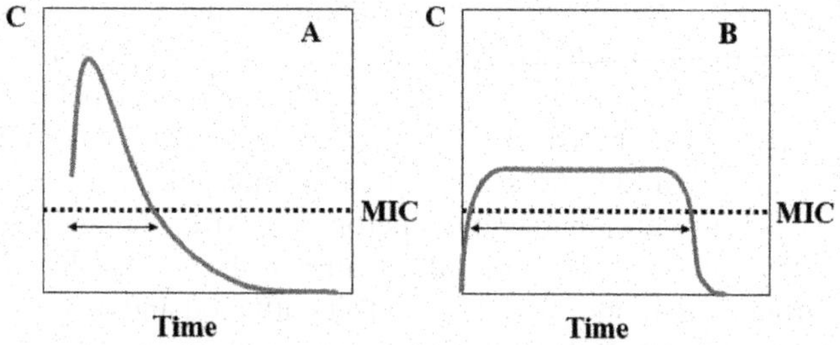

FIGURE 13.3 Concentration profile at the surface of foods. (A) Unconstrained free diffusion system and (B) monolithic system dashed line indicates MIC.

the migration curve of antibacterial agents should be thoroughly understood and controlled in order to achieve the desired antibacterial effect throughout the shelf life.

The design of antimicrobial packaging systems needs to incorporate controlled-release technology and microbial growth kinetics. If the migration rate of an antimicrobial agent is faster than the growth rate of the target microorganisms, the antimicrobial agent will be depleted before the end of the expected storage period. Because the volume of packaged food is large relative to the concentration of the antimicrobial agent, once the antimicrobial agent is depleted the packaging system will lose its antimicrobial activity and microorganisms will begin to grow. In contrast, if the release rate of the antimicrobial agent is too slow to control the growth of microorganisms, they will rapidly proliferate before the antimicrobial agent is released. Therefore, the release rate of antimicrobial agents from the packaging material into the food must be specifically controlled to match the growth kinetics of the target microorganisms. The importance of controlling the release rate of antimicrobial agents is demonstrated in Figure 13.3. In order to inhibit the growth of microorganisms on food surfaces, antimicrobial agents must be able to migrate from the packaging material to the food. When the concentration of antimicrobial agents on the food surface (C) is maintained above the MIC, the system will have effective antimicrobial activity. The effective antimicrobial period of system B is longer than that of system A because the concentration curve of system B remains above the MIC throughout the entire storage period.

In addition to antibacterial effects and migration rates, attention must be paid to the compatibility between EOs and packaging materials when EOs are used in antibacterial packaging systems. Specifically, whether the addition of EOs will affect the normal physical properties of the packaging material (e.g., barrier and mechanical properties, thickness, etc.) should be considered.

13.4 ESSENTIAL OIL–ACTIVE PACKAGING

13.4.1 Active Packaging

Active packaging mainly refers to a kind of packaging technology that adds active agents to the original packaging system in order to inhibit food corruption and oxidation, ensure food quality and safety, and extend the shelf life of food (Zanetti et al., 2018; Sharma et al., 2020; Mukurumbira et al., 2022). It is a new method to protect food from internal and external environmental factors and prolong the shelf life of food. Common food additives and their functions for active food packaging are shown in Figure 13.4.

The interaction between food and active ingredients in active packaging can absorb or release active ingredients, thus prolonging the shelf life of food and maintaining or improving the original quality of food in packaging. In addition to the long-term protection of food, these packages can be made of biodegradable and environmentally friendly materials.

Common food AP can be divided into two types, as shown in Figure 13.5. The first type of packaging is suitable for solid or liquid foods and directly comes in contact with food without any gaps. The antimicrobial agents in this type of antimicrobial packaging can be directly added to the food surface or combined with the packaging materials and then spread to the food surface. In the second type of packaging, food is placed in containers that are then packed with flexible packaging materials, leaving a gap between the packaging material and the food. This type of antimicrobial packaging is suitable for volatile antimicrobial agents, which spread

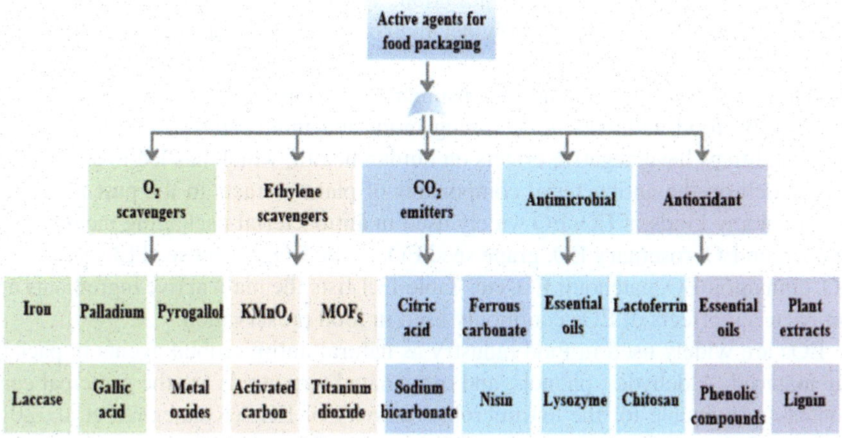

FIGURE 13.4 The common food additives and their functions for active food packaging.

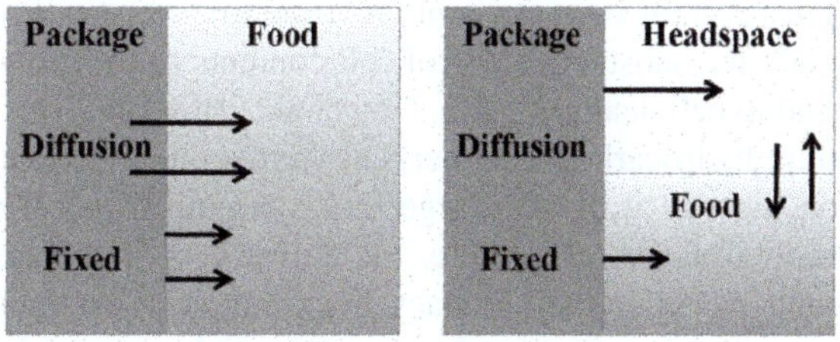

FIGURE 13.5 The common active food packaging types.

through the gap to the food surface (Ogunsona et al., 2020). For example, loading oregano essential oil with β-cyclodextrin as an active antibacterial ingredient in polylactic acid packaging can significantly ($P < 0.05$) prolong the shelf life of BlackBerry. At the same time, the active packaging showed significant antibacterial and antifungal ability. The biosafety of the packaging material was proved by coculture with zebrafish (Shi et al., 2022). In addition, the poly(hydroxybutyric acid–hydroxyvalerate) nanocomposites containing oregano essential oil had good inhibitory effect on both Gram-positive and Gram-negative bacteria, and the active packaging could prolong the action time of essential oil (Costa et al., 2020). Similarly, some researchers prepared active films containing clove essential oil and ZnO hybrid nanoparticles using chitosan and amylopectin as substrates and evaluated their antibacterial and antioxidant activities. Finally, the author confirmed that the new active film could be used to prolong the shelf life of chicken (Gasti et al., 2022).

At present, with the increase of world population and the increasing attention of industrial manufacturers and consumers to limited resources and environmental protection, the use of renewable resources and biodegradable materials has received special attention (Soares et al., 2017). For example, carbohydrates, proteins, and lipids are used as polymeric substrates to develop food packaging instead of synthetic packaging materials. In active packaging, antimicrobial packaging combined with EOs has attracted more and more attention. These packages are mainly composed of different kinds of EOs and polymer matrix, which can protect food from food pathogens.

With the increasing demand of consumers, the trend of looking for natural, safe, and healthy food additives is becoming more and more intense. In addition, synthetic additives have negative effects on human health, which has led to a great deal of research on the antibacterial components of plant extracts in the past few years. Among many kinds of EO, EO widely used in antibacterial packaging include clove EO, lemon EO, rosemary EO, grape seed EO, garlic EO, cumin seed EO, straw fruit EO, cinnamon EO, patchouli EO, etc. Table 13.1 lists the main active ingredients and antibacterial effects of EOs commonly used in food preservation.

EOs are widely used in food industry as natural antimicrobial agents in packaging material. Aldehydes, phenols, and oxygenated terpenoids are the principal components responsible for the antimicrobial activity of EOs (Khaneghah et al., 2018; Ju et al., 2018a). An important characteristic of EOs and their constituents is their hydrophobicity, which allows them to interact with the lipids of the microbial cell membrane and mitochondria, making the structures less organized and thus more permeable. This increased permeability allows the outflow of ions and other cell contents. Although a certain amount of outflow from microbial cells may be tolerated with no loss of viability, substantial loss of cell contents or the loss of vital ions and molecules will lead to cell death (Ju, Xie, Guo et al., 2018b; Ju, Xie, Yu et al., 2022).

EOs that contain hydrocarbon monoterpenes can simply interact with the active agents present on the cell wall of a target microorganism to enhance the uptake of other antimicrobial agents. Therefore, the delivery of EOs together with other antimicrobial agents increases their efficacy. For example, the combination of essential oil and nisin, sodium caseinate and thyme essential oil, or sodium alginate and citronella essential oil can exert synergistic antibacterial effect (Xue et al., 2016;

TABLE 13.1
Application of Active Food Packaging Containing Essential Oil in Food Preservation

Packaging type	Name of essential oil	Concentration	Product	Conclusion	Reference
Polylactic acid	Oregano essential oil	0.6 ml	Blackberry	The active packaging delayed the postharvest decay, deterioration, and storage quality loss of blackberry	Shi et al. (2022)
Chitosan and starch	Clove essential oil	2 ml	Chicken	The shelf life of chicken wrapped with composite membrane was extended to the 5th day at 8°C ± 2°C	Gasti et al. (2022)
Gelatin	Lemon peel essential oil	1–7% (w/w)	Cheese	LPO-loaded gelatin centrifugal spun positively affected the shelf life by suppressing the growth of aerobic mesophilic bacteria and yeast molds that cause spoilage in cheese	Doğan et al. (2022)
Polylactic acid/ polyhydroxyalkanoates	Oregano essential oil	4 wt%	Fish fillet	The slow-release composite films could extend the shelf life of pufferfish fillets by 2–3 days at 4°C ± 1°C	Zheng et al. (2022)
Poly(butylene adipate-*co*-terephthalate)	Cinnamon essential oil	Cellulose nanofibers: oil 2:1	Strawberry	The results indicated that the developed films improved the strawberries' qualities and have antimicrobial properties against *Salmonella* and *Listeria monocytogenes*, offering a potential alternative to synthetic materials as food packaging	Montero et al. (2021)
Poly(lactic acid) (PLA)/nanofibrillated cellulose (NFC)	Thymol	5%, 10%, and 15%	Cherries and tomatoes	The PLA/NFC films with thymol and anthocyanin formulation could inhibit fungus growth better in the cherry tomato sample than the PLA/NFC films with curry and anthocyanin	Nazri et al. (2022)
Polylactic acid	Fenugreek essential oil	5 wt%	Strawberry	The PLA-based composite film exhibited good antibacterial and antioxidant properties. In addition, a food quality test was performed on strawberry, and the results were compared to the commercial (polyethylene) film	Subbuv and Kavan (2022)

(Continued)

TABLE 13.1
(Continued)

Packaging type	Name of essential oil	Concentration	Product	Conclusion	Reference
Chitosan	*Cinnamodendron dinisii* Schwanke essential oil	200 µl	Beef	The active packaging containing zein nanoparticles was efficient in the conservation of ground beef, stabilizing the deterioration reactions, and preserving the color	Xavier et al. (2020)
Sodium alginate	Tea tree essential oil	10% (v/v)	Banana	The addition of TEON significantly improved the antifungal and antioxidant properties of the bilayer film	Yang et al. (2022)
Poly(3-hydroxybutyrate-*co*-4-hydroxybutyrate)	Thyme essential oil	0% (control), 10%, 20%, 30%, 40% (v/w)	White bread	The antimicrobial activity of oil films was checked against total bread molds and observed that when bread was sealed in P(3HB-*co*-4HB) films containing 30% v/w thyme oil, its shelf life extended to at least 5 days compared to 1–4 days for the neat film	Sharma et al. (2022)
Arabic gum	Cinnamon essential oil	4, 8 and 10% (v/w)	Cheese	Shelf life of cheese samples, packed with the self-stick membranes loaded with cinnamon extract, has extended from 3 to 8 weeks	Nada et al. (2022)
Gelatin	Cinnamon essential oil	Cinnamon oil and surfactant at different ratios of 70:30, 60:40, and 50:50	Apple	Cinnamon oil emulsion-based films could enhance and modify the functionality of protein-based films creating a biodegradable, sustainable, and cost-effective food packaging materials	Ganeson et al. (2022)
Gelatin, carboxymethyl cellulose	*Mentha longifolia* L. essential oil	0.5, 1, and 2%	Shrimp	CMC-GE-MEO 1% and CMC-GE-MEO 2% nanofibrous films successfully improved the microbial population and chemical property of peeled prawns for 12 and 14 days, respectively	Yasser et al. (2022)
Poly(butylene adipate terephthalate) and poly(lactic acid) (PBAT/PLA)	Citral	6%	Shrimp	Findings indicated high potential of EO compounded films as functional active packaging to preserve seafood qualities	Yeyen et al. (2021)

Polylactic acid, chitosan, and caseinate	Rosemary essential oil	1% and 2%	Chicken breast	Overall, this study has demonstrated that PLA films coated with an active coating are a promising delivery method for providing antioxidant effects in packaging for fresh meat products	Afberto et al. (2021)
Chitosan	Garlic essential oil	2:3 v/v	Chicken	According to the current findings, the chitosan film used for chicken fillet prevented the factors affecting its chemical and microbial spoilage. In particular, the addition of different levels of nano-liposome garlic essential oil to the chitosan film had a synergistic and dose-dependent effect. The shelf life of the chicken fillet samples was at least two to three times more than the usual shelf life which has been regulated for 3 days at 4°C	Ablofa et al. (2021)
Polyvinyl alcohol	*Rosmarinus officinalis*	10 g	Chicken breast	These active packaging coatings containing LEO and REO enhanced the shelf life of chicken breast fillets, reducing the lipid oxidation process and reducing *Listeria* counts during cold storage	Ggarn et al. (2021)
Potato starch/apple peel pectin	*Zataria multiflora* essential oil	0.5 g	Quail meat	Examination of the chemical properties of quail meat packaged with active films indicates the positive effect of encapsulated essential oil and ZrO_2 nanoparticles in increasing the shelf life of quail meat	Sani et al. (2021)
Chitosan	Cinnamon essential oil	0.60%	Roasted duck	Taken together, CH incorporated with cinnamon EO could be developed as prospective edible packaging materials to preserve roast duck meat	Xue et al. (2021)
Chitosan	Rosemary and ginger essential oil	2%	Fresh poultry meat	Films have shown to be effective by extending the shelf life of the fresh poultry meat	Afonso et al. (2018)
Gelatin	Thyme essential oil	TEO:β-CD (w:w) = 1:8	Chicken	The TEGNs packaged chicken samples possessed of lower aerobic bacterial count, TBA, TVB-N, and pH values without adverse impact on color, texture, and sensory evaluation, which indicated TEGNs had a promising prospect in meat preservation	Lin et al. (2018)
Tamarind starch/whey protein	Thyme essential oil	5, 10, 15, and 20%	Tomato	With the incorporation of WPC and TEO into film, water gain, and solubility of the films improved significantly to enhance the shelf life of tomato till 14 days	He et al. (2021)

Salvia-Trujillo et al., 2015; Zhang et al., 2021). Not only these, but also related researchers have proved that the combination of nanoemulsions and EO can significantly ($P < 0.05$) improve the antimicrobial activity of EO. This may be due to the influence of nanoemulsions on the transport of EO to cell membranes and the interaction of nanoemulsions with multiple molecular sites on microbial cell membranes (Donsì et al., 2016; Barradas et al., 2021). Interestingly, however, it is reported that the interaction of some macromolecules as emulsifiers and EO can reduce the antimicrobial activity of EO to a certain extent. For example, the coating of whey protein and maltose dextrin combined with lilac EO shows lower antimicrobial activity than free EO. The reason for this phenomenon may be that some polymer matrices are made into EO–edible coatings after the interaction with EOs, which can produce electrostatic interactions with corresponding cell membranes. Or because the proportion of the water phase contained in the coating is large, which greatly reduces the interaction between the EO and the cell membrane (Ju et al., 2018b).

The main indicators related to the antibacterial mechanism of EOs were shown in Figure 13.6. At present, relevant research reports mainly focus on the research of

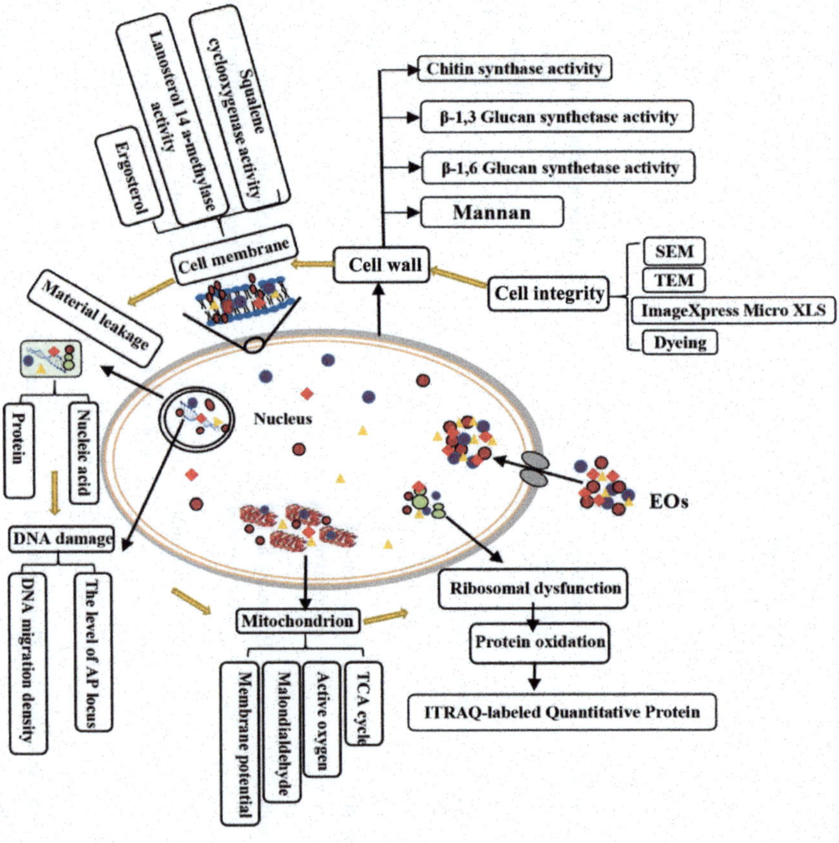

FIGURE 13.6 Main test indicators and means.

cell wall, cell membrane, nucleus, mitochondria, ribosome, and cell contents leakage. The most popular view is that plant EOs are lipophilic and can interact with cell walls and cell membranes, leading to cell membrane rupture, EOs entering the cell and then interacting with organelles, causing cell contents to leak out and eventually cells to be killed.

The process mainly goes through five stages:

(1) Primary contact of EOs with cell walls and cell membranes.
(2) Changes in cell wall and cell membrane permeability.
(3) Interaction of EOs with organelles.
(4) Leakage of cell contents.
(5) Cell death.

13.4.2 Realization of Essential Oil–Active Packaging

The incorporation of active compounds into natural and synthetic polymers and their application in the formulations of coatings are valuable strategies for increasing the shelf lives of packaged food products. From a theoretical perspective, antibacterial agents should be delivered at a controlled rate. Additionally, the concentration of released antibacterial agent should be neither too high nor too low to avoid adverse effects on sensorial and toxicological properties. In other words, a balance between the microbial growth kinetics and the controlled release rate should be established to guarantee the proper protective function during the expected shelf life. Therefore, one of the most interesting challenges in the field of antibacterial systems is controlling the release rate of the antibacterial agents from the packaging and their subsequent transfer into the food products.

The realization way of EO as an additive in antibacterial packaging mainly includes the following five ways. (1) Be directly mixed into the packaging matrix material. (2) Be coated on packaging materials or adsorbed by packaging materials. (3) Be made into small antibacterial package. (4) Be added to packaging in gaseous form. (5) Be microencapsulated and then packaged. Figure 13.7 showed the realization way of antibacterial packaging containing EO.

13.4.2.1 Be Directly Mixed Into the Packaging Matrix Material

Because the technology is relatively simple, directly mixed into packaging substrates is the commonly studied type of EO antibacterial packaging. According to the application mode, there are two kinds of adding ways. One is that the EO is directly added to the packaging materials, and the antibacterial material is made by blending or solvent casting technology (Figure 13.7a). For example, Suppakul et al. (2011) prepared polypropylene (PP) and polyethylene packaging film containing cinnamon EO, oregano EO, and clove EO. The results showed that polypropylene film had better antifungal activity than polyethylene film. Kuorwel et al. (2014) added different contents of EO components such as carvol, linalool, and thymol into thermoplastic starch film. The results showed that the tensile strength of the film decreased with the increase of the concentration of EO, while the thermal properties, water vapor permeability, and transparency of the film did not change

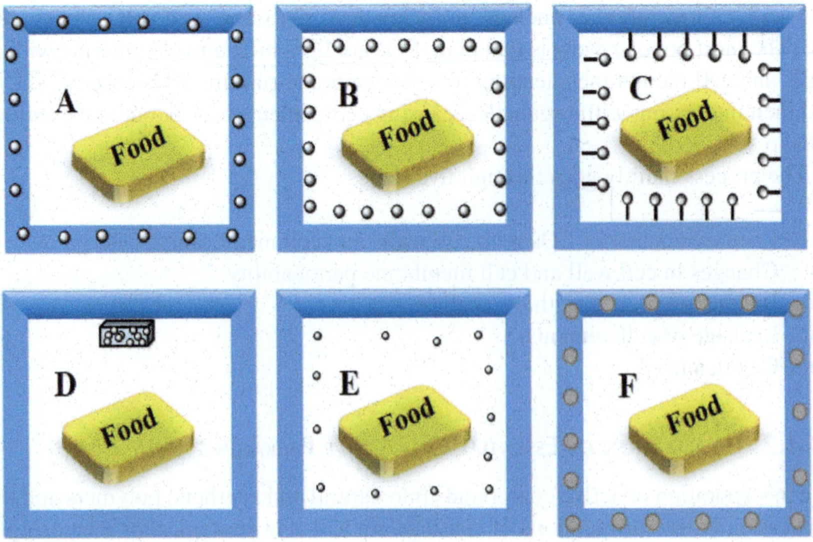

FIGURE 13.7 The realization way of EO as an additive in antibacterial packaging.

much. Another way of application is to mix EOs into the membrane matrix and cover the surface or interior of food by wrapping, impregnating, coating, and spraying, so as to reduce or prevent moisture and gas (O_2, CO_2) from entering. The main material used is edible film. For example, Gómez-Estaca et al. (2010) found that when cod was coated with gelatin-chitosan solution containing clove essential oil, the Gram-negative bacteria in fish could effectively be inhibited. The shelf life of cod was up to 20 days at 2°C.

13.4.2.2 Be Coated on Packaging Materials or Adsorbed by Packaging Materials

EOs can be coated on packaging materials directly or absorbed by it because of its heat-resistant and volatile nature (Figure 13.7B and C). At present, the main packaging systems are polymer packaging and paper packaging. For example, Rodriguez et al. (2008) found that paper packages coated with cinnamon EO can prevent water vapor from seeping into them and prevent fungal contamination. Some researchers also prepared compound coatings with oregano EO, beta-cyclodextrin and chitosan, which were coated on the inner surface of the carton. The results showed that the carton could significantly ($P < 0.05$) prolong the shelf life of strawberries. Bekele et al. (2017) coated the compound preparation of lemon grass EO and polyethylene on LDPE film to explore the volatilization of EO under high-humidity conditions and the fresh-keeping effect on grapes at room temperature. Recently, Ju et al. (2017) found that applying cinnamon or lemon EO to bread bags could prolong the shelf life of bread better. This method can not only prolong the shelf life of bread, but also have no significant effect on the sensory organs of bread.

13.4.2.3 Be Made Into Small Antibacterial Package

Be made into small antibacterial package is that the EOs are adsorbed or embedded in carrier material to make small antibacterial packaging bags and then put into food packaging (Figure 13.7D). During the whole storage process, EOs will slowly and continuously release into the packaging gap. For example, Espitia et al. (2012) made antimicrobial bags from polymer resins containing oregano, cinnamon, and lemon grass EOs and placed them in cartons containing papaya. The results showed that the method could prolong the shelf life of papaya better. Relevant researchers also put antimicrobial bags containing oregano EO into sliced bread bags. The results also showed that this method could prolong the shelf life of bread and significantly ($P < 0.05$) inhibit the growth of *E. coli*, *Salmonella*, mold, and yeast in bread (Passarinho et al., 2014).

13.4.2.4 Be Added to Packaging in Gaseous Form

Eos can be used in modified atmosphere packaging because of its strong volatility. When the Eos fill the whole package in the form of gas, it forms a bacteriostatic environment (Figure 13.7E). In the past few years, Nielsen et al. (2000) have shown that allyl-isothiocyanate (the active ingredient in mustard EO) can be added to the packaging bag in the form of gas. When the mass concentration in the packaging bag reached 3.5 mg/ml, it could completely kill *Penicillium* and *Aspergillus flavus*. Serrano et al. found that the total bacterial count of cherries treated with eugenol, thymol, and menthol decreased significantly ($P < 0.05$) after 16 days storage under certain atmospheric and low temperature conditions, and the number of fungi and yeasts decreased from 2.1 log cfu/g to 1.5 log cfu/g, while the number of fungi and yeasts increased to 4.9 log cfu/g in the control group (Serrano et al., 2005).

13.4.2.5 Be Microencapsulated and Then Packaged

The so-called microencapsulation is to use the semipermeability or the holes of microcapsules to make EOs embedded in a relatively sealed environment and release at a controlled rate under certain conditions (Figure 13.7F). The size of microcapsules is usually between 1 and 1,000 μm. The main advantage of microencapsulation technology containing EOs is achieving controlled release of EOs. Second, it can conceal the bad flavor of EOs, isolate the reacting components, and stabilize them in a system. At present, carbohydrates are widely used as wall materials of microcapsules. For example, Hao et al. (2012) prepared the microcapsules containing clove EO and then successfully added them to the polyvinyl alcohol film. The results showed that the antibacterial film not only had good air permeability, but also could inhibit the growth and reproduction of *Aspergillus niger*. da Silva Barbosa et al. (2021) used gelatin and chitosan as wall materials of microcapsules to prepare microcapsules containing garlic and cinnamon EO by composite coacervation method and then successfully applied them to the preservation of rice. The results showed that this method not only had better bacteriostatic effect, but also can significantly ($P < 0.05$) delay the aging of rice. At present, with the continuous development of related science and technology, nano-EO capsules with particle sizes ranging from 1 to 1,000 nm have emerged. The sustained-release effect of nano-EO capsules is more obvious than that of microcapsules, and the antibacterial effect has also been significantly

($P < 0.05$) improved. For example, Liolios et al. (2009) found that the antimicrobial activity of EOs in micrometer-sized capsules did not increase significantly ($P < 0.05$) generally, while nano-microcapsules could enhance the passive absorption mechanism of microbial cells, reduce the resistance of substance transport, and thus improve the antimicrobial effect of EOs. Donsì et al. (2011) prepared nano-microcapsules of terpenoids extracted from tea tree EO. The results showed that when the concentration of terpenoids was low (1.0 g/l), the microcapsules could delay the growth of microorganisms, and when the concentration increased to 5.0 g/l, it could completely inhibit the reproduction of microorganisms.

13.5 RELEASE MECHANISM OF ACTIVE COMPOUNDS

It is the most common type of active packaging that active compounds are directly mixed into polymer matrix. In this type of system, the activator combines with the polymer to form a composite matrix. As this type of active packaging is the most studied food packaging system in recent years, this part focuses on analyzing and describing the main release mechanism of this type of packaging system. There are three main types of release mechanisms recommended for directly integrated food active packaging systems.

1. *Diffusion-Induced Release Mechanism:* In this release mechanism, the active compounds are diffused through the micropore or macroporous structure of the polymer matrix and transported from the membrane surface to the food. The chemical composition, porosity and permeability of the polymer are the most important parameters in this kind of release rate. This mechanism is usually used to release active compounds in synthetic and waterproof polymers.
2. *Disintegration-Induced Release Mechanism:* The main reason for this release is the degradation, cracking, or deformation of the polymer. The rate of deformation and degradation is caused by the absorption of fluids in the polymer matrix, usually in aqueous solution. This type of release occurs in biodegradable or reactive nonbiodegradable polymers such as polyanhydrides, poly(lactide), and poly(lactide-*co*-glycolide).
3. *Swelling-Induced Release Mechanism:* In this system, the active compound cannot be diffused in the polymer matrix because of its low diffusion coefficient. When the polymer matrix is placed in a compatible liquid medium, the polymer swells by penetrating fluid to its matrix. In the swollen state of polymer, the active agent's diffusion coefficient increases, and then it diffuses out. Since most foods are wet and contain water, water is the most common penetrant. This type of release often occurs in moisture-sensitive packaging materials, such as proteins or polysaccharide films.

13.6 APPLICATION OF EOs–ACTIVE PACKAGING IN FOOD PRESERVATION

With the continuous improvement of people's living standards, consumers' requirements for the quality of food are also increasing. They expect not only to eat local fresh food, but also to taste exotic food. Unfortunately, the long-term storage and transportation conditions have a negative impact on the taste and quality of these products. Although active packaging could reduce these negative effects to some extent, with the potential risks of synthetic food additives to human health and the increasing demands on environment and food safety, the development of natural, friendly, and healthy active food packaging has become a research and development hotspot in the field of food. EOs-AP has great application potential.

EOs-AP technology, as a method of food preservation, has been recognized by the international community and listed as a key research and development hi-tech in the 21st century (Ribeiro-Santos et al., 2017). Although the development history of EOs-AP technology is relatively short, its emergence has solved many problems in food, showing great superiority especially in the application of food preservation industry. Many products waste a lot of resources because the preservation technology is not guaranteed, but the packaging technology of EOs can better solve this problem. Especially that it not only conforms to the concept of green environmental protection, but also has less impact on human health with EO as active agent. At present, we can see from the applied food that the application of EOs-AP mainly concentrates on meat products, fruit, and vegetable products. This may be mainly related to the structural characteristics of EOs-AP itself. EOs may have poorer odor and mechanical and permeability properties than general synthetic packaging, which may limit their application in some foods. However, with the continuous improvement and exploration of processing technology, these limitations will be overcome.

13.7 CONCLUDING REMARKS

At present, although the research on antibacterial packaging of essential oils has achieved some results, there are still many problems to be solved. (1) Because of the unique odor of EO, it may affect the sensation of food when used in antibacterial packaging of food. (2) There are relatively few studies on the migration characteristics of antibacterial agents in packaging materials, especially the correlation analysis of numerical analysis and finite element method. (3) Although many EO components have been listed as harmless components or approved as food flavor additives, some studies have shown that some of them have certain irritation or toxicity. In addition, some components of EOs are regulated differently in different countries. The regulations of different countries should be paid attention to when exporting products. (4) The antimicrobial activity and physicochemical properties of EOs-AP are affected by many factors, such as the type of EO, the concentration of active ingredients, and the composition of packaging materials. Finally, it is necessary to make further in-depth investigation on the acceptability of consumers.

ACKNOWLEDGEMENTS

This work was supported by the National Natural Science Foundation of China (32202192), Special fund for Taishan Scholars Project.

REFERENCES

Ablofa, K., Ebrahim, B., Khanjari, A., Afshin, A., Behshad, N., and Nsb, C. 2021. Nanocomposite active packaging based on chitosan biopolymer loaded with nano-liposomal essential oil: Its characterizations and effects on microbial, and chemical properties of refrigerated chicken breast fillet. *Int. J. Food Microbiol.* 342: 109071.

Afberto, A., Stevrnpa, C., Sv, B., Elebat, D., and Pmb, D. 2021. Active packaging based on PLA and chitosan-caseinate enriched rosemary essential oil coating for fresh minced chicken breast application. *Food Packag. Shelf Life.* 29: 100708.

Afonso, P., Lauriano, D., and Luisa, F. A. 2018. Chitosan/montmorillonite bionanocomposites incorporated with rosemary and ginger essential oil as packaging for fresh poultry meat. *Food Packag. Shelf Life.* 17: 142–149.

Alamri, M. S., Qasem, A. A., Mohamed, A. A., Hussain, S., Ibraheem, M. A., Shamlan, G., and Qasha, A. S. 2021. Food packaging's materials: A food safety perspective. *Saudi J. Biol. Sci.* 28(8): 4490–4499.

Almasi, H., Jahanbakhsh Oskouie, M., and Saleh, A. 2021. A review on techniques utilized for design of controlled release food active packaging. *Crit. Rev. Food Sci. Nutr.* 61(15): 2601–2621.

Azman, N. H., Khairul, W. M., and Sarbon, N. M. 2022. A comprehensive review on biocompatible film sensor containing natural extract: Active/intelligent food packaging. *Food Control.* 109189.

Barradas, T. N., and de Holanda e Silva, K. G. 2021. Nanoemulsions of essential oils to improve solubility, stability and permeability: A review. *Environ. Chem. Lett.* 19(2): 1153–1171.

Bekele, L. D., Zhang, W., Liu, Y., Duns, G. J., Yu, C., Jin, L., and Chen, J. 2017. Preparation and characterization of lemongrass fiber (*Cymbopogon* species) for reinforcing application in thermoplastic composites. *BioResources.* 12(3): 5664–5681.

Costa, R., Daitx, T. S., Mauler, R. S., Silva, N., and Carli, L. N. 2020. Poly(hydroxybutyrate-co-hydroxyvalerate)-based nanocomposites for antimicrobial active food packaging containing oregano essential oil. *Food Packag. Shelf Life.* 26: 100602.

da Silva Barbosa, R. F., Yudice, E. D. C., Mitra, S. K., and dos Santos Rosa, D. 2021. Characterization of Rosewood and *Cinnamon cassia* essential oil polymeric capsules: Stability, loading efficiency, release rate and antimicrobial properties. *Food Control.* 121: 107605.

Doğan, N., Doğan, C., Eticha, A. K., Gungor, M., and Akgul, Y. 2022. Centrifugally spun micro-nanofibers based on lemon peel oil/gelatin as novel edible active food packaging: Fabrication, characterization, and application to prevent foodborne pathogens *E. coli* and S. aureus in cheese. *Food Control.* 139: 109081.

Donsì, F., Annunziata, M., Sessa, M., and Ferrari, G. 2011. Nanoencapsulation of essential oils to enhance their antimicrobial activity in foods. *LWT-Food Sci. Technol.* 44(9): 1908–1914.

Donsì, F., and Ferrari, G. 2016. Essential oil nanoemulsions as antimicrobial agents in food. *J. Biotechnol.* 233: 106–120.

Espitia, P. J. P., Soares, N. D. F. F., Botti, L. C. M., Melo, N. R. D., Pereira, O. L., and Silva, W. A. D. 2012. Assessment of the efficiency of essential oils in the preservation of post-harvest papaya in an antimicrobial packaging system. *Braz. J. Food Technol.* 15(4): 333–342.

Ganeson, K., Razifah, M. R., Mubarak, A., Kam, A., Vigneswari, S., and Ramakrishna, S. 2022. Improved functionality of cinnamon oil emulsion-based gelatin films as potential edible packaging film for wax apple. *Food Biosci.* 47: 101638.

Gasti, T., Dixit, S., Hiremani, V. D., Chougale, R. B., Masti, S. P., Vootla, S. K., and Mudigoudra, B. S. 2022. Chitosan/pullulan-based films incorporated with clove essential oil loaded chitosan-ZnO hybrid nanoparticles for active food packaging. *Carbohydr. Polym.* 277: 118866.

Ggarn, B., Mjfa, C., Pc, A., Hie, B., Gs, A., and Lra, C. 2021. Biodegradable active food packaging structures based on hybrid cross-linked electrospun polyvinyl alcohol fibers containing essential oils and their application in the preservation of chicken breast fillets. *Food Packag. Shelf Life*. 27: 100613.

Gómez-Estaca, J., De Lacey, A. L., López-Caballero, M. E., Gómez-Guillén, M. C., and Montero, P. 2010. Biodegradable gelatin—chitosan films incorporated with essential oils as antimicrobial agents for fish preservation. *Food Microbiol.* 27(7): 889–896.

Hao, X. H., Sun, M., and Deng, J. 2012. Clove EO microcapsule antibacterial packaging film. *Plastic*. 41(1): 64–66.

He, Q., Guo, M., Jin, T. Z., Arabi, S. A., and Liu, D. 2021. Ultrasound improves the decontamination effect of thyme essential oil nanoemulsions against *Escherichia coli* O157: H7 on cherry tomatoes. *Int. J. Food Microbiol.* 337: 108936.

Ju, J., Xie, Y., Guo, Y., Cheng, Y., Qian, H., and Yao, W. 2018a. The inhibitory effect of plant essential oils on foodborne pathogenic bacteria in food. *Crit. Rev. Food Sci. Nutr.* 3281–3292.

Ju, J., Xie, Y., Guo, Y., Cheng, Y., Qian, H., and Yao, W. 2018b. Application of starch microcapsules containing essential oil in food preservation. *Crit. Rev. Food Sci. Nutr.* 2825–2836.

Ju, J., Xie, Y., Yu, H., Guo, Y., Cheng, Y., Qian, H., and Yao, W. 2022. Synergistic interactions of plant essential oils with antimicrobial agents: A new antimicrobial therapy. *Crit. Rev. Food Sci. Nutr.* 62(7): 1740–1751.

Ju, J., Xu, X., Xie, Y., Guo, Y., Cheng, Y., Qian, H., and Yao, W. 2017. Inhibitory effects of cinnamon and clove essential oils on mold growth on baked foods. *Food Chem.* 240: 850–855.

Khaneghah, A. M., Hashemi, S. M. B., and Limbo, S. 2018. Antimicrobial agents and packaging systems in antimicrobial active food packaging: An overview of approaches and interactions. *Food Bioprod. Process*. 1–19.

Kuorwel, K. K., Cran, M. J., Sonneveld, K., Miltz, J., and Bigger, S. W. 2013. Migration of antimicrobial agents from starch-based films into a food simulant. *LWT-Food Sci. Technol.* 50(2): 432–438.

Kuorwel, K. K., Cran, M. J., Sonneveld, K., Miltz, J., and Bigger, S. W. 2014. Physico-mechanical properties of starch-based films containing naturally derived antimicrobial agents. *Packag. Technol. Sci.* 27(2): 149–159.

Li, J., Ye, F., Lei, L., and Zhao, G. 2018. Combined effects of octenylsuccination and oregano essential oil on sweet potato starch films with an emphasis on water resistance. *Int. J. Biol. Macromol.* 115: 547–553.

Lim, Z. Q. J., Tong, S. Y., Wang, K., Lim, P. N., and San Thian, E. 2022. Cinnamon oil incorporated polymeric films for active food packaging. *Materials Letters*. 313: 131744.

Lin, L., Zhu, Y., and Cui, H. 2018. Electrospun thyme essential oil/gelatin nanofibers for active packaging against campylobacter jejuni in chicken. *LWT-Food Sci. Technol.* 97: 711–718.

Liolios, C. C., Gortzi, O., Lalas, S., Tsaknis, J., and Chinou, I. 2009. Liposomal incorporation of carvacrol and thymol isolated from the essential oil of *Origanum dictamnus* L. and in vitro antimicrobial activity. *Food Chem.* 112(1): 77–83.

Mascheroni, E., Guillard, V., Gastaldi, E., Gontard, N., and Chalier, P. 2011. Anti-microbial effectiveness of relative humidity-controlled carvacrol release from wheat gluten/montmorillonite coated papers. *Food Control.* 22(10): 1582–1591.

Montero, Y., Souza, A., Oliveira, D., and Rosa, D. 2021. Nanocellulose functionalized with cinnamon essential oil: A potential application in active biodegradable packaging for strawberry. *Sustainable Mater. Technol.* 29: 00289.

Mukurumbira, A. R., Shellie, R. A., Keast, R., Palombo, E. A., and Jadhav, S. R. 2022. Encapsulation of essential oils and their application in antimicrobial active packaging. *Food Control.* 108883.

Nada, A. A., Eckstein Andicsová, A., and Mosnáček, J. 2022. Irreversible and self-healing electrically conductive hydrogels made of bio-based polymers. *Int. J. Mol. Sci.* 23(2): 842–851.

Nazri, F., Tawakkal, I. S. M. A., Basri, M. S. M., Basha, R. K., and Othman, S. H. 2022. Characterization of active and pH-sensitive poly (lactic acid) (PLA)/nanofibrillated cellulose (NFC) films containing essential oils and anthocyanin for food packaging application. *Int. J. Biol. Macromol.* 212: 220–231.

Nielsen, P. V., and Rios, R. 2000. Inhibition of fungal growth on bread by volatile components from spices and herbs, and the possible application in active packaging, with special emphasis on mustard essential oil. *Int. J. Food Microbiol.* 60(2–3): 219–229.

Ogunsona, E. O., Muthuraj, R., Ojogbo, E., Valerio, O., and Mekonnen, T. H. 2020. Engineered nanomaterials for antimicrobial applications: A review. *Applied Materials Today*. 18: 100473.

Passarinho, A. T. P., Dias, N. F., Camilloto, G. P., Cruz, R. S., Otoni, C. G., Moraes, A. R. F., and Soares, N. D. F. 2014. Sliced bread preservation through oregano essential oil-containing sachet. *J. Food Process Eng.* 37(1): 53–62.

Ribeiro-Santos, R., Andrade, M., and Sanches-Silva, A. 2017. Application of encapsulated essential oils as antimicrobial agents in food packaging. *Curr. Opin. Food Sci.* 14: 78–84.

Rodriguez, A., Nerin, C., and Batlle, R. 2008. New cinnamon-based active paper packaging against Rhizopusstolonifer food spoilage. *J. Agric. Food Chem.* 56(15): 6364–6369.

Salvia-Trujillo, L., Rojas-Graü, M. A., Soliva-Fortuny, R., and Martín-Belloso, O. 2015. Physicochemical characterization and antimicrobial activity of food-grade emulsions and nanoemulsions incorporating essential oils. *Food Hydrocolloids.* 43: 547–556.

Sani, I. K., Geshlaghi, S. P., Pirsa, S., and Asdagh, A. 2021. Composite film based on potato starch/apple peel pectin/ZrO_2 nanoparticles/microencapsulated *Zataria multiflora* essential oil; investigation of physicochemical properties and use in quail meat packaging. *Food Hydrocolloids.* 117: 106719.

Serrano, M., Martinez-Romero, D., Castillo, S., Guillén, F., and Valero, D. 2005. The use of natural antifungal compounds improves the beneficial effect of MAP in sweet cherry storage. *Innovative Food Sci. Emerging Technol.* 6(1): 115–123.

Sharma, P., Ahuja, A., Izrayeel, A. M. D., Samyn, P., and Rastogi, V. K. 2022. Physicochemical and thermal characterization of poly (3-hydroxybutyrate-co-4-hydroxybutyrate) films incorporating thyme essential oil for active packaging of white bread. *Food Control.* 133: 108688.

Sharma, S., Barkauskaite, S., Jaiswal, A. K., and Jaiswal, S. 2020. Essential oils as additives in active food packaging. *Food Chem.* 343(8): 128403.

Shi, C., Zhou, A., Fang, D., Lu, T., Wang, J., Song, Y., and Li, W. 2022. Oregano essential oil/β-cyclodextrin inclusion compound polylactic acid/polycaprolactone electrospun nanofibers for active food packaging. *Chem. Eng. J.* 445: 136746.

Soares, M. J., Dannecker, P. K., Vilela, C., Bastos, J., Meier, M. A., and Sousa, A. F. 2017. Poly (1, 20-eicosanediyl 2, 5-furandicarboxylate), a biodegradable polyester from renewable resources. *Eur. Polym. J.* 90: 301–311.

Subbuvel, M., and Kavan, P. 2022. Preparation and characterization of polylactic acid/fenugreek essential oil/curcumin composite films for food packaging applications. *Int. J. Biol. Macromol.* 194: 470–483.

Suppakul, P., Sonneveld, K., Bigger, S. W., and Miltz, J. 2011. Loss of AM additives from antimicrobial films during storage. *J. Food Eng.* 105(2): 270–276.

Uz, M., and Altınkaya, S. A. 2011. Development of mono and multilayer antimicrobial food packaging materials for controlled release of potassium sorbate. *LWT-Food Sci. Technol.* 44(10): 2302–2309.

Xavier, L. O., Sganzerla, W. G., Rosa, G. B., Rosa, C., and Nunes, M. R. 2020. Chitosan packaging functionalized with *Cinnamodendron dinisii* essential oil loaded zein: A proposal for meat conservation. *Int. J. Biol. Macromol.* 169: 183–193.

Xue, C. A., Wc, A., Xiao, L. A., Ym, A., Xin, L. A., and Gl, B. 2021. Effect of chitosan coating incorporated with oregano or cinnamon essential oil on the bacterial diversity and shelf life of roast duck in modified atmosphere packaging. *Food Res. Int.* 147: 110491.

Xue, Q., Liu, Y. J., Xiang, X. W., and Wu, R. M. 2016. Microencapsulation of Cinnamon EO and its application in preserving fruits and vegetables. *Packag. Eng.* 37(5): 50–56.

Yang, Z., Li, M., Zhai, X., Zhao, L., Tahir, H. E., Shi, J., and Xiao, J. 2022. Development and characterization of sodium alginate/tea tree essential oil nanoemulsion active film containing TiO_2 nanoparticles for banana packaging. *Int. J. Biol. Macromol.* 213: 145–154.

Yasser, A., Nassim, A., Negin, B., Rl, C., and Farzad, D. 2022. Electrospun carboxymethyl cellulose-gelatin nanofibrous films encapsulated with mentha longifolia l. essential oil for active packaging of peeled giant freshwater prawn. *LWT-Food Sci. Technol.* 152: 112322.

Yeyen, A., and Nha, B. 2021. Carvacrol, citral and α-terpineol essential oil incorporated biodegradable films for functional active packaging of pacific white shrimp. *Food Chem.* 363: 130252.

Zanetti, M., Carniel, T. K., DalcaCnton, F., dos Anjos, R. S., Riella, H. G., de Araujo, P. H., and Fiori, M. A. 2018. Use of encapsulated natural compounds as antimicrobial additives in food packaging: A brief review. *Trends Food Sci. Technol.* 51–60.

Zhang, J., Li, Y., Yang, X., Liu, X., and Luo, Y. 2021. Effects of oregano essential oil and nisin on the shelf life of modified atmosphere packed grass carp (*Ctenopharyngodon idellus*). *LWT-Food Sci. Technol.* 147(11): 111609.

Zheng, H., Tang, H., Yang, C., Chen, J., Wang, L., Dong, Q., and Liu, Y. 2022. Evaluation of the slow-release polylactic acid/polyhydroxyalkanoates active film containing oregano essential oil on the quality and flavor of chilled pufferfish (*Takifugu obscurus*) fillets. *Food Chem.* 385: 132693.

14 In Vivo Experiment, Stability, and Safety of Essential Oils

14.1 INTRODUCTION

In nature, essential oils play an important role in plant protection as antimicrobial agents, antiviral drugs, antifungal agents, and pesticides. In addition, essential oils can fight plants by reducing their appetite for herbivores. At the same time, essential oils may also attract some insects to spread pollen and seeds or repel other unwanted insects. In recent years, the applications of EOs as food additives have rapidly developed as well, owing to their strong antimicrobial and antioxidant properties (Reis et al., 2022; Ju et al., 2020, 2022). The antimicrobial effect of cinnamon EO on spores of *Bacillus anthracis* was first reported in the 1980s. Typically, EOs are a complex mixture of various secondary metabolites produced by plants, including aromatic and terpenoid compounds and their derivatives (e.g., aldehydes, alcohols, ketones, phenols, esters, and ethers). Because EOs are a mixture of these complex active ingredients, they can effectively control pathogens in food, including bacteria, fungi, and related toxins, without promoting drug resistance (Wardana et al., 2022; Hou et al., 2022; Álvarez-Martínez et al., 2021). In addition, because EOs are derived from plant extracts, they are recognized as being relatively safe, healthy, and environment-friendly. Table 14.1 lists traditional EOs that are classified as generally recognized as safe (GRAS) by the US Food and Drug Administration (FDA). Although many EOs are natural plant fragrance components and are considered GRAS by various countries and international food additive regulatory agencies, some components are regulated differently in different countries, such as *Artemisia annua* cerebrum, which is approved by the FDA but is specifically prohibited in the European Union (Chiriac et al., 2021; Bafort et al., 2021).

Generally speaking, it is the main component of essential oil that determine the biological characteristics of essential oil, including beneficial effects (antibacterial, antifungal, antioxidation, anti-inflammatory, etc.) and the toxicological characteristics of these substances. Therefore, in order to ensure people's safety, the pros and cons of EOs should be considered comprehensively. It should also be noted that different subspecies, cultivation time, growth stage, and delivery time will cause great differences in the composition of essential oils. Between different species, the carvacrol content may change considerably from 22.0% in *Origanum compactum* to 64.3% in *Origanum onites* or even exhibit different chemical composition, such as the case for β-caryophyllene and α-thuyene, which are present in *O. compactum* and not detected in *O. onites*. For these reasons, the toxicological characteristics

DOI: 10.1201/9781003329268-14

TABLE 14.1
The Traditionally Used Essential Oils Categorized as Generally Recognized in Safe (GRAS) Category by US Food and Drug Administration (Prakash et al., 2018)

S. No.	Essential oils	Active ingredients (GC-MS analysis)	Experimental design	Bioactivity reported
1	*Cinnamomum zeylanicum* Nees.	(*E*)-Cinnamaldehyde (78.95%); phenol, 2-methoxy-3-(2-propenyl)- (74.65%) and caryophyllene (8.46%)	Antimicrobial activity of EO was determined using potato dextrose agar (PDA) medium. Requisite amounts of EO was dissolved separately in Petri plates containing PDA to obtain desired concentrations (0.1–1.0 ml/ml). Observation was made after 1 week	Exhibited strong activity against foodborne molds, viz., *Aspergillus flavus*, *Aspergillus niger*, *Aspergillus candidus*, *Aspergillus sydowi*, *Aspergillus fumigatus*, *Cladosporium cladosporoides*, *Curvularia lunata*, *Alternaria alternata*, *Penicillium species*, *Mucor* sp., and aflatoxin B1 contamination
2	*Cuminum cyminum* L.	Cymene (47.08%), gamma-terpinene (19.36%), cuminaldehyde (14.92%)	Antimicrobial activity of EO was determined via poisoned food assay. Requisite amount of EO was dissolved separately in Petri plates containing potato dextrose broth (PDB) medium to obtain desired concentrations (0.1–1.0 ml/ml). Observation was made after 1 week	Exhibited strong antimicrobial activity against molds isolated from chickpea seed, aflatoxin B1 contamination, and lipid peroxidation
3	*Zanthoxylum alatum* Roxb.	Linalool (56.10); methyl cinnamate (19.73%)	Antimicrobial activity of EO was determined using PDA medium. Requisite amounts of EO was dissolved separately in Petri plates containing PDA medium to obtain desired concentrations (0.25–5.0 m l/ml). Observation was made after 1 week	Exhibited strong efficacy against molds, viz., *Aspergillus flavus*, *Aspergillus niger*, *Aspergillus terreus*, *Aspergillus candidus*, *Aspergillus sydowi*, *Aspergillus fumigatu*, *Alternaria alternata*, *Cladosporium cladosporioides*, *Curvularia lunata*, *Fusarium nivale*, *Penicillum italicum*, and *Trichoderma viride*. In addition, it shows remarkable antioxidant and aflatoxin B1 inhibitory potential

4	*Rosmarinus officinalis* L.	1,8-Cineole (18.20%), α-pinene (22.25%) and camphor (12.35%)	Antimicrobial activity of EO was determined using Czapek-Dox agar (CDA). Requisite amount of EO was dissolved separately in Petri plates containing CDA medium to obtain desired concentrations (0.25–5.0 ml/ml)	Exhibited strong antifungal, antiaflatoxin and antioxidant activity
5	*Cananga odorata* Hook. f. and Thoms.	Linalool (24.5%), benzyl acetate (9.77%), and benzyl benzoate (33.61%)	The discs diffusion technique was used to determine the antimicrobial activity of the EO. Requisite amounts of EO were dissolved separately in DMSO to obtain desired concentrations (0.01–0.75 mg/ml)	Exhibited strong antimicrobial property against food spoiling microbes, viz., *Candida albicans* ATCC 48274, *Rhodotorula glutinis* ATCC 16740, *Schizosaccharomyces pombe* ATCC 60232, *Saccharomyces cerevisiae* ATCC 2365, *Yarrowia lypolitica* ATCC 16617
6	*Curcuma longa* L.	α-Turmerone (19.8%), α-ephellandrene (20.42%), and 1,8-cineole (10.3%)	Observation was made after 24 h	Exhibited strong antimicrobial property against foodborne microbes
7	*Curcuma zedoria* Rosc.	epi-Curzerenone (19.0%), Zingiberene (12.0%), and ar-Curcumene (12.1%)	The Petri plate-paper disk method was performed to determine the antimicrobial activity of the EO. Observation was made after 24 h for bacterium and 48 h for fungus	Exhibited strong antioxidant and antimicrobial activity against *Staphylococcus aureus* (IFO14462), *Corynebacterium amycolatum* (IFO 15207), *Escherichia coli* (IFO 15034), *Candida albicans* (IFO 1594), and *Aspergillus ochraceus* (IFO 31221)
8	*Mentha spicata* L.	Carvone D (60.07%) and limonene (19.91%)	Antimicrobial activity of EO was determined by Broth micro-dilution method. Requisite amounts of EO were dissolved separately in mineral oil to obtain desired concentrations (9.7–2,500 mg/ml). Observation was made after 24 h for bacterium and 48 h for fungus	Exhibited strong antimicrobial and anticholinesterase activity
9	*Pimenta officinalis* lindl.	Eugenol (64.29%) and methyl eugenol (20.55%)	The agar dilution method was performed to determine the antimicrobial activity of the EO at 1 ml/ml. Observation was made after 1 week	Exhibited strong antifungal activity against *Fusarium oxysporum*, *Fusarium verticillioides*, *Penicillium expansum*, *Penicillium Brevicompactum*, *Aspergillus flavus*, and *Aspergillus fumigatus*

(*Continued*)

TABLE 14.1
(Continued)

S. No.	Essential oils	Active ingredients (GC-MS analysis)	Experimental design	Bioactivity reported
10	*Cymbopogon martini* Stapf.	Geraniol (79.7%) and Nerol (8.9%)	Antimicrobial activity of EO was determined by Broth micro-dilution method. Observation was made after 24 h for bacterium and 48 h for fungus	Exhibited strong bioactivity against foodborne bacteria, viz., *Staphylococcus aureus* and *Enterococcus faecalis*), two Gram-negative (*Escherichia coli* and *Moraxella catarrhalis*), and two yeast species (*Candida albicans* and *Candida tropicalis*)
11	*Coriandrum sativum* L.	2E-decenal (15.9%), decanal (14.3%), 2E-decen-1-ol (14.2%), and *n*-decanol (13.6%)	Antimicrobial activity of EO was determined using disc diffusion method. Requisite amounts of EO were dissolved in sterile distilled water to obtain desired concentrations (20%, 25%, 33%, and 50%). Observation was made after 24 h	Exhibited strong antimicrobial activity against Gram-positive (*Staphylococcu saureus*, *Bacillus* spp.) and Gram-negative (*Escherichia coli*, *Salmonella typhi*, *Klebsiella pneumonia*, *Proteus mirabilis*, *Pseudomonas aeruginosae*) bacteria and a pathogenic fungus, *Candida albicans*
12	*Artemisia dracunculus* L.	Terpinolene (19.1%), methyl chavicol (16.2%), and methyl eugenol (35.8%)	The drop agar diffusion method was performed to determine the antimicrobial activity of EO Observation was made after 1 week	Exhibited strong antioxidant and antimicrobial activity against foodborne mold *Aspergillus niger* and bacteria *Escherichia coli*, *Staphylococcus aureus*, and *Staphylococcus epidermidis*, yeasts (*Candida albicans* and *Cryptococcus neoformans*)
13	*Cymbopogan nardus* Rendle.	Geranial (42.1%) and Neral (30.5%)	Antimicrobial activity of EO was determined by broth micro-dilution method. Observation was made after 24 h for bacterium and 48 h for fungus	Exhibited strong bioactivity against bacterial species *Staphylococcus aureus* and *Enterococcus faecalis*, two Gram-negative *Escherichia coli* and *Moraxella catarrhalis*, and two yeast species (*Candida albicans* and *Candida tropicalis*)
14	*Citrus sinensis* (L.) Osbeck.	Limonene (85.50%)	The Petri plate-paper disk method was performed to determine the antimicrobial activity of the EO. Observation was made after 24 h for bacterium and 48 h for fungus	Exhibited strong antibacterial activity against *Enterococcus faecium* (ATCC 19434), *Staphylococcus aureus* (ATCC 6538), *Pseudomonas aeruginosa* (ATCC 10145), and *Salmonella enterica* subsp. enterica ser. *Enteritidis* (ATCC 49214).
15	*Thymus vulgaris* L.	*p*-Cymene (47.9%), thymol (43.1%), and 1,8-cineole (11.3%)	Antimicrobial activity of EO was determined by agar disk diffusion assay. Observation was made after 24 h	Exhibited strong activity against foodborne bacteria, viz., *Staphylococcus aureus*, *Listeria monocytogenes*, *Salmonella enteritidis*, and *Campylobacter jejuni*

of essential oils will also vary with the aforementioned differences, which is very important for risk assessors. It should be emphasized that the quantitative and qualitative properties of EO extracts can be modified according to pretreatment, processing, and extraction processes.

14.2 THE RELATIONSHIP BETWEEN ESSENTIAL OIL AND FOOD

The close relationship between EOs and their active components and food itself makes it more difficult to evaluate its safety. The Federal Food, Drug and Cosmetics Act (FFDCA) acknowledges that naturally occurring substances in food must apply different and lower safety standards than ingredients deliberately added to food. For naturally occurring substances, the Act adopts a realistic standard, that is, the substance must not harm health. For the added substances, the applicable standards are much higher. If the added substance "extra ingredients" Then the food is adulterated, which may make the food harmful to human health. However, EOs are in the middle because they are obtained from natural plants. Many of these EOs are deliberately added to food as separate chemicals. Because they are neither considered as direct food additives nor food itself, the current standards cannot be easily applied to the safety evaluation of EOs.

The safety assessment of documented EOs with a long history of use in food first assumes that they are safe. Because of their long-term human use, there are no known adverse effects reported. With a high degree of confidence, people may think that EOs extracted from plants or spices are likely to be safe. The history of the use of many EOs is recorded in the annual survey of condiment use in the United States. On the contrary, confidence in the safety of natural plant extracts that have undergone significant changes in their use patterns has declined by the time they enter the food supply. In addition, some natural plant extracts or spices may be consumed not only as condiments. On the contrary, they may often be consumed as so-called functional dietary supplements. These changes in consumption patterns have renewed interest in the safety evaluation of natural plant extracts (including EOs). Although the safety assessment of natural plant extracts must still depend to a large extent on the understanding of the history of use, a flexible science-based approach will allow for rigorous safety assessments for different uses of the same EOs.

14.3 HOW TO ENSURE THE SAFETY OF PLANT ESSENTIAL OIL ITSELF

The safety of essential oil must be considered in the fields of food processing, agricultural production, cosmetics, and biomedicine. Several main links affecting the safety of plant essential oils include planting, processing, storage, and so on. First of all, the residues caused by the excessive use of pesticides and fertilizers in the planting process of plant essential oils. The second is the extraction and processing process of plant essential oil. For example, in the process of extracting essential oils with organic solvents, organic solvents will be mixed with essential oils, which may lead to contamination of essential oils. The third is the pollution and oxidation of essential oil during storage. Conditions such as strong light, humidity, and high temperature can easily lead to the deterioration of the quality of essential oils. Therefore, how to ensure the safety of plant essential oil itself is an important prerequisite for its wide application. During the period of plant cultivation, the use of pesticides and fertilizers should be controlled, and the use of organic solvents should be avoided in the process of extracting plant essential oils, so as to control the safety of essential oils from the source. Usually, putting the essential oil in the cold storage is an effective way to prolong the shelf life of the essential oil. This method can prolong the shelf life of essential oil to more than twice that of the original. In addition, it is scientific to store essential oils in brown glass bottles. It is recommended that the capacity of each bottle of pure essential oil should not be larger than 20 ml. The purpose of this is to prevent children from misusing it and cause more harm. The name, chemical type, place of production, purity, extraction method, extraction site, capacity, manufacturer, production batch number, main chemical composition, shelf life, and place of production should be clearly indicated on the label. Because the essential oil is easy to oxidize and decompose, the essential oil after opening the bottle is safer to use within 1 year.

14.4 STABILITY OF PLANT ESSENTIAL OILS

The application of plant essential oil is limited because of its volatile and poor stability. The study on the stability of plant essential oil has a very important guiding significance for its application in various fields. However, it is regrettable that the current research information related to the stability of essential oils is still very scarce. Here we systematically summarize and analyze some published studies on the stability of essential oils. At the same time, in Figure 14.1, we summarized the factors affecting the stability of plant essential oils in food preservation.

Artemisia argyi essential oil can not only significantly inhibit *Staphylococcus aureus*, *Bacillus subtilis*, and *Escherichia coli*, but also has good UV stability.

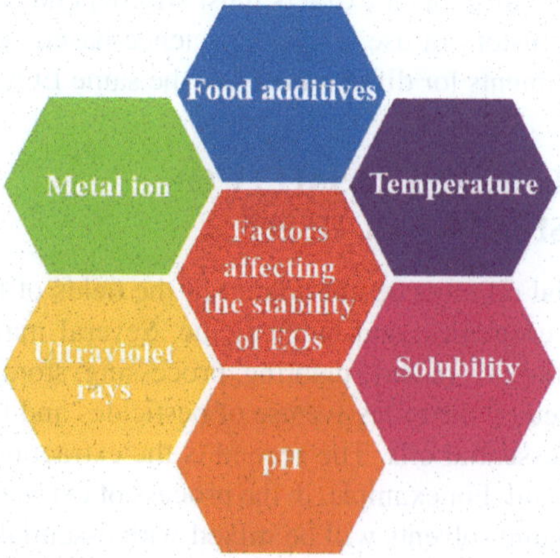

FIGURE 14.1 Factors affecting the stability of plant essential oils in food preservation.

Temperature and pH have great influence on the bacteriostatic effect of *Artemisia annua* essential oil. The antibacterial activity of *A. argyi* essential oil is better when the temperature is lower than 40°C, and decreases gradually when the temperature is higher than 40°C. Its bacteriostatic activity can be enhanced under acidic conditions. Chitosan, $CaCl_2$, and NaCl can all increase the bacteriostatic effect of *Artemisia* essential oil, but the effect of chitosan and $CaCl_2$ on the bacteriostatic effect of *A. annua* essential oil is greater than that of NaCl (Duan et al., 2015). Similarly, litchi wood essential oil has good UV stability. Temperature and pH value have great influence on the bacteriostatic effect of litchi wood essential oil. The bacteriostatic effect of the essential oil had no significant change when the temperature was lower than 100°C, but decreased when the temperature was higher than 100°C. Acidic conditions can enhance the antibacterial activity of litchi wood essential oil (Dong et al., 2021). The essential oil of *Zingiber officinale* showed strong antibacterial activity under acidic conditions. In the range of pH 6–8, the bacteriostatic zone of the five tested bacteria decreased at first and then increased. In addition, the essential oil of *Z. officinale* has good thermal stability, and the effect of temperature on its antibacterial activity is not significant, but NaCl can increase the antibacterial effect of the essential oil (Mahboubi, 2019). In the study of oregano essential oil, we found that strong alkali conditions could affect the UV absorbance of carvacrol and destroy the antibacterial activity of oregano essential oil. Natural light, humidity, and 60°C treatment had no significant effect on the stability of carvacrol (Zhao et al., 2020). *Alpinia officinalis* essential oil has good thermal stability and good stability to ultraviolet treatment. Although the antibacterial activity of *A. officinalis* essential oil decreased under 121°C and 100 min UV treatment, it still had moderate bacteriostatic effect on *Escherichia coli*, *Bacillus subtilis*, *Staphylococcus aureus*, and yeast (Ai et al., 2020). In investigating the effect of food preservatives on the stability of essential oil, we found that sodium benzoate and sodium carboxymethyl cellulose (CMC) had little effect on coriander essential oil. NaCl and ascorbic acid can promote its stability. In the reductant experiment, the resistance of cilantro flavonoids to Na_2SO_3 with strong reducibility was weak. Therefore, coriander essential oil should avoid contact with reducing agents such as sulfite or sulfur dioxide in the process of production, storage, and use. In addition, the citric acid has a destructive effect on many plant flavonoids, which should be avoided in the process of food processing (Burdock and Carabin, 2009).

14.5 TOXICOLOGICAL STUDIES

Food is a complex system, which is composed of many interrelated micro-environments. Even edible grade essential oils may change the microsystem of food itself. Under the same environmental conditions, the in vitro effects of essential oils are not exactly the same. The use of essential oils in food preservation may replace some chemically synthesized preservatives to extend the shelf life of food. Relevant researchers have also found that some essential oils can be added to various drugs to treat various diseases (Parker et al., 2022). However, whether plant essential oils and their active ingredients can be safely used in food industry needs to be carefully evaluated. Even though many essential oils have been listed as harmless components or approved as food flavor additives, some studies have shown that EOs may have potential toxic effects, mainly including genotoxicity and mutagenicity. Table 14.2 summarizes some commonly used toxicity assessment methods and the corresponding toxicity levels of essential oils.

14.5.1 Genotoxicity and Mutagenicity

There are two main factors affecting the results of genotoxicity and mutagenicity experiments. First, it is experimental model. Some researchers found that under the action of thyme essential oil, DNA damage occurred in lymphocytes, while no genotoxicity was found in human colon cancer Caco-2 cells. The second is the concentration of essential oil. For example, thymol showed no cytotoxicity at low doses, but showed low toxicity at high doses (Stammati et al., 1999). In addition, the genotoxicity of EOs may also depend on its growth conditions and extraction methods. *Salmonella typhi* and *E. coli* are commonly used to evaluate mutagenicity of essential oils. The possible changes and frameshift mutation of all base pairs would be detected (Di Sotto et al., 2013; Llana-Ruiz-Cabello et al., 2014b; Zegura et al., 2011).

The genetic toxicity and mutagenicity of essential oils may be partly due to oxidative stress, but the current research reports on this aspect are very limited, and there are more reports on oxidation promotion action of EOs. For example, studies have shown that rosemary essential oil can induce protein oxidation in sausages (Soncu et al., 2020). Other studies have also shown that polyphenols can promote oxidation, and high doses of eugenol and thymol can increase the level of malondialdehyde (MDA), resulting in membrane damage and DNA damage in different cell lines (Jiang et al., 2022; Ranjbar et al., 2022). In this regard, the current research is more about essential oils that can cause ROS outbreak, induce cell apoptosis, and lead to oxidative stress. These studies indicates that high concentrations of single compounds can induce oxidative damage and, in some cases, play a role of oxidant (Wu et al., 2016; Ji et al., 2018). Therefore, when EOs and their compounds are used in food preservation, pre-oxidation experiments should be carried out to determine the safe dose range.

14.5.2 Cytotoxicity

Because EOs contain a large number of components, they do not seem to have specific cellular targets. They can pass through the cell wall and plasma membrane, destroying the internal structure of the cell. Cytotoxicity seems to include this membrane damage.

In bacteria, the permeability of the membrane is related to the loss of ions and the decrease of membrane potential, the collapse of proton pump, and the depletion of ATP. The damage of EOs to cell wall and membrane can lead to the leakage and dissolution of macromolecules (Pathania et al., 2022; Wang et al., 2022; de Souza et al., 2022). In eukaryotic cells, EOs can depolarize the mitochondrial membrane by reducing the membrane potential, affect the ion Ca++ cycle and other ion channels, and reduce the pH gradient, affecting the proton pump and ATP. EOs change the fluidity of the membrane, causing the membrane to become unusually permeable,

TABLE 14.2
Toxicity Evaluation Methods of Commonly Used EOs and Corresponding Toxicity Levels

Essential oil/ active ingredient	Family/ source	Evaluation method	Experimental object	LD50/ (g/kg)	Toxicity grade	Reference
Lemon balm	*Melissa officinalis* L.	Acute toxicity by oral administration	Little mouse	2.57	Low toxicity	Stojanović et al. (2019)
Cinnamaldehyde	Cinnamon EO	Acute toxicity by oral administration	Little mouse	2.220	Low toxicity	Jakhetia et al. (2010)
Eugenol	Clove EO	Acute toxicity by oral administration	Little mouse	3.000	Low toxicity	Zin et al. (2011)
1,8-Cineole	Tea tree EO	Acute toxicity by oral administration	Little mouse	7.220	Actual nontoxic	Niu et al. (2019)
Verbenone	Rosemary EO	Acute toxicity by oral administration	Little mouse	8.088	Actual nontoxic	Wu et al. (2010)
Azadirachtin	Azadirachtin EO	Acute toxicity by oral administration	Little mouse	13.000	Actual nontoxic	Rajashekar et al. (2012)
Mentha haplocalyx	Mentha	Acute toxicity by oral administration	Little mouse	2.000	Low toxicity	Zhang et al. (2013)
Cymbopogon citratus	Citronella Gramineae	Acute toxicity by oral administration	Little mouse	3.500	Low toxicity	Costa et al. (2011)
Litsea cubeba	Litsea	Acute toxicity by oral administration	Little mouse	4.000	Low toxicity	Luo et al. (2005)
Cinnamomum cassia	*Cinnamomum camphora*	Acute toxicity by oral administration	Little mouse	5.038	Actual nontoxic	Liu et al. (2010)
Eugenia caryophyllata	Syringa of Myrtle Family	Acute toxicity by oral administration	Little mouse	5.523	Actual nontoxic	Ma et al. (2010)
Thymus vulgaris	Thymus of Labiatae	Acute toxicity by oral administration	Little mouse	17.500	Actual nontoxic	Zhang et al. (2011)
Aegle marmelos	Rutaceae	Acute toxicity by oral administration	Little mouse	23.660	Actual nontoxic	Singh et al. (2009)

leading to the leakage of free radicals, cytochrome C, calcium, and proteins, as in the case of oxidative stress and bioenergy exhaustion. The increase of permeability inside and outside the mitochondria leads to apoptosis and necrosis.

Scanning and transmission electron microscopy showed that ultrastructural changes occurred in several compartments of the cells, such as cytoplasmic leakage and nuclear dissolution. The cell membrane components of several bacteria treated with some EOs were analyzed by gas chromatography and scanning electron microscope. the results showed that the unsaturated fatty acids decreased significantly, and the saturated fatty acids increased significantly. It can also be observed by electron microscope that essential oil destroys the membrane of HSV virus to prevent host cell infection (Schnitzler et al., 2007).

Recent studies on *Saccharomyces cerevisiae* have shown that the cytotoxicity of some EOs varies greatly depending on their chemical composition (Ridaoui et al., 2022). In particular, recent work in the yeast *Saccharomyces cerevisiae* has shown that the cytotoxicity of some EOs, based on colony forming ability differed considerably depending on their chemical composition; EOs treated cells in stationary phase of growth showed 50% lethality at 0.45 µl/ml of *O. compactum* EOs, 1.6 µl/ml of *Coriandrum sativum* EOs, >8 µl/ml of *Cinnamomum camphora*, *Artemisia herba-alba*, and *Helichrysum italicum* EOs (Bakkali et al., 2005). In addition, the cytotoxicity of EOs also depends on the state of cell growth, and cells are usually more sensitive to EOs during cell division, which may be due to the fact that EOs permeate more effectively in the sprouts. Generally speaking, the cytotoxic activity of EOs is mainly caused by phenols, aldehydes and alcohols.

This cytotoxic property is very important in the application of EOs, not only for pathogens or parasites of some people or animals, but also for the preservation of agricultural or marine products. EOs or some of their components are indeed effective against a variety of organisms, including bacteria, viruses, fungi, mites, insects, and molluscs.

Cytotoxic activities of EOs or their major components, sometimes activated by light, were also demonstrated in mammalian cells in vitro by short-term viability assays using specific cell staining or fluorescent dyes, including NRU (Neutral Red Uptake) test, MTT (3-(4,5-dimethylthiazol-2-yl)-2,5-diphenyl-tetrazolium bromide) test, Alamar Blue (resazurin) test, Trypan Blue exclusion testor Hoechst 33342, and propidium iodide test. Essential oil cytotoxicity in mammalian cells is caused by induction of apoptosis and necrosis.

The cytotoxicity of EOs to mammalian cells is caused by the induction of apoptosis and necrosis. Unscheduled DNA synthesis tests were also carried out in mammalian cells to detect the presence and removal of adducts in DNA. For example, eugenol can induce cytotoxicity and genotoxicity in rat and mouse hepatocytes through lactate dehydrogenase. Basil EOs and its main component estradiol can also induce hamster fibroblast V79. So far, because their mode of action affects several targets at the same time, there is usually no description of the special resistance or adaptability of EOs. However, *Bacillus cereus* is resistant to carvanol at sublethal concentrations. The pretreatment of carvanol reduced the fluidity of the membrane by changing the fatty acid ratio and composition of the membrane (Pasqua et al., 2006). Table 14.3 lists the cell toxicity test cases of EOs or its active components.

TABLE 14.3
Cell Toxicity Test Cases of EOs or Its Active Components

Essential oil or components	Microorganism	Dosage	Reference
Juniperus squamata root	Breast cancer cells	20 μg/ml	Rana et al. (2022)
Cymbopogon citratus	Human leukocytes and erythrocytes	2.5 μl	Hacke et al. (2022)
Rosemary essential oil	T1 cell	100 μl	Amani et al. (2022)
Psidium brownianum	Nctc929 mouse fibroblasts	200 μl	Bezerra et al. (2022)
Cymbopogon citratus	Haemaphysalis longicornis	10, 15, 20, and 25 mg/ml	Agwunobi et al. (2021)
Gallesia integrifolia	Human tumor cell lines MCF-7, NCI-H460, HeLa, and HepG2 and nontumor plp2	MCF-7 (66 μg/ml), NCI-H-460 (147 μg/ml), HeLa (182 μg/ml), and HepG2 (240 μg/ml)	Bortolucci et al. (2021)
Cedrus atlantica	MCF-7 breast cancer cells	143 μg/ml	Belkacem et al. (2021)
C. austroarabica	Eggs of Artemia salina	50 mg/ml	Weli et al. (2021)
Acmella oleracea	Human breast cancer cells (mda-mb 231)	1.95–500 μg/ml	Spinozzi et al. (2021)
Bauhinia cheilantha	Myeloid leukemia (HL-60) and breast adenocarcinoma (MCF-7) cells	8.6 and 18.3 μg/ml	Silva et al. (2020a)
Artemisia arborescens	Cervical adenocarcinoma (Hela)	10, 50, 100, 500 and 700 μg/ml	Jaradat et al. (2022)
Lavender	Human epithelial colorectal adenocarcinoma (LoVo cells) and human epithelial hepatocytes	10, 25, 50 and 100 μg/ml	Oueslati et al. (2020)
Gautheria procumbens	Neuroblastoma cell viability	1.82–58.34 mg/ml	Verdi et al. (2022)
Cinnamomum stenophyllum	Colon adenocarcinoma (HCT-116) and breast cancer (MCF-7)	50 μg/ml	Silva et al. (2020b)
Philippine calamansi	Breast cancer cell line and normal epithelial cell line	7.98 ± 1.77 and 16.15 ± 8.35 μg/ml	Palma et al. (2019)
Foeniculum vulgare Mill	Gastric cancer cell line (MGC-803), breast cancer cell line (MCF-7), cervical cancer cell line (Hela), lung cancer cell line (A549), liver cancer cell lines (Huh-7 and HepG2), and colon cancer cell line (SW620)	100 μl	Chen et al. (2020)

(*Continued*)

TABLE 14.3 (Continued)

Essential oil or components	Microorganism	Dosage	Reference
Mentha pulegium, *Mentha spicata*, Pulegone, Menthone, Carvone *Dracocephalum foetidum*	*Drosophila melanogaster*	0.2–2.1 µl on paper disk	Gopanraj et al. (2005)
Lippia sidoides	*Saccharomyces cerevisiae*	MIC 0.625–10 mg/ml	Botelho et al. (2007)
Pulicaria odora	*Bacillus cereus* *Streptococcus* *Proteus vulgaris* *Enterococcus faecalis* *Escherichia coli* *Pseudomonas aeruginosa*	5, 10 µg/filter paper disk	Hanbali et al. (2005)
Chenopodium ambrosioides	*Aspergillus niger* *Aspergillus fumigatus* *Botryodiplodia theobromae* *Fusarium oxysporum* *Sclerotium rolfsii* *Macrophomina phaseolina* *Cladosporium cladosporioides* *Helminthosporium oryzae* *Pythium debaryanum*	100 µg/ml	Kumar et al. (2007)
Actinidia macrosperma	*Staphylococcus aureus* *Bacillus subtilis* *Escherichia coli*	MIC 0.78–25.50 µl/ml	Lu et al. (2007)
Hippomarathrum microcarpum	Eight bacteria, nine fungi, one yeast	MIC 62.50–125 µl/ml	Ozer et al. (2007)
Foeniculum vulgare (FE) (FE1, FE2, FE3)	*Alternaria alternata* *Aspergillus niger*	MIC 1.0–3.0 µl/ml MIC 1.0–2.8 µl/ml	Mimicadukic et al. (2003)
Pinus densiflora *Pinus koraiensis* *Chamaecyparis obtusa*	*Salmonella typhimurium* *Listeria monocytogenesis* *Escherichia coli*	50 ml of dilutions 1/2, 1/4, 1/8, 1/16 on filter paper disks	Hong et al. (2004)
Tamarix boveana	*Staphylococcus aureus* *Staphylococcus epidermidis* *Escherichia coli* *Pseudomonas aeruginosa* *Micrococcus luteus* *Salmonella typhimurium* *Fusarium oxysporum* *Aspergillus niger*	1, 0.8, 0.5, 0.3 mg/ml 0.5, 2, 4 mg/filter paper disk MICs 0.3, 0.5, 0.8 mg/ml 80, 200, 500 µg/disk (no antifungal activity)	Saidana et al. (2008)
Eucalyptus robusta *Eucalyptus saligna*	*Staphylococcus aureus* *Escherichia coli* *Candida albicans*	0.25–1% (v/v)	Hammer et al. (2004)
Melaleuca alternifolia	*Filamentous fungi* *Dermatophytes*	MFC 0.03–8% MIC 0.004–0.25%	Hammer et al. (2002)

TABLE 14.3
(Continued)

Essential oil or components	Microorganism	Dosage	Reference
Salvia fruticosa	*Fusarium oxysporum*	50–2000 μl/l	Pitarokili et al. (2003)
1,8-cineole	*Fusarium solani*	20–500 μl/l	
Camphor	*Fusarium proliferatum*	MIC > ou = 2,000 μl/l	
Grammosciadium platycarpum	*Bacillus subtilis*	MIC 0.5–1.9 mg/ml	Sonboli et al. (2005)
	Enterococcus faecalis	MIC 7.5–15 mg/ml	
Linalool	*Staphylococcus aureus*	MIC 0.2–2.5 mg/ml	
Limonene	*Staphylococcus epidermidis*	MIC 0.6–5 mg/ml	
Ziziphora clinopodioides	*Staphylococcus epidermidis*	10 ll/filter paper disk	Sonboli et al. (2006)
	Staphylococcus aureus	MIC 3.75 to >15 mg/ml	
Pulegone	*Escherichia coli*		
1,8-cineole	*Bacillus subtilis*	MIC 1.8–7.2 mg/ml	
		MIC 0.9–7.2 mg/ml	
Pimpinella anisum	*Candida albicans*	MIC 0.1–1.56% v/v	Kosalec et al. (2005)
	Candida parapsilosis		
	Candida tropicalis		
	Candida pseudotropicalis		
	Candida krusei		
	Candida glabrata		
	Trichophyton rubrum		
	Trichophyton mentagrophytes		
	Microsporum canis		
	Microsporum gypseum		
Thyme, Basil,	*Shigella sonnei*	0.1–10%	Bagamboula et al. (2004)
Thymol, Estragol,	*Shigella flexneri*	0.05% (lettuce)	
Linalool, carvacrol,	*Escherichia coli*		
Coriandrum sativum	*Salmonella choleraesuis*		
Coriandrum sativum volatiles	*Salmonella choleraesuis*	6.25 μg/ml	Kubo et al. (2004)
		12.5 μg/ml	

14.5.3 Phototoxicity

Some EOs contain photoactive molecules. For instance, *Citrus bergamia* (= *Citrus aurantium* spp. bergamia) EOs contain psoralens which bind to DNA under ultraviolet A light exposure producing mono- and biadducts that are cytotoxic and highly mutagenic. In the dark, however, the EOs itself is not cytotoxic or mutagenic. In the presence of cytotoxicity, EOs destroy the membranes of cells and organelles, and act as promoters of proteins and DNA to produce reactive oxygen species (ROS), while light exposure does not increase the overall reaction much. In the presence of phototoxicity, EOs can penetrate cells without damaging cell membranes or proteins and DNA. When cells are exposed to activated light, the free radical reactions excited by some molecules and the energy transfer during the production of oxygen singlet will change. This may cause damage to cellular macromolecules and in some cases form covalent adducts with DNA, proteins, and membrane lipids. Obviously, cytotoxicity or phototoxicity mainly depends on the molecular types present in essential oils. However, such confrontations are not very strict rules. Abelan et al. (2022) showed that citrus and citronella EOs have phototoxicity and cytotoxicity. Therefore, it may be meaningful to systematically determine the cytotoxicity and possible phototoxicity of a certain EOs when studying it.

14.6 ANTIMUTAGENICITY OF ESSENTIAL OIL

So far, most studies have shown that the antimutagenic properties of EOs may be due to the inhibition of the penetration of mutagens into cells. Antioxidants can deactivate mutagens by direct scavenging, capture free radicals produced by mutagens, or activate cellular antioxidant enzymes. The enzyme detoxification ability of mutagens is activated by plant extracts. However, little is known about the antimutagenic interference to the DNA repair system after genotoxic damage is induced. Some antimutagenic drugs can inhibit error-prone DNA repair or promote error-free DNA repair.

The biological effect of antimutagenesis interfering with protometabolism to prevent mutation has been well-known and has been well-documented. In recent years, there are many studies on the effects of reactive oxygen species scavengers such as glutathione, superoxide dismutase, catalase, *N*-acetylcysteine, retinoic acid, carotenoids and tocopherols, flavonoids, and other polyphenols.

In the experiment of *E. coli*, some researchers have shown that lavender EOs has a strong inhibitory effect on the mutagenesis of 2-nitrofluorene (Evandri et al., 2005). Similarly, 1,8-cineole and limonene, the main components of *Salvia miltiorrhiza* extract, could also inhibit the mutagenic effect of UVC on *S. typhimurium*, *E. coli*, and *S. cerevisiae*. In addition, it has been proved that menthol, α-pinene, (+)-pinene, α-terpinene, α-terpineol, cineole, D-limonene, camphor, citronellal, and citral can regulate the activity of liver monooxygenase (Du et al., 2022), for example, the interaction of CYP1A1 and CYP2B1 with mutagens or carcinogen precursors. In a recent study, it was found that oregano EOs and some of its subcomponents had antimutagenic effects on indirect mutagenic agent ethyl carbamate and direct mutagenic agent methyl methanesulfonate (De-Oliveira et al., 1999). It is now recognized that the activity of oxidant can induce late apoptosis and necrosis of cells (Hadi et al., 2000). Oxidant promoters may damage the cell membrane, especially the mitochondrial membrane, thus promoting the release of Ca-++, cytochrome C, and ROS. This can lead to cell death, at least in mammalian cells. But yeast cells can survive mitochondrial damage (Martemucci et al., 2022). The antimutagenic effect has nothing to do with the type of mutation. The extent of this antimutagenic effect depends on the concentration of mutagens and EOs. However, surprisingly, the mechanism of reduced mutagenicity depends not on the type of EOs, but on the type of mutagen.

14.7 SAFETY OF SOME COMMON PLANT ESSENTIAL OILS

Since genotoxicity and mutagenicity are the core toxicological studies needed for the application of EOs in food preservation, it is very important to introduce the genotoxicity and mutagenicity potential of relevant EOs and their main compounds.

The main component of tea tree EOs is terpineol, which also belongs to terpenes, and its content is up to 40%. Relevant scholars have selected healthy rats, rabbits, and guinea pigs without specific pathogens to curry out the oral, percutaneous toxicity, skin irritation, and eye irritation tests. The results showed that the oral and transdermal toxicity of terpineol to SD rats was low, and the acute inhalation was slightly toxic. It was mild to moderate irritation to rabbit eyes, but no irritation after 4 s of rinsing, and had certain sensitization to guinea pigs (Liao et al., 2018). At present, it is known that tea tree essential oil can produce certain cytotoxicity and cause some neurological symptoms, but it has not been found that tea tree essential oil can cause poisoning or even death. Through the clinical trial, it was found that 15.0% tea tree oil could cause some irritation, among which 1,8-cineole was the main irritant component.

The main component of citrus EOs is limonene, which is a monoterpene compound. There are three isomers of limonene, namely, D-limonene, L-limonene, and DL-limonene, which generally exist in the form of D-limonene. The safety and anticancer properties of D-limonene were evaluated comprehensively by IARC (International Agency for Research on Cancer) and WHO (World Health Organization) in the 1990s. It was found that limonene had no genetic, immune, and reproductive toxicity, and only had carcinogenic effects on adult male rats, but no carcinogenicity on other animals and humans (Chen et al., 2017; Huang et al., 2015). Similarly, Mendanha et al. (2013) compared and studied the cytotoxicity and hemolysis of major terpenes to fibroblasts by methyl thiazolyl tetrazolium (MTT) method. The results showed that compared with other terpenes, limonene had very low cytotoxicity, cell membrane aggression, and hemolysis. They believed that terpenes such as limonene and eucalyptus brains could promote the penetration of polar and nonpolar drugs into skin.

The main component of clove EOs is eugenol, which has strong clove aroma and is insoluble in water. According to the National Toxicology Program, eugenol whose contents were 0, 0.6%, and 25%, respectively, was fed to female F344/N rates and male F344/N rates for 103 weeks. There was no evidence that eugenol was carcinogenic to rats. Female B6C3F1 mice fed the same way showed a significant increase in hepatocellular carcinoma cells and adenomas. Male B6C3F1 mice in low-dose group showed an increased incidence of cancer cells and adenomas (Jin et al., 2017). Some researchers also carried out acute toxicity test of eugenol by gastric perfusion in rats and irritation test of intact skin in rabbits. The results showed that the LD50 of eugenol in rats was 3,189.9 mg/kg, which was moderately irritating to rabbit skin, and decreased with the decrease of concentration (Jiang et al., 2015). Recently, eugenol was predicted to be a strong embryotoxic compound based on the experimental model of embryonic stem cells. This provides research data for the further evaluating of eugenol safety (Li et al., 2015). Consistency between microbial and mammalian outcomes is important because genotoxicity experiments need to be conducted not only at the microbial cell level, but also in mammals.

Linalool is the main component of coriander EOs, the content of which reaches 55–74% and is considered nonmutagenic in several different genotoxicity tests. For example, Higashimoto et al. (1993) conducted Ames test with *S. typhimurium* TA98 and TA100 to study the mutagenicity of water, methanol, and *n*-hexane extracts of coriander. The results showed that the extracts had no mutagenic effect on both strains. Gaworski et al. (1994) also reached a similar conclusion that coriander was negative in *Salmonella microsome* test. The researchers also used the single-cell gel electrophoresis assay to determine the genotoxicity of coriander and its ethanol extracts. The results showed that there was no genotoxicity to embryonic fibroblasts of rat at concentrations of 1,019 and 1,020 mg/l, respectively. At present, coriander essential oil has been approved for food use by FDA, FEMA (Federal Emergency Management Agency), and the European Commission (Heibatullah et al., 2008).

The main compounds of rosemary EOs are camphor (14.6%), 1,8-eucalyptol (7.26%), and alpha-pinene (6.65%). At present, rosemary EOs has also been widely used as seasoning food additives. And none of the aforementioned compounds showed mutagenicity in the detection of *Salmonella* and *E. coli* (Di Sotto et al., 2013). The genotoxicity of camphor was assayed in the Vero cell line by the alkaline Comet assay and nogenotoxicityupto100 μM was found, although higher tested concentrations induced DNA strand breaks (Nikolic et al., 2011)

The main volatile component of cinnamon EOs is cinnamaldehyde and it is the only α,β-unsaturated aldehyde approved for food use by FDA. Experiments on genotoxicity in *E. coli* have shown negative results. And the DNA repair tests of *E. coli* and *B. subtilis* showed positive results (Stammati et al., 1999). According to the results of genetic toxicology, *trans*-cinnamaldehyde mutation occurred in *S. typhimurium* strain TA100 in the presence of S9 activator in mouse liver (Jo et al., 2015). However, it also has antimutagenic effects. For example, cinnamaldehyde has an antimutagenic effect on spontaneous mutation of *A. flavus* TA104 (Shaughnessy et al., 2001).

Pinene is the main component of thuja EOs. The skin irritation and sensitization experiments of thuja EOs showed it was irritating to the skin. Acute oral tests showed that it was nontoxic and nonirritating. However, studies have shown that thuja EOs may have epileptic effects (Peng et al., 2016). In the past few years, the authors evaluated the genotoxicity of thuja EOs by *S. typhimurium* recovery mutation (AMES) test and mammalian micronucleus test. They eventually concluded that there was no obvious genetic toxicity or mutagenicity of thuja EOs (Puškárová et al., 2017).

Garlic is a kind of medicinal and edible plant, which can be used as condiment or medicine. The main component of garlic EOs is sulfide, and the content of allyl sulfide is the highest. At present, studies on mutagenicity/genotoxicity of garlic EOs are almost nonexistent. The only one was that researchers explored the toxicity and mutagenicity of garlic EOs soft capsules in 2005. The result showed that garlic EOs soft capsule is a kind of nontoxic and nonmutagenic effect substance with negative mutagenic test result (Delinour et al., 2005).

14.8 CONCLUDING REMARKS

Essential oils have been widely used in food preservation research and have a long history of use in both medicine and food with positive outcomes. However, given the complexity of the components of EOs and the conclusions of previous toxicological evaluations, it is necessary to evaluate the safety of EOs and their components before they are formally used in food preservation. Considering the enormous potential of EOs in food preservation, we hope that the relevant regulatory authorities will develop specific ISO standards that define EOs, provide general rules for their use, and specify labeling requirements that ensure the highest standard for EOs in the near future.

ACKNOWLEDGEMENTS

This work was supported by the National Natural Science Foundation of China (32202192), Special fund for Taishan Scholars Project.

REFERENCES

Abelan, U. S., de Oliveira, A. C., Cacoci, É. S. P., Martins, T. E. A., Giacon, V. M., Velasco, M. V. R., and Lima, C. R. R. D. C. 2022. Potential use of essential oils in cosmetic and dermatological hair products: A review. *J. Cosmet. Dermatol.* 21(4): 1407–1418.

Agwunobi, D. O., Zhang, M., Zhang, X., Wang, T., Yu, Z., and Liu, J. 2021. Transcriptome profile of Haemaphysalis longicornis (Acari: Ixodidae) exposed to *Cymbopogon citratus* essential oil and citronellal suggest a cytotoxic mode of action involving mitochondrial Ca^{2+} overload and depolarization. *Pestic. Biochem. Physiol.* 179: 104971.

Bafort, F., Perraudin, J. P., Jijakli, H., Bohlouli, M., Yin, T., Hammami, H., and Daklaoui, S. 2021. Potential for the use of lactoperoxidase against postharvest diseases on fruit. *Acta Horticulturae*. 1323.

Ai, W., Zhang, X., and Feng, L. Y. 2020. Study on antibacterial activity and stability of essential oil of *Alpinia officinalis*. *J. Food Addit.* 31(9): 8–13.

Álvarez-Martínez, F. J., Barrajón-Catalán, E., Herranz-López, M., and Micol, V. 2021. Antibacterial plant compounds, extracts and essential oils: An updated review on their effects and putative mechanisms of action. *Phytomedicine*. 90: 153626.

Amani, F., Rezaei, A., Damavandi, M. S., Doost, A. S., and Jafari, S. M. 2022. Colloidal carriers of almond gum/gelatin coacervates for rosemary essential oil: Characterization and in-vitro cytotoxicity. *Food Chem.* 377: 131998.

Bagamboula, C., Uyttendaele, M., and Debevere, J. 2004. Inhibitory effect of thyme and basil essential oils, carvacrol, thymol, estragol, linalool and p-cymene towards *Shigella sonnei* and *S. flexneri*. *Food Microbiol.* 21(1): 33–42.

Bakkali, F., Averbeck, S., Averbeck, D., Zhiri, A., and Idaomar, M. 2005. Cytotoxicity and gene induction by some essential oils in the yeast *Saccharomyces cerevisiae*. *Mutat. Res. Genet. Toxicol. Environ. Mutagen.* 585(1): 1–13.

Belkacem, N., Khettal, B., Hudaib, M., Bustanji, Y., Abu-Irmaileh, B., and Amrine, C. S. M. 2021. Antioxidant, antibacterial, and cytotoxic activities of *Cedrus atlantica* organic extracts and essential oil. *Eur. J. Integr. Med.* 42: 101292.

Bezerra, J. N., Gomez, M. C. V., Rolón, M., Coronel, C., Almeida-Bezerra, J. W., Fidelis, K. R., and Bezerra, C. F. 2022. Chemical composition, evaluation of antiparasitary and cytotoxic activity of the essential oil of *Psidium brownianum* MART EX. DC. *Biocatal. Agric. Biotechnol.* 39: 102247.

Bortolucci, W. D. C., Raimundo, K. F., Fernandez, C. M. M., Calhelha, R. C., Ferreira, I. C., Barros, L., and Gazim, Z. C. 2021. Cytotoxicity and anti-inflammatory activities of *Gallesia integrifolia* (Phytolaccaceae) fruit essential oil. *Nat. Prod. Res.* 1–6.

Botelho, M. A., Nogueira, N. A., Bastos, G. M., Fonseca, S. G., Lemos, T. L., Matos, F. J., and Brito, G. A. 2007. Antimicrobial activity of the essential oil from *Lippia sidoides*, carvacrol and thymol against oral pathogens. *Braz. J. Med. Biol. Res.* 40(3): 349–356.

Burdock, G. A., and Carabin, I. G. 2009. Safety assessment of coriander (*Coriandrum sativum* l.) essential oil as a food ingredient. *Food Chem. Toxicol.* 47(1): 22–34.

Chen, C., Sheng, Y. H., Tang, L. M., and Chen, G. L. 2017. Advances in pharmacology, toxicology and clinical application of dextran. *Indian J. Pharm.* 48(12): 1698–1703.

Chen, F., Guo, Y., Kang, J., Yang, X., Zhao, Z., Liu, S., and Luo, D. 2020. Insight into the essential oil isolation from *Foeniculum vulgare* mill. fruits using double-condensed microwave-assisted hydrodistillation and evaluation of its antioxidant, antifungal and cytotoxic activity. *Ind. Crops Prod.* 144: 112052.

Chiriac, A. P., Rusu, A. G., Nita, L. E., Chiriac, V. M., Neamtu, I., and Sandu, A. 2021. Polymeric carriers designed for encapsulation of essential oils with biological activity. *Pharmaceutics*. 13(5): 631.

Costa, C. A., Bidinotto, L. T., Takahira, R. K., Salvadori, D. M., Barbisan, L. F., and Costa, M. 2011. Cholesterol reduction and lack of genotoxic or toxic effects in mice after repeated 21-day oral intake of lemongrass (*Cymbopogon citratus*) essential oil. *Food Chem. Toxicol.* 49(9): 2268–2272.

De-Oliveira, A. C. A. X., Fidalgo-Neto, A. A., and Paumgartten, F. J. R. 1999. In vitro inhibition of liver monooxygenases by beta-ionone, 1,8-cineole, (-)-menthol and terpineol. *Toxicology*. 135(1): 33–41.

de Souza, D. P., de Carvalho Gonçalves, J. F., de Carvalho, J. C., da Silva, K. K. G., Fernandes, A. V., de Oliveira Nascimento, G., and Santos, A. S. 2022. Untargeted metabolomics used to describe the chemical composition and antimicrobial effects of the essential oil from the leaves of *Guatteria citriodora* Ducke. *Ind. Crops Prod.* 186: 115180.

Delinour, S. L., Kang, J. Z., and Mu, B. 2005. Experimental study on toxicity and mutagenicity of garlic essential oil soft capsules. *Disease Prevention and Control B*. 20(4): 10–19.

Di Sotto, A., Maffei, F., Hrelia, P., Castelli, F., Sarpietro, M. G., and Mazzanti, G. 2013. Genotoxicity assessment of β-caryophyllene oxide. *Regul. Toxicol. Pharmacol.* 66: 264–268.

Dong, L., Zheng, Y. X., Han, M., and Jia, Q. 2021. Study on antibacterial activity and stability of litchi wood essential oil in vitro. *Food Res. Dev.* 42(9): 43–48.

Du, Y., Zhou, H., Yang, L., Jiang, L., Chen, D., Qiu, D., and Yang, Y. 2022. Advances in Biosynthesis and Pharmacological Effects of *Cinnamomum camphora* (L.) Presl Essential Oil. *Forests*. 13(7): 1020.

Duan, W. L., Liu, Y. Q., and Bao, H. Y. 2015. Antibacterial activity and stability of Artemisia essential oil. *J. Food Chem. Biotechnol.* 34(12): 54–66.

Estévez, M., and Cava, R. 2006. Effectiveness of rosemary essential oil as an inhibitor of lipid and protein oxidation: Contradictory effects in different types of frankfurters. *Meat Sci.* 72: 348–355.

Evandri, M. G., Battinelli, L., Daniele, C., Mastrangelo, S., Bolle, P., and Mazzanti, G. 2005. The antimutagenic activity of *Lavandula angustifolia* (lavender) essential oil in the bacterial reverse mutation assay. *Food Chem. Toxicol.* 43(9): 1380–1387.

Gaworski, C. L., Vollmuth, T. A., Dozier, M. M., Heck, J. D., Dunn, L. T., Ratajczak, H. V., and Thomas, P. T. 1994. An immunotoxicity assessment of food flavouring ingredients. *Food Chem. Toxicol.* 32(5): 409–415.

Gopanraj, Dan M., Shiburaj, S., Sethuraman, M. G., and George, V. 2005. Chemical composition and antibacterial activity of the rhizome oil of *Hedychium larsenii*. *Acta Pharmaceutica*. 55(3): 315–320.

Hacke, A. C. M., da Silva, F. D. A., Lima, D., Vellosa, J. C. R., Rocha, J. B. T., Marques, J. A., and Pereira, R. P. 2022. Cytotoxicity of *Cymbopogon citratus* (DC) Stapf fractions, essential oil, citral, and geraniol in human leukocytes and erythrocytes. *J. Ethnopharmacol.* 291: 115147.

Hadi, S. M., Asad, S. F., Singh, S., and Ahmad, A. 2000. Putative mechanism for anticancer and apoptosis-inducing properties of plant-derived polyphenolic compounds. *Iubmb Life*. 50(3): 167–171.

Hammer, K. A., Carson, C. F., and Riley, T. V. 2002. In vitro activity of *Melaleuca alternifolia* (tea tree) oil against dermatophytes and other filamentous fungi. *J. Antimicrob. Chemother*. 50(2): 195–199.

Hammer, K. A., Carson, C. F., and Riley, T. V. 2004. Antifungal effects of *Melaleuca alternifolia* (tea tree) oil and its components on *Candida albicans*, *Candida glabrata* and *Saccharomyces cerevisiae*. *J. Antimicrob. Chemother*. 53(6): 1081–1085.

Hanbali, F. E. L., Akssira, M., Ezoubeiri, A., Gadhi, C. E. A., Mellouki, F., and Benherraf, A. 2005. Chemical composition and antibacterial activity of essential oil of Pulicaria odoral. *J. Ethnopharmacol.* 99(3): 399–401.

Heibatullah, K., Marzieh, P., Arefeh, I., and Ebrahim, M. 2008. Genotoxicity determinations of coriander drop and extract of coriander sativum cultured fibroblast of rat embryo by comet assay. *Saudi Pharm. J.* 16(1): 85–88.

Higashimoto, M., Purintrapiban, J., Kataoka, K., Kinouchi, T., Vinitketkumnuen, U., and Akimoto, S. 1993. Mutagenicity and antimutagenicity of extracts of three spices and a medicinal plant in Thailand. *Mutat. Res. Lett.* 303(3): 135–142.

Hong, E., Na, K., Choi, I., Choi, K., and Jeung, E. 2004. Antibacterial and antifungal effects of essential oils from coniferous trees. *Biol. Pharm. Bull.* 27(6): 863–866.

Hou, T., Sana, S. S., Li, H., Xing, Y., Nanda, A., Netala, V. R., and Zhang, Z. 2022. Essential oils and its antibacterial, antifungal and anti-oxidant activity applications: A review. *Food Biosci.* 101716.

Huang, Q. J., Huang, L. H., Sun, Z. G., Hao, L. M., and Guo, L. 2015. Advances in safety studies of limonene. *J. Food Sci.* 36(15): 277–281.

Jakhetia, V., Patel, R., Khatri, P., Pahuja, N., Garg, S., Pandey, A., and Sharma, S. 2010. Cinnamon: A pharmacological review. *J. Adv. Sci. Res.* 1(2): 19–23.

Jaradat, N., Qneibi, M., Hawash, M., Al-Maharik, N., Qadi, M., Abualhasan, M. N., and Bdir, S. 2022. Assessing *Artemisia arborescens* essential oil compositions, antimicrobial, cytotoxic, anti-inflammatory, and neuroprotective effects gathered from two geographic locations in Palestine. *Ind. Crops Prod.* 176: 114360.

Ji, D., Chen, T., Ma, D., Liu, J., Xu, Y., and Tian, S. 2018. Inhibitory effects of methyl thujate on mycelial growth of *Botrytis cinerea* and possible mechanisms. *Postharvest Biol. Technol.* 142: 46–54.

Jiang, N., Wang, L., Jiang, D., Wang, M., Liu, H., Yu, H., and Yao, W. 2022. Transcriptomic analysis of inhibition by eugenol of ochratoxin A biosynthesis and growth of *Aspergillus carbonarius*. *Food Control.* 135: 108788.

Jiang, W. Y., Ling, J., Zhong, Z. D., Wang, H. X., Wen, Z. P., and Liu, R. 2015. Acute toxicity and skin irritation test of eugenol. *J. Physiol. Sci.* 2: 49–51.

Jin, Y., He, Y. J., Chen, B., Song, J. L., Ke, C. L., Liu, Y. T., and Li, L. 2017. Discussion on the safety of eugenol anesthetics. *J. Food Saf. Food Qual.* 1: 33–40.

Jo, Y. J., Chun, J. Y., Kwon, Y. J., Min, S. G., Hong, G. P., and Choi, M. J. 2015. Physical and antimicrobial properties of trans-cinnamaldehyde nanoemulsions in water melon juice. *LWT-Food Sci. Technol.* 60(1): 444–451.

Ju, J., Xie, Y., Yu, H., Guo, Y., Cheng, Y., Qian, H., and Yao, W. 2022. Synergistic interactions of plant essential oils with antimicrobial agents: A new antimicrobial therapy. *Crit. Rev. Food Sci. Nutr.* 62(7): 1740–1751.

Ju, J., Xie, Y., Yu, H., Guo, Y., Cheng, Y., Zhang, R., and Yao, W. 2020. Synergistic inhibition effect of Citral and eugenol against *Aspergillus niger* and their application in bread preservation. *Food Chem.* 310: 125974.

Kosalec, I., Pepeljnjak, S., and Kustrak, D. 2005. Antifungal activity of fluid extract and essential oil from anise fruits (*Pimpinella anisum* L., Apiaceae). *Acta Pharmaceutica.* 55(4): 377–385.

Kubo, I., Fujita, K., Kubo, A., Nihei, K., and Ogura, T. 2004. Antibacterial activity of coriander volatile compounds against *Salmonella choleraesuis*. *J. Agric. Food Chem.* 52: 3329–3332.

Kumar, R., Mishra, A. K., Dubey, N. K., and Tripathi, Y. B. 2007. Evaluation of *Chenopodium ambrosioides* oil as a potential source of antifungal, antiaflatoxigenic and antioxidant activity. *J. Food Microbiol.* 115(2): 159–164.

Li, F. G., Chen, J., Chen, W. M., and Ji, M. F. 2015. Evaluation of eugenol embryo toxicity: An experimental model using embryonic stem cells. *Crter.* 19(19): 3017–3021.

Liao, M., Pan, F., Ma, J., Xiao, J. J., Yang, Q. Q., and Hua, R. M. 2018. Acute toxicity of tea tree essential oil to higher animals. *Afr. J. Agric. Sci.* 45(04): 162–166.

Liu, D. L., Ma, S. T., Zeng, R. Y., Pu, Y. X., and Meng, L. J. 2010. Determination of median lethal dose of *Cinnamomum cassia* volatile oil in mice. *Med. J. Southwest. Natl. Def.* 20(5): 209–215.

Llana-Ruiz-Cabello, M., Maisanaba, S., Puerto, M., Prieto, A. I., Pichardo, S., and Jos, A. 2014b. Evaluation of the mutagenicity and genotoxic potential of carvacrol and thymol using the Ames *Salmonella* test and alkaline, Endo III-and FPG-modified comet assays with the human cell line Caco-2. *Food Chem. Toxicol.* 72: 123–128.

Lu, Y., Zhao, Y., Wang, Z. C., Chen, S., and Fu, C. 2007. Composition and antimicrobial activity of the essential oil of *Actinidia macrosperma* from China. *Nat. Prod. Chem. Res.* 21(3): 227–233.

Luo, M., Jiang, L. K., and Zou, G. L. 2005. Acute and genetic toxicity of essential oil extracted from *Litsea cubeba* (Lour.) Pers. *J. Food Prot.* 68(3): 581–588.

Ma, S. T., Liu, D. L., Lan, X. P., Cai, Y., and Bai, P. 2010. Determination of lethal dose of essential oil from clove in mice. *World J. Tradit. Chin. Med.* 5: 67–68.

Mahboubi, M. 2019. *Zingiber officinale* Rosc. essential oil, a review on its composition and bioactivity. *Clinical Phytoscience.* 5(1): 1–12.

Martemucci, G., Portincasa, P., Di Ciaula, A., Mariano, M., Centonze, V., and D'Alessandro, A. G. 2022. Oxidative stress, aging, antioxidant supplementation and their impact on human health: An overview. *Mech. Ageing Dev.* 206: 111707.

Mendanha, S. A., Moura, S. S., Anjos, J. L., Valadares, M. C., and Alonso, A. 2013. Toxicity of terpenes on fibroblast cells compared to their hemolytic potential and increase in erythrocyte membrane fluidity. *Toxicology in Vitro.* 27(1): 323–329.

Mimicadukic, N., Kujundžic, S., Sokovic, M., and Couladis, M. 2003. Essential oil composition and antifungal activity of *Foeniculum vulgare* Mill. obtained by different distillation conditions. *Phytother. Res.* 17(4): 368–371.

Nikolic, B., Mitic-Culafic, D., Vukovic-Gacic, B., and Knezevic-Vukcevic, J. 2011. Modulation of genotoxicity and DNA repair by plant monoterpenes camphor, eucalyptol and thujone in *Escherichia coli* and mammalian cells. *Food Chem. Toxicol.* 49: 2035–2045.

Niu, B., Liang, Y., Liang, J. P., and Liu, Y. 2019. Enrichment of characteristic components of tea tree oil by molecular distillation. *Appl. Food Res.* (4): 25–31.

Oueslati, M. H., Abutaha, N., Al-Ghamdi, F., Nehdi, I. A., Nasr, F. A., Mansour, L., and Harrath, A. H. 2020. Analysis of the chemical composition and in vitro cytotoxic activities of the essential oil of the aerial parts of *Lavandula atriplicifolia* Benth. *J. King Saud Univ. Sci.* 32(2): 1476–1481.

Ozer, H., Sokmen, M., Gulluce, M., Adiguzel, A., Sahin, F., Sokmen, A., and Baris, O. 2007. Chemical composition and antimicrobial and antioxidant activities of the essential oil and methanol extract of Hippomarathrum microcarpum (Bieb.) from Turkey. *J. Agric. Food Chem.* 55: 937–942.

Palma, C. E., Cruz, P. S., Cruz, D. T. C., Bugayong, A. M. S., and Castillo, A. L. 2019. Chemical composition and cytotoxicity of *Philippine calamansi* essential oil. *Ind. Crops Prod.* 128: 108–114.

Parker, R. A., Gabriel, K. T., Graham, K. D., Butts, B. K., and Cornelison, C. T. 2022. Antifungal activity of select essential oils against *Candida auris* and their interactions with antifungal drugs. *Pathogens.* 11(8): 821.

Pasqua, R. D., Hoskins, N., Betts, A. G., and Mauriello, G. 2006. Changes in membrane fatty acids composition of microbial cells induced by addiction of thymol, carvacrol, limonene, cinnamaldehyde, and eugenol in the growing media. *J. Agric. Food Chem.* 54(7): 2745–2749.

Pathania, D., Kumar, S., Thakur, P., Chaudhary, V., Kaushik, A., Varma, R. S., and Khosla, A. 2022. Essential oil-mediated biocompatible magnesium nanoparticles with enhanced antibacterial, antifungal, and photocatalytic efficacies. *Sci. Rep.* 12(1): 1–13.

Peng, J., Tang, L., Xiao, Y., Chen, Y., and Bao, M. H. 2016. Acute transoral toxicity, skin irritation and sensitization of arborvitae essential oil. *China Pharmacist.* 19(7): 1420–1422.

Pitarokili, D., Tzakou, O., Loukis, A. A., and Harvala, C. 2003. Volatile metabolites from *Salvia fruticosa* as antifungal agents in soilborne pathogens. *J. Agric. Food Chem.* 51(11): 3294–3301.

Puškárová, A., Bučková, M., Kraková, L., Pangallo, D., and Kozics, K. 2017. The antibacterial and antifungal activity of six essential oils and their cyto/genotoxicity to human HEL 12469 cells. *Sci. Rep.* 7(1): 1–11.

Rajashekar, Y., Bakthavatsalam, N., and Shivanandappa, T. 2012. Botanicals as grain protectants. *Acad. J. Entomol.* 1–13.

Rana, A., Matiyani, M., Tewari, C., Negi, P. B., Arya, M. C., Das, V., and Sahoo, N. G. 2022. Functionalized graphene oxide based nanocarrier for enhanced cytotoxicity of Juniperus squamata root essential oil against breast cancer cells. *J. Drug Delivery Sci. Technol.* 72: 103370.

Ranjbar, A., Ramezanian, A., Shekarforoush, S., Niakousari, M., and Eshghi, S. 2022. Antifungal activity of thymol against the main fungi causing pomegranate fruit rot by suppressing the activity of cell wall degrading enzymes. *LWT-Food Sci. Technol.* 161: 113303.

Reis, D. R., Ambrosi, A., and Di Luccio, M. 2022. Encapsulated essential oils: A perspective in food preservation. *Future Foods.* 100126.

Ridaoui, K., Guenaou, I., Taouam, I., Cherki, M., Bourhim, N., Elamrani, A., and Kabine, M. 2022. Comparative study of the antioxidant activity of the essential oils of five plants against the H2O2 induced stress in *Saccharomyces cerevisiae. Saudi J. Biol. Sci.* 29(3): 1842–1852.

Saidana, D., Mahjoub, M. A., Boussaada, O., Chriaa, J., Cheraif, I., Daami, M., and Helal, A. N. 2008. Chemical composition and antimicrobial activity of volatile compounds of Tamarix boveana (Tamaricaceae). *Microbiol. Res.* 163(4): 445–455.

Schnitzler, P., Koch, C., and Reichling, J. 2007. Susceptibility of drug-resistant clinical herpes simplex virus type 1 strains to essential oils of ginger, thyme, hyssop, and sandalwood. *Antimicrob. Agents Chemother.* 51(5): 1859–1862.

Shaughnessy, D. T., Setzer, R. W., and De Marin, D. M. 2001. The antimutagenic effect of vanillin and cinnamaldehyde on spontaneous mutation in Salmonella TA104 is due to a reduction in mutations at GC but not AT sites. *Mutat. Res.* 480–481.

Silva, A. M. A., da Silva, H. C., Monteiro, A. O., Lemos, T. L. G., de Souza, S. M., Militão, G. C. G., and Santiago, G. M. P. 2020a. Chemical composition, larvicidal and cytotoxic activities of the leaf essential oil of *Bauhinia cheilantha* (Bong.) Steud. *S. Afr. J. Bot.* 131: 369–373.

Silva, F. L., Silva, R. V., Branco, P. C., Costa-Lotufo, L. V., Murakami, C., Young, M. C., and Moreno, P. R. 2020b. Chemical composition of the Brazilian native *Cinnamomum stenophyllum* (Meisn.) Vattimo-Gil essential oil by GC-qMS and GC× GC-TOFMS, and its cytotoxic activity. *Arabian J. Chem.* 13(4): 4926–4935.

Singh, P., Kumar, A., Dubey, N. K., and Gupta, R. 2009. Essential oil of *Aegle marmelos* as a safe plant-based antimicrobial against postharvest microbial infestations and aflatoxin contamination of food commodities. *J. Food Sci.* 74(6): M302–M307.

Soncu, E. D., Özdemir, N., Arslan, B., Küçükkaya, S., and Soyer, A. 2020. Contribution of surface application of chitosan–thyme and chitosan–rosemary essential oils to the volatile composition, microbial profile, and physicochemical and sensory quality of dry-fermented sausages during storage. *Meat Science.* 166: 108127.

Sonboli, A., Babakhani, B., and Mehrabian, A. 2006. Antimicrobial activity of six constituents of essential oil from aalvia. *Zeitschrift für Naturforschung C.* 160–164.

Sonboli, A., Salehi, P., Kanani, M. R., and Ebrahimi, S. N. 2005. Antibacterial and antioxidant activity and essential oil composition of *Grammosciadium scabridum* Boiss. from iran. *Zeitschrift für Naturforschung C.* 60(7–8): 534–538.

Spinozzi, E., Pavela, R., Bonacucina, G., Perinelli, D. R., Cespi, M., Petrelli, R., and Maggi, F. 2021. Spilanthol-rich essential oil obtained by microwave-assisted extraction from *Acmella oleracea* (L.) RK Jansen and its nanoemulsion: Insecticidal, cytotoxic and anti-inflammatory activities. *Ind. Crops Prod.* 172: 114027.

Stammati, A., Bonsi, P., Zucco, F., Moezelaar, R., Alakomi, H. L., and Von Wright, A. 1999. Toxicity of selected plant volatiles in microbial and mammalian short-term assays. *Food Chem. Toxicol.* 37(8): 813–823.

Stojanović, N. M., Randjelović, P. J., Mladenović, M. Z., Ilić, I. R., Petrović, V., Stojiljković, N., and Radulović, N. S. 2019. Toxic essential oils, part VI: Acute oral toxicity of lemon balm (Melissa officinalis L.) essential oil in BALB/c mice. *Food Chem. Toxicol.* 133: 110794.

Verdi, C. M., Machado, V. S., Machado, A. K., Klein, B., Bonez, P. C., de Andrade, E. N. C., and Santos, R. C. V. 2022. Phytochemical characterization, genotoxicity, cytotoxicity, and antimicrobial activity of *Gautheria procumbens* essential oil. *Nat. Prod. Res.* 36(5): 1327–1331.

Wang, B., Li, P., Yang, J., Yong, X., Yin, M., Chen, Y., and Wang, Q. 2022. Inhibition efficacy of *Tetradium glabrifolium* fruit essential oil against *Phytophthora capsici* and potential mechanism. *Ind. Crops Prod.* 176: 114310.

Wardana, A. A., Kingwascharapong, P., Wigati, L. P., Tanaka, F., and Tanaka, F. 2022. The antifungal effect against *Penicillium italicum* and characterization of fruit coating from chitosan/ZnO nanoparticle/Indonesian sandalwood essential oil composites. *Food Packag. Shelf Life.* 32: 100849.

Weli, A. M., Al-Omar, W. I., Al-Sabahi, J. N., Gilani, S. A., Alam, T., Philip, A., and Al Touby, S. S. 2021. Biomarker profiling of essential oil and its antibacterial and cytotoxic activities of *Cleome austroarabica. ABST.* 3: 1–7.

Wu, Y. N., OuYang, Q., and Tao, N. 2016. Plasma membrane damage contributes to antifungal activity of citronellal against *Penicillium digitatum. J. Food Sci. Technol.* 53(10): 3853–3858.

Wu, Y. N., Wang, Y., Huang, J., Wang, R., and Yao, L. 2010. Safety analysis of two chemical rosemary essential oils. *J. Agric. Sci.* 28(2): 147–150.

Zegura, B., Dobnik, D., Niderl, M. H., and Filipic, M. 2011. Antioxidant and antigenotoxic effects of rosemary (*Rosmarinus officinalis* L.) extracts in *Salmonella typhimurium* TA98 and HepG2 cells. *Environ. Toxicol. Pharmacol.* 32: 296–305.

Zhang, F. Z., Wang, Y., and Yan, W. J. 2013. Determination of LD50 in menthol oil by up-down method and traditional acute toxicity method. *Medi Sci.* 30(4): 268–270.

Zhang, Y. L., Zhang, R. G., and Zhong, Y. 2011. Chemical constituents, antimicrobial activity, antioxidant activity and toxicological properties of thyme essential oil. *Chin. Agric. Sci.* 44(9): 1888–1897.

Zhao, Y. Z., Hao, X. X., Meng, J., and Gao, X. 2020. Stability of antibacterial active components of oregano essential oil and its slow-release characteristics in antibacterial cartons. *Allergen Manage. Food Ind.* 46(20): 114–119.

Zin, W. A., Silva, A. G., Magalhães, C. B., Carvalho, G. M., Riva, D. R., Lima, C. C., and Faffe, D. S. 2011. Eugenol attenuates pulmonary damage induced by diesel exhaust particles. *J. Appl. Physiol.* 112(5): 911–917.

Index

Note: Page numbers in *italics* indicate a figure and page numbers in **bold** indicate a table on the corresponding page.

A

acetylcholinesterase (AChE), 140–141
acidity, 90, 219
active food packaging
  antibacterial properties of, 1–2
  antimicrobial activity and, 243–244
  common food additives and, 247, *247*, 248
  common packaging types, *247*, 248
  definition of, 227
  diffusion-induced release mechanisms, 256
  disintegration-induced release mechanisms, 256
  electrospinning and, 228, 235
  EOs-antibacterial packaging and, 243–248, **249–251**, 252–256
  main types of, *235*
  release mechanisms of active compounds, 256
  sustained-release antimicrobial, 243, 246
  swelling-induced release mechanisms, 256
adventitia, 21
Aeromonas, 4
aflatoxin, **47**, 54
agar plug method, 75
aldehydes
  antibacterial properties of, 14, 17, 87
  antimicrobial activity and, 71, 218, 248
  cytotoxicity and, 272
  fatty acids in cell membranes and, 25
  ligands and, 153
  nanoemulsion and, 211
Alisma plantago-aquatica essential oil, 214
alkaline phosphatase, 23
alkalinity, 90
alkaloids, 136–137, **138**, 139–140
allicin, 29, 72, 124
Alpinia guilinensis essential oil, 26
Alpinia officinalis essential oil, 269
Alternaria alternata, 50
amino acids, 23, 104, 149
antibacterial activity of EOs
  active food packaging and, 1, 243–244
  checkerboard test, 19
  chemical components and, 17
  combined with antibiotics, 87, 94–95, *96*
  combined with food additives, 93–94
  damage effect of essential oil on cell membrane, 23
  damage effect of essential oils on cell wall, 20–21, 23
  dilution method, 18
  disc diffusion method, 18
  edible coatings and, 183–184
  effect of temperature on, 90
  effect on cell morphology, 26
  effect on energy metabolism, 25
  effect on fatty acids in cell membrane, 24–25
  effect on genetic material, 25–26
  effect on protein in cell membrane, 23–24
  evaluation methods of, 17–20
  experimental methods for evaluating, 43–44, *44*, 45–46
  food industry applications, 1–5, 7
  food preservation and, 13–14, 30
  inhibitory mechanisms of, 20–21, 23–27, 29
  nanoemulsions and, 217
  phenols and, 71
  sources of, **70**
  starch microcapsules and, 158, 162
  time sterilization method, 19
  vapor phase and, 76
antibiotics
  accumulation in fish culture, 4
  bacterial disease treatment and, 4
  combined with EOs, 87, 89–90, 93–95, *96*, 220
  livestock and poultry breeding industry applications, 5
  resistance of bacteria to, 53, 87, 89, 94–95
  tetracycline, 94
antifungal activity of EOs
  chemical constituents and, 17
  effect on biofilms, 54–56
  effect on cell membrane, 46, 50
  effect on cell wall, 46, 50
  effect on membrane pumps, 52–53
  effect on mitochondria, 51, *52*
  effect on mycotoxins, 54, **55**
  effect on nucleic acids and proteins, 50–51
  generation of ROS in mitochondria, 53, *53*, 54
  main targets/components of, 46, **47–50**
  minimum inhibitory concentration (MIC) and, **47–50**
  possible action mode, *51*

terpenes and terpenoids, 46
vapor phase of EOs and, 76
anti-inflammatory effects of EOs
cosmetics industry applications, 7, 13
electrospinning and, 238
food preservation and, 219
medical applications, 13, 39
sources and main extraction methods, 14
toxicological characteristics and, 263
antimicrobial activity of EOs, *see also* EOs-antibacterial packaging (EOs-AP)
acidity and alkalinity, 90
active food packaging and, 243
carvacrol and, 17
checkerboard test, 19
chemical constituents and, 17, 40, 70
cultivation conditions for, 89–90
dilution method, 18
disc diffusion method, 18
dispersants and emulsifiers of EOs, 89
edible coatings and, 184, **185–190**, 191
effect of nanoemulsion embedding on, 207, 211–212
evaluation methods of, 18–19, **91–92**
experimental methods for evaluating, 43–46
exposure time and, 89–90
factors affecting, 88–90, 217–218
food preservation and, *211*, 219, 248
lauric acid and, 191
low water solubility and, 30, 207, 219, 227
nanoemulsions and, 201, *206*, 207, **208–210**, 211, 215–217, 252
natural food additives and, 39
number of microorganisms and, 88–89
oxidability of, 30, 207
oxygen content and, 90
packaging theory and, 244
phenols and, 17, 70–71, 74
species of microorganism and, 88
starch microcapsules and, 158, 162–163
stress response of microorganisms and, 89
synergistic antibacterial effects and, 30, 87–88
synergistic effect of drug combinations, 90, **91**, 92, **92**, 93
temperature and, 90
terpenes and, 70–71
time sterilization method, 19
vapor phase of EOs and, 69–70, 74–76, **77–78**
antimicrobial coatings, 182–183, *183*
antimutagenesis, 276
antioxidant activity of EOs
antioxidant capacity and, 105
antioxidant mechanisms and, 108–111
antioxidant system and, 106
antioxidative components and, 106–107, *108*
chelation with metal ions, 109, *109*, 110
direct antioxidant effects, 108–110
extraction methods, 112
factors affecting, 111–113
food industry applications, 94, 103–104
free radicals and, 105–106
free radical scavenging, 108–110
indirect antioxidant effects, 110–111
initiator of oxidation, 113
natural food additives and, 39, 103
oxidative stress and, 104–105
phenolic compounds and, 103
receptor type, 113
regulating level of antioxidant enzymes, 111, *111*
solubility and, 112
sources of, 14
storage condition and, 113
synergism of EO combinations, 30
thyme essential oil and, 6–7
antioxidants
active food packaging and, 227
antimutagenicity and, 276
chemical synthesis of, 103
enzyme systems, 106
eugenol and, 107
exogenous, 105
fat-soluble, 106
food industry and, 103
food protection and, 182
free radicals and, 105, *106*, 110
microencapsulation technology and, 149
nonenzyme systems, 106
oil-soluble, 112
phenylpropyl esters and, 107
protein-based, 106
receptor type and, 113
rosemary essential oil and, 107
solubility and, 112
thyme essential oil and, 7, 107
water-soluble, 106, 112
antiseptic properties, 13, 183, 243
antiviral activity of EOs
active components of, 123–124, **125–126**, 127–129
antiseptic role of, 13
commercial use and, 30
electrospinning and, 238
immunity and, 129
inhibition of plant viruses, 124
mechanisms of resistance, 126–128, **128**, 129
virus plaque reduction method and, 127
virus types and, 122
A. parasitica, 54
Artemisia annua essential oil, 13, 269
Artemisia argyi essential oil, 93, 268–269
Artemisia herba-alba essential oil, 51
Aspergillus flavus, 6, 14, 50, 53, 278

Aspergillus niger
antibacterial properties of EOs and, 14
antifungal activity of EOs and, 46
cell membrane permeability and, 44, 46, 50
effect of EOs on nucleic acids and proteins, 51
effect of EOs on ROS production, 53
inhibitory effects of EOs on, 78, 92, 94
vapor phase of EOs and, 76
ATP (adenosine triphosphate)
cell membrane permeability and, 45, 218, 270
effect of terpenoids on, 51
essential oils and synthesis, 24–25
fungal cells and, 52
ATPase, 25, 51–52, 141

B

Bacillus anthracis, 263
Bacillus cereus, 72, 93, 272
Bacillus lester, 72
Bacillus subtilis, 2, 93, 268–269, 278
bacteria, *see also* antibacterial activity of EOs; Gram-negative bacteria; Gram-positive bacteria
animal disease and, 4–5
antibiotics and, 4
drug resistance in, 53, 87, 89, 93–95
effect of EOs on toxins, 29, *30*
growth morphology of, 26
pathogenic microorganisms, 4
permeability of membrane and, 270
quorum sensing (QS) and, 27, 29
bacterial nitric oxide synthase (BNOS), 53
baked goods, 3–4
basil essential oil, 140, 219
Beer's law, 213
Bergamot essential oil, 7
Betula platyphylla essential oils, 54
biofilm formation
characteristics of, *28*
filamentous fungal, 55–56, *56*
impact of fungal, 55
inhibitory effects of essential oils on, 27, 54–56
sessile cells and, 54–55
three-dimensional, 27
biogenic diseases, 4
biological activity
antimicrobial activity and, 207
checkerboard test and, 19
chemical composition of EOs and, 76
encapsulation and, 219
food preservation and EOs, 13, 220, 243
microorganism concentrations and, 88
nanoemulsions and, 201, 211
natural pesticides and, 136–137
starch microcapsules and, 165
botanical pesticides, 6, 135, 137–138
Botrytis cinerea, 53
browning inhibitors, 182

C

calcium, 25, 109, 182, 194, 272
Campylobacter coli, 25
Campylobacter jejuni, 25
Candida albicans, 46, 50, 53, 56, 71, 95
Candida tropicalis, 56
cardamom essential oil, 25
carvacrol
antifungal activity and, 52–53
antimicrobial activity and, 17, 71, 244
antioxidant activity of, 107
cell membrane fluidity and, 24
damage effect on cells, 26, 53
differences in content of, 263, 267
inclusion complexes and, 164–165
pesticides industry applications, 6
phenolic compounds and, 71
stability of, 269
thyme essential oil and, 52
UV absorbance of, 269
carvanol
antibacterial activity of, 4–5
antimicrobial activity and, 89
antiviral activity of, 123–124
cytotoxicity and, 272
damage to cells, 71
Gram-positive bacteria and, 26
mycotoxins and, 54
carvol, 5, 7, 23, 87, 92, 253
C. austroindica leaf oil, 136
cell membrane
antifungal activity and, 46, 50
carvacrol and, 24
cytotoxicity and, 272
damage effect of carvacrol on, 26, 53
damage effect of essential oil on, 23, *24*, 71
effect of EOs on fatty acids in, 24–25
effect of EOs on microbial cells, 43–46
effect of EOs on protein in, 23–24
EO nanoemulsions and, 216, *216*, 217, 252
experimental methods for evaluating, *44*
integrity of, 45
interactions between EOs-edible coatings, 184, 191, *191*
permeability of, 43–45
phototoxicity and, 275
release of cellular components, 45–46
cell morphology, 26–27, 45–46
cell wall
active food packaging and, 249, 253
antifungal activity and, 46
cytotoxicity and, 270

damage effect of citral on, 50–51
damage effect of EOs on, 20–21, 23, 26
edible coatings and, 191
fungal growth and, 46, 50
Gram-positive/negative bacteria and, 184
nanoemulsions and, 217
pectin and, 177
terpenoids and, 23
cell wall synthase, 20
chamomile essential oil, 7
checkerboard test, 19, *19*
chelation of phenols with metal ions, 109, *109*, 110, *110*
chemical composition of EOs
acids, 16
alcohols, 14, 16
aldehydes, 14
antimicrobial activity and, 70, 76
biological activity and, 76
biological characteristics and, 263
carvacrol content and, 263, 267
concentrations of, 70
distillation time, 42–43
drying conditions and, 42
esters, 16
extraction methods and, 41, 69
factors affecting, 40–43
functional groups of, 14, 16–17
geographical position and, 41
harvest time of plants, 40–41
hydro-carbons, 16
insecticidal effects and, 136–137, *137*
interactions between, 72, 74, 92–93
ketones, 16
molecular formulas in food preservation, 218, *218*, 219
oxides, 16
phenols, 16
structure of, *16*, *123*
terpenes, 16
toxicological characteristics and, 263, 267, 269–270
weather conditions and, 41
chemical ecology, 135
chemical synthetic preservatives, 20
Chenopodium ambrosioides essential oils, 6
chitin, 46, 177
chitosan, 94, 176–178, 193, 211, 215, 255
chrysanthemum essential oil, 23
cinnamaldehyde
anti-biofilm effect of, 27
antifungal activity and, 46
ATP synthesis and, 25
chemical component interactions, 74
food preservation and, 239, 278
inhibitory effects of, 27, 93
livestock and poultry breeding industry applications, 5
nanoemulsion and, 211
vapor phase of, 74
cinnamic acid, 93
Cinnamomum camphora essential oil, 51
cinnamon essential oil
active food packaging and, 254–255
antibacterial properties of, 14, 26, 72
antifungal activity and, 17
antimicrobial activity and, 263
antiviral activity of, 129
cinnamaldehyde and, 278
edible coatings and, 176, 178, 180–181
elecrospun mats and, 239
food industry applications, 3
food packaging and, 76
release of cellular components, 45–46
citral
antifungal activity and, 50–51
antimicrobial activity and, 88
cell membrane permeability and, 44
food packaging and, 76, 78
release of cellular components, 46
citronella essential oil, 17, 128, 162, 211, 276
citronellal, 14, 40, 42, 54, 276
Citrus bergamia essential oil, 275
citrus essential oil, 276–277
citrus peel essential oil, 5
clove essential oil
active food packaging and, 248, 255
antibacterial properties of, 14, 72
antimicrobial activity and, 71, 211
edible coatings and, 181
effect on protein in cell membrane, 23
eugenol and, 277
fruit preservation and, 1
nanoemulsions and, 217
conifer essential oil, 93
Conyza sumatrensis essential oil, 6
core materials, 148–149
coriander essential oil, 269, 277–278
Coriandrum sativum essential oil, 50
Coriaria nepalensis essential oil, 50
cosmetics industry, 1, *2*, 6–7
cumin essential oil, 14, 25, 176, 248
cyclodextrin (CDs), 163–164, *164*, 165
cytotoxicity, 270, 272, 275–277

D

dairy products, 3
Daucus Sahariensis Murb. essential oil, 40
diffusion-induced release mechanisms, 256
dill essential oil, 25
dilution method, 18

disc diffusion method, 18
disintegration-induced release mechanisms, 256
distillation time, 42–43
diterpenes, 138
diterpenoids, 40, 71
DNA (deoxyribonucleic acid)
effect of citral on, 50–51
effect of essential oil on, 25–26, 104
EO inhibition of biosynthesis of, 50
PI dye and, 44
replication of, 25–26
sensitivity to ROS, 104–105
viruses and, 122
drug resistance
bacteria and, 26, 29, 53, 87, 89, 93–95
causes of bacterial, 93
combined EOs and antibiotics, 87, 90, 94–95
combined EOs and food additives, 93–94
EO nanoemulsions and, 220
foodborne pathogens and, 87
microbial, 90, 94
pests/weeds and, 6
plant pests and microorganisms, 141
drugs
antibacterial, 217
antifungal, 55–56
antimutagenicity and, 276
antiviral, 122, **128**, 129
cell walls and sensitivity of, 20
chemical synthetic, 56
core materials and, 149
dilution range of, 18
evaluation methods of, **91–92**
nanoemulsions and, 202–204, 213
synergistic antibacterial effects and, 87
synergistic effects of, 90, 96
traditional medicine and, 121
drying conditions, 42
duodenal villi, 5

E

E. canadensis essential oils, 6
E. coli (*Escherichia coli*)
Alpinia officinalis essential oil and, 269
antimutagenicity of EOs and, 276
Artemisia argyi essential oil and, 268
biofilm formation in, 27
cell membrane permeability and, 44, 71
chrysanthemum essential oil and, 23
cinnamaldehyde and, 24
combined EOs and food additives, 94
effect of essential oil on, 29, *30*, 72
electrospun mats with EOs and, 239
essential oils and, 2–3
foodborne bacteria and, 14
food poisoning and, 87
genotoxicity and, 278
inhibitory effects of EO combinations, 92–93
mutagenicity and, 270
oregano essential oil and, 26
peppermint essential oil and, 24
release of cellular components, 46
synergistic antibacterial effects and, 92
tea tree essential oil and, 89
edible coatings with EOs
antibacterial properties of, 183–184
antimicrobial activity and, 184, **185–190**, 191
composite, 178–179
definition of, 171, 179
development of, 171
dipping method, 179–180
food packaging and, 171–172, 179, 192–193, 195
food preservation and, 176–182, 194
food product appearance and, 192–193
food product texture and, 193–194
food sensory properties related to, 191–194
interactions between microbial cell membranes and, 184, 191, *191*
lipid-based, 172, 178
materials used for, 172, **173–175**
modified starch and, 177
physical and chemical stability of, 194
polysaccharide-based, 172, 176–177
potential food applications of, *176*
production method of, 179–180, *180*, 181, **181**, 182
protein-based, 172, 177–178
requirements for use of, 172
shellac coating and, 193–194
spraying method, 180
spreading method, 181
structure of, *172*
taste and smell of products, 194
thin film hydration method, 181–182
edible starch microcapsules with EOs, *see* EOs–starch microcapsules
Edwardsiella, 4
electron flow system, 108
electrospinning
active food packaging and, 228, 235
development of, 227–228
entrapment of bioactive substances in, 234–235
environmental parameters and, 234
factors affecting quality of, 230, **231–232**, 233
food industry applications, 235, *236*, 239
food preservation and, 235
forces acting on, *229*, 230
manufacturing process, 238, *238*

mats containing EOs and, 238–239
nanofibers and, 230, 233, 235, *236*, *238*
optimization of active-component loading, 236–237
principles of, 228–229, *229*, 230
process parameters and, 233–234, *234*
simulation of release behavior, 237
solution parameters and, 230, *232*, 233
Taylor cone and, 228, 230, 233
embedding techniques, 207
endogenous pathways, 105
energy metabolism, 20, 25
enveloped viruses, 122, 128
enzyme antioxidant systems, 106
EO, *see* essential oils (EO)
EOs-antibacterial packaging (EOs-AP)
active food packaging and, 243–248, **249–251**, 252–256
added to packaging in gaseous form, 255
coated on/adsorbed by packaging materials, 254
controlled-release technology and, 246, *246*
directly mixed in packaging matrix material, 253–254
film-forming materials and, 245
food preservation and, 256–257
made into small antibacterial packages, 255
main indicators related to antibacterial mechanisms, 252, *252*, 253
microencapsulation and, 255–256
migration curve of antimicrobial agents, 245, *245*, 246
process of, 253
realization of, 253–254, *254*, 255–256
research status of antimicrobial packaging theory, 244
tape casting technology and, 244
EOs–starch microcapsules
action mechanisms and, 158
antibacterial properties of, 158, 162
antimicrobial activity and, 158, 162–163
cyclodextrin (CDs) and, 163–165
evaluation index for, **154–157**
food preservation and, 147, 162–165
main materials in, **154–157**
modified starch and, 150
porous starch and, 153, 158
potential food applications of, 163, *163*
preparation formulas for, **159–161**
preparation methods and, 152–153, **154–157**, 158
research purposes of, **159–161**
schematic diagram of mechanisms, *162*
Erigeron canadensis essential oils, 6
essential oils (EO)
absorption mechanism of small molecules, *22*
antianxiety effects of, 7
antimutagenicity of, 276
antiseptic properties of, 13, 183, 243
baked goods and, 3–4
binding constants of, 164
bioactive components in, 122
cell toxicity test cases of, **273–275**
chemical components of, 14, 16, *16*, 17, 40–41
combined with antibiotics, 94–95, 220
combined with food additives, 93–94
cosmetics industry applications, *2*, 6–7
cytotoxicity and, 270, 272–**275**
dairy products and, 3
fish culture and, 2, *2*, 3–5
foodborne bacteria and, 14
food industry applications, 1–2, *2*, 3, 13–14
food preservation and, 13–14, 20, 227, 269–270, 276, 278
food safety and, 267–268, 276–278
generally recognized as safe (GRAS), 263, **264–266**
genotoxicity and, 270
hydrophobicity and, 23–24, 248
livestock and poultry breeding industry applications, *2*, 5
main compounds in, 39–40
medicinal properties of, 7–8, 39
mutagenicity and, 270
odor threshold of, 228, 238
pesticides industry applications, *2*, 5–6
pharmaceutical industry applications, *2*
phototoxicity and, 275–276
possible action pathways of, *21*
schematic representation of possible mechanisms of actions, *21*
sources and main extraction methods, 14, **15–16**, 39, 72
stability of, 268, *268*, 269
thermal sensitivity of, 212
toxicity and, 269–270, **271**, 276–277
volatility of, 72, 76, 135, 165, 212, 243, 255
E. stipitate essential oil, 93
eucalyptus essential oil, 72, 76, 123
eugenol
active food packaging and, 255
antibacterial properties of, 23
antifungal activity and, 46, 51–52, 54
antimicrobial activity and, 71, 87–88
ATP synthesis and, 25
cell membrane permeability and, 44
food packaging and, 76, 78
inhibition of plant viruses, 124
inhibitory effects of, 5
livestock and poultry breeding industry applications, 5
release of cellular components, 46

synergistic inhibition and, 72, 74
toxicity and, 277
Eupatorium buniifolium essential oils, 6
European Regulation (ER), 227
exercise ability, 27
exogenous pathways, 105
extraction methods
antioxidant components and, 112
chemical composition of EOs and, 40, 71–72
chemical composition of essential oils and, 40–42
cold pressing, 72
distillation and, 72
freeze pressing, 72
high-pressure liquid-phase, 112
immersion, 72
main types of, 14, **15–16**
microwave-assisted, 72, 112
oil separation, 72
solid-phase extraction, 112
solvent extraction, 72, 112
solvents for, 112
source and, 14, **15**, 16, **16**, 17
steam distillation, 112
supercritical extraction, 112
supercritical fluid extraction, 72
ultrasonic-assisted, 72
water distillation, 72

F

F1F0-ATP enzyme, 25
fatty acids
aquatic products and, 2
cell membrane and, 45, 104
cytotoxicity and, 272
damage effect of EOs on, 23–25
food additives and, 182
free radicals and, 108
lipid coating and, 178
Federal Food, Drug and Cosmetics Act (FFDCA), 267
fennel essential oil, 5, 8, 13, 123, 137, 215
Feynman, Richard, 201
filamentous fungal biofilms, 55–56, *56*
finger citron essential oil, 217
fish culture, 2, *2*, 3–5
flame ionization detector (FID), 214
food additives
active food packaging and, 247, *247*
antibacterial properties of EOs and, 183, 248
antioxidant activity of EOs and, 39, 103
bioactive release-rate and, 237
combined with EOs, 93–94
consumer preference for natural, 20, 39, 69, 103, 183, 227, 243, 248
edible coatings and, 172, 182–183
essential oils and, 13, 263, 267, 278
microencapsulated, 149
risks of synthetic, 257
foodborne pathogens
antibacterial properties of EOs and, 14, 30, 87
drug resistance and, 87
essential oils and, 14, 20
food poisoning and, 87
inhibitory effects of EOs on, 20, 30
multidrug-resistant, 87
food industry
antioxidant activity of EOs and, 39, 103–104
edible coatings and, 171–172
EO nanoemulsions and, 219–220
essential oils in, 1–3, 13–14, 69, 248
foodborne pathogens and, 14, 20, 87
fresh-keeping technology and, 3, 163, 165, 182
microencapsulation technology in, 149–150
nanotechnology and, 3, 201, 205–206, *206*
starch microcapsules with EOs and, 147, 162–163
food packaging, *see also* active food packaging; electrospinning
antimicrobial, 30, 69
antioxidant, 30, 103
Crank's prediction model and, 244
edible coatings and, 171–172, 179, 192–193, 195
electrospinning and, 227–228, 234–235, 239–240
environmental impact of, 171, 212
essential oils and, 76, 78, 248
inert materials and, 243
microencapsulation technology and, 150, 162
nanoemulsions and, 219
vapor phase of EOs and, 69, 76, 78
food poisoning, 87
food preservation
antimicrobial coatings, 182–183, *183*
antioxidants and, 39, 103, 182
browning inhibitors and, 182, 192–193
chemical additives and, 20, 182, 269
cinnamaldehyde and, 239
combined EOs and food additives, 94
consumer preference for natural, 1, 20, 39, 69, 103, 162, 165, 183, 220, 227, 239
cyclodextrin (CDs) and EOs in, 163–165
edible coatings and, 176–182, 194
EOs-antibacterial packaging and, 256–257
essential oils in, 13–14, 20, 227, 269–270, 276, 278
molecular formulas of EO active components, *218*
nanoemulsions with EOs and, 201, 205–207, *211*, 218–220
protection mechanisms for, 182–183, *183*

stability of EOs and, 269
starch microcapsules with EOs and, 147, 162–163
thymol and, 239
food quality
active food packaging and, 246
decline during storage, 182
effect of essential oil on, 69, 219
enzymatic browning and, 182, 192–193
EOs-edible coatings and, 191–192
lipid-based coatings and, 178
lipid oxidation and, 110
microbial contamination and, 20
nanomaterials and, 201
product appearance and, 192–193
product texture and, 193–194
protection mechanisms for, 182–183
sensory properties of, 191–194, 219
starch microcapsules and, 162–163
food safety
active food packaging and, 243
drug-resistant pathogens and, 87
electrospun packaging and, 240
environmental impact of packaging and, 162, 235
essential oils and, 267–268, 276–278
foodborne bacteria and, 29–30
nanomaterials and, 220
free radicals
effect on lipids, 104
endogenous and exogenous pathways, 105
EO scavenging and, 108–110
exogenous scavengers and, 108
oxidative damage and, 103–104
production of, *106*
RNS and, **107**, 108
ROS and, **107**, 108
scavenging rate of EOs to, 109
thyme essential oil and, 6–7
types of, **107**
fruit preservation
browning inhibitors and, 182
edible coatings and, 171, 176–177, 179–180, 184, 192–194
essential oils in, 1–2, 14
vapor phase of EOs and, 76
F. solani, 50
Fucus spiralis brown algae essential oil, 137
fungal infections, 40, 46, 53–54, *see also* antifungal activity of EOs
Fusarium graminearum, 54
Fusarium moniliforme, 50

G

garlic essential oil
active food packaging and, 255
allicin in, 72
antifungal activity and, 17
antioxidant activity of, 113
antiviral activity of, 124
combined with nisin, 94
starch microcapsules and, 162
sulfide and, 278
gas chromatography (GC), 72, 214
gas chromatography and mass spectrometry (GC-MS), 72
genetic material, 25–26, 104–105, *see also* DNA (deoxyribonucleic acid)
genotoxicity, 270, 278
geographical position, 41
ginger essential oil, 25, 140
ginger extract, 13
glucose, 25, 39, 163
Gram-negative bacteria
absorption mechanism of small molecules, *22*
antimicrobial efficacy of edible coatings and, 184
damage effect of EOs on cell morphology, 20–21, 26
EOs-antibacterial packaging and, 248
inhibitory effects of EOs on, 88
lipopolysaccharide and, 21, 88
Gram-positive bacteria
absorption mechanism of small molecules, *22*
antimicrobial efficacy of edible coatings and, 184
damage effect of EOs on cell morphology, 20–21, 26
EOs-antibacterial packaging and, 248
inhibitory effects of EOs on, 88
grass fruit essential oil, 26

H

headspace (HS), 214
headspace sampling (HS), 72
hemp essential oil, 3
herbal medicine, 121
high-performance liquid chromatography (HPLC), 214–215
high-pressure liquid-phase extraction, 112
hydrophobicity
antibacterial properties of EOs and, 17, 183–184
bioactive release-rate and, 237
biofilm formation and, 56
cyclodextrin (CDs) and, 163–165
essential oils and, 23–24, 90, 153, 201, 215, 239, 248
Gram-positive bacteria and, 20–21
lipid-based coatings and, 178
lipids and, 153
hydroxypropyl methylcellulose (HPMC) coatings, 193
Hymenaea rubriflora essential oil, 93

I

imidacloprid, 143
indirect quantization, 215
Indonesian wood essential oil, 141
insecticidal effects of EOs, *see also* pesticides industry
- active components and, 136–137, *137*
- alkaloids and, 136, 139
- chemical ecology and, 135
- development of resistance to, 141
- inhibitory effects on reproductive ability, 139
- mechanism of, 141
- mechanisms of, 140, **142**
- phenolic compounds and, 138–139
- plants with insect repellents, *136*
- repellent activity, 139–140
- terpenes and, 137–138
- terpenoids and, 137–138

inverted Petri dish method, 75
isoprenes, 39–40

K

Kaempferia galanga Linn essential oil, 109
Kundmannia anatolica Hub.-Mor essential oils, 41

L

Labiatae essential oils, 107
lactic acid bacteria, 3, 94
Lactobacillus campylobacter, 93
Lauraceae essential oil, 107, 136
lauric acid, 182, 191
Lavandula angustifolia Mill essential oil, 111
lavender essential oil, 7–8, 42, 76, 129, 276
lemon essential oil, 45, 176, 178, 182, 254
lemon grass essential oil, 129, 177, 184, 217, 254–255
lilac essential oil, 184, 252
limonene, 277
linalool
- antibacterial properties of, 89, 93
- antimicrobial activity and, 54
- antimicrobial packaging and, 244, 253
- antiviral activity of, 123
- application in cosmetics industry, 7
- distillation time and, 42
- insecticidal effects of, 136, 140–141
- isoprenes and, 40
- mutagenicity and, 277

lipids
- edible coatings and, 172, 178
- effect of citral on, 51
- effect of EOs on, 13, 24–25, 44, 110
- effect of free radicals on, 104
- effect of oxidative stress on, 25, 104
- effect of terpenoids on, 23
- hydrophobicity and, 153
- inhibition of peroxidation, 110
- in livestock and poultry meat, 2, 13
- oxidation of, 110
- peroxidation process, 25, 104, *105*

lipopolysaccharide, 21, 23, 88
lipoteichoic acid, 20
Listeria monocytogenes
- antibacterial properties of EOs and, 24, 29
- antimicrobial activity of EOs and, 89
- antimicrobial efficacy of edible coatings against, 178–179, 184
- combined EOs and food additives, 94
- effect of EOs on cells, 23, 25–26
- effect of EOs on energy metabolism, 25
- effect of EOs on exercise ability, 27
- food poisoning and, 87
- inhibitory effects of EO combinations, 92–93
- inhibitory effects of EOs on, 14
- vapor phase of EOs and, 76

litchi wood essential oil, 269
Litsea cubeba essential oil, 29, 93
livestock and poultry breeding industry, 2, 2, 5, 248

M

maltodextrin, 149, 162, 217
marjoram essential oil, 13
medicinal properties, 7–8, 39, *see also* traditional medicine
melanin synthesis pathway, 6
MERS-CoV, 124
metal ions, 109, *109*, 110, *110*
methicillin-resistant *S. aureus* (MRSA), 93
methylparaben, 94
microbial drug resistance, 39, 90, 94
microemulsion (ME), 202
microencapsulation technology, *see also* EOs–starch microcapsules
- active food packaging and, 255–256
- binding characteristics of, 152–153
- core materials and, 148–149
- development of, 147–148
- direct adsorption method, 152
- food industry applications, 149–150, 163–165
- modified starch and, 150
- porous starch and, 150–151, *151*, 152
- preparation methods and, 152–153, 158
- spray drying method, 152
- structure of, *148*
- wall materials and, 148–149

microorganisms
- antibacterial effects of EOs on, 88–90
- bacteria and, 4

biofilm formation and, 27
chemical synthetic preservatives and, 20
cultivation conditions for, 89–90
dispersants and emulsifiers, 89
drug resistance in, 93–94
effect of essential oil on, 18, 23, 25, 29
exercise ability and, 27
human disease and, 87
livestock and poultry products, 2, 5
number of, 88–89
species of, 88
stress response of, 89
synergism of EO combinations and, 92–93
vapor phase of EOs and, 69
volatile components in, 25
microwave steam diffusion gravity (MHG), 42
Middle East respiratory syndrome (MERS), 121
minimum antimicrobial concentration (MIC), 18
minimum bactericidal concentration (MBC), 18
minimum inhibitory concentration (MIC)
antifungal activity and, 46, **47–50**
antimicrobial activity and natural products, 87
EOs-antibacterial packaging and, 246
EOs in the food matrix and, 238
essential oils and pathogenic organisms, 227–228
vapor phase of EOs and, 75–76
mitochondria, 51, *52*, 53, *53*
modified starch, 150, 177
Monilinia fructicola, 44
monolayer peptidoglycan, 20–21
monoterpenes, 39–41, 123, 138, 141, 248
monoterpenoids, 40, 71
Moringa oleifera essential oil, 29
M. piperita L. oil, 139
mustard essential oil, 76
mutagenicity, 270, 276–278
mycelia, 54, 56
mycotoxins, 54, **55**
Myrtle essential oil, 107

N

Nandina domestica essential oil, 110
nanoemulsion (NEN)
advantages of, 204
antimicrobial activity and, *206*
bottom-up preparation of, 204
characteristics of, 203
development of, 201–202
drug delivery systems and, 202
embedding techniques and, 207
food industry applications, 3
food systems and, *206*
formation mechanism of, 204–205
negative interfacial tension and, 204–205
oil-in-water (O/W), 202–203, *203*
preparation methods and, 202
schematic diagram of add-order, 202, *202*
top-down preparation of, 204
nanoemulsion containing essential oils (EOs-NE)
antibacterial properties of, 217
antibiofilm activity of, 201
antimicrobial activity and, 201, 207, **208–210**, 211, 252
antimicrobial mechanisms and, 215–218
application prospects and challenges, 219–220
average droplet size and, 217–218
chromatographic methods and, 214–215
effect of embedding on antimicrobial activity, 207, 211–212
encapsulation efficiency and, 212, **213**, 215
encapsulation materials and, 205, *207*
entrapment efficiency of, 212–215
fish culture and, 4
food preservation and, 205–207, *211*, 218–220
formulation of, 217
gas chromatography and, 214
high-performance liquid chromatography (HPLC), 214–215
indirect quantization and, 215
interactions with microbial cell membranes and, 216, *216*, 217
nanocarriers and, 205–207
preparation methods and, 204, *205*, *206*, 212
prolonged release times and, 211
quantification of, 212–213, **213**, 214–215
sensory properties of food and, 219
spectrophotometry and, 212–213
surface charge and, 217
Nectandra grandiflora Nees essential oil (NGN-EO), 4
negative interfacial tension, 204–205
nettle essential oil, 177
nisin, 94
nitric oxide (NO), 53
nonenzyme antioxidant system, 106
nuclei acids
effect of EOs on cell membrane and, 23–24, 45–46
effect of EOs on proteins, 50–51
free radical scavenging and, 108
viruses and, 122
nutmeg essential oil, 7

O

Ocimum basilicum essential oil, 93
Oenanthecrocata essential oil, 128
oil-in-water (O/W) nanoemulsions, 202–203, *203*

onion essential oil, 124
orange essential oil, 26
oregano essential oil
  active food packaging and, 248, 254–255
  antibacterial properties of, 26
  antifungal activity and, 53
  antimicrobial activity and, 211
  antioxidant activity of, 107
  antiviral activity of, 128
  carvacrol and, 17
  combined with nisin, 94
  dairy products and, 3
  edible coatings and, 184
  effect on bacterial toxins, 29
  food industry applications, 2
  nanoemulsions and, 213
  poultry industry and, 5
  stability of, 269
Origanum compactum essential oil, 51
oxidant, 276
oxidation, 108, 110, 112–113
oxidative stress, 25, 53, 104–105, 270, 272
oxygen
  antimicrobial activity of EOs and, 90
  antioxidant systems and, 106
  edible coatings and, 179, 192
  electron flow system and, 108
  electrospinning and, 234, 238
  free radicals and, 108–109
  lipid-based coatings and, 178
  microencapsulation technology and, 148
  microwave steam diffusion gravity (MHG) and, 42
  monoterpenes and, 39, 41, 141
  terpenoids and, 40, 248

P

Paracelsus, 39
parsley essential oil, 211
penicillin binding protein (PBP2a), 93
Penicillium roqueforti, 44, 46, 51, 53, 74
peony essential oil, 7
pepper essential oil, 139
peppermint essential oil, 24, 26, 56, 94, 139, 217
perilla essential oil, 94, 129, 180
peroxidation, 25, 104, *105*, 108, 110
pesticides industry, *see also* insecticidal effects of EOs
  botanical pesticides and, 6, 135, 137–138
  chemical pesticides and, 5–6, 135–136, 141, 143
  drug resistance and, 141
  environmental impact of, 135–136, 141, 143
  imidacloprid and, 143
  plant essential oils in, 2, 5–6, 135–143
pharmaceutical industry, 2, 20, 57, 151
phenols
  antibacterial properties of, 14, 16, 87
  antifungal activity and, 76
  antimicrobial activity and, 17, 70–71, 74, 218,248
  chelation with metal ions, *110*
  cytotoxicity and, 272
  effect on genetic material, 26
  effect on protein in cell membrane, 23
  inhibitory effects of, 26
  insecticidal effects of, 138–139
  interactions with cell membrane lipids, 183, 215
  lipopolysaccharide release and, 23
  starch microcapsules and, 153
phenylpropanoid, 123
phototoxicity, 275–276
Pimpinella anisum essential oil, 215
pinene, 278
pine needle essential oil, 107
pine nut essential oil, 162
plant harvest time, 40
pneumolysin (PLY), 29
Pogostemon cablin Benth essential oil, 110
polyphenols, 94, 108–109, 124, 138, 270, 276
polysaccharides, 172, 176–177
porous starch, 150–151, *151*, 152–153, 158
proteins, *see also* whey protein
  antifungal activity of EOs and, 46, 50–51
  antioxidants, 106, 111
  antiviral activity of EOs and, 124, **125**, 126
  cell content leakage, 45–46
  cosmetics industry and, 6
  edible coatings and, 172, 177–179, 194
  effect of EOs on cells, 23–24
  effect of oxidative stress on, 104
  hydrophobic, 56
  phototoxicity and, 275
Pseudomonas aeruginosa, 27, 29, 71, 94
Pseudomonas fluorescens, 27
Pseudomonas tolaasii, 94
PV (peroxide value), 3

Q

quorum sensing (QS), 27, 29, *29*

R

reactive nitrogen species (RNS), 104, **107**, 108
reactive oxygen species (ROS)
  antimicrobial activity of EOs and, 51–53
  cytotoxicity and, 275
  EO alteration of mitochondrial function and, 51

free radicals and, **107**, 108
proteins as targets of, 104
sensitivity of DNA to, 104–105
respiratory syncytial virus (RSV) infection, 128
Rhizopus solani, 74
Rhizopus stolonifer, 44, 50, 53
Ridolfia segetum essential oil, 128
RNA, 122, 124
rose essential oil, 8
rosemary essential oil
antifungal activity and, 56
antioxidant activity of, 107
dairy products and, 3
edible coatings and, 178
food industry applications, 2
food preservation and, 13
protein oxidation and, 270
starch microcapsules and, 162
Rutaceae essential oil, 109–110

S

Saccharomyces cerevisiae, 51, 92–93, 272
sage essential oil, 26, 107
salicylic acid, 94
Salmonella, 71, 87
Salmonella enteritis, 24, 94
Salmonella microsome, 277
Salmonella typhi, 270
Salmonella typhimurium, 23, 44
Salvia miltiorrhiza essential oil, 217
Sandalwood essential oil, 128
Santolina insularis essential oil, 213
SARS-CoV-2, 121–122, *122*, 124, 126, *127*
scanning electron microscopy (SEM), 45
Schinus molle essential oil, 2
sesquiterpenes, 40, 123
sesquiterpenoids, 40, 71, 123
severe acute respiratory syndrome (SARS), 121–122
shellac coating, 193–194
Shigella, 87
solid-phase extraction, 112
solid-phase microextraction (SPME), 72
solubility
antibacterial properties of EOs and, 245
antioxidant activity of EOs and, 112
core materials and, 149
cyclodextrin (CDs) and, 164
EOs and, 7, 17, 30, 89, 153, 219, 227, 238–239
modified starch and, 150, 152
nanoemulsions and, 204, 206–207, 216
terpenoids and, 71
solvent extraction, 72, 112
spectrophotometry, 212–213
Staphylococcus aureus (*S. aureus*)
Alpinia officinalis essential oil and, 269
Artemisia argyi essential oil and, 268
cell membrane permeability and, 44–45, 71
chrysanthemum essential oil and, 23
combined EOs and antibiotics, 93–94
combined EOs and food additives, 94
electrospun mats with EOs and, 239
essential oils and, 2, 14, 26, 29, *30*
food poisoning and, 87
inhibitory effects of EO combinations, 92
release of cellular components, 46
tea tree essential oil and, 89
vapor phase of EOs and, 76
Staphylococcus epidermidis, 89
Staphylococcus typhimurium, 278
steam distillation, 112
streptococcal hemolysin (SLO), 29
Streptococcus, 4, 29, *30*
Streptococcus haemolyticus, 87
Streptococcus pyogenes, 27
supercritical extraction, 112
swelling-induced release mechanisms, 256
synergism of EO combinations
antibacterial properties and, 20, 72, 74, 87
antifungal activity and, 46, 53
antimicrobial effects and, 78, 90, **91–92**, 92–93
checkerboard test and, 19
combined application, 93
combined EOs and antibiotics, 94–95, *96*
combined EOs and food additives, 93–94
effect on ROS production, 53
methods for evaluating, 90, **91–92**
microorganism types and, 92–93
synthetic antimicrobial agents, 39, 76
synthetic fungicides, 40

T

Taniguchi, Nario, 201
tape casting technology, 244
Taylor cone, 228, 230, 233
TBA (thiobarbituric acid), 3
tea bark essential oil, 5
tea leaf essential oil, 5
tea tree essential oil
active films and, 1
antibacterial properties of, 89
antifungal activity and, 46, 56
food packaging and, 76
nanoemulsions and, 215
terpineol and, 276–277
teichoic acid, 20–21
temperature
antimicrobial activity of EOs and, 90
antimicrobial packaging and, 244–245

checkerboard test and, 19
edible coatings and, 193
electrospinning and, 234, 238–239
EO stability and, 269
inverted Petri dish method and, 75
microencapsulation technology and, 150, 158
mycotoxin inhibition and, 54
nanoemulsions and, 4, 205
storage condition and, 113
terpenes
antibacterial properties of, 7, 14, 16, 87
antifungal activity and, 46
antimicrobial activity and, 70–71
antioxidant activity of, 111
essential oils and, 39
inhibitory effects of, 26
insecticidal effects of, 136–138
toxicity and, 277
terpenoids
antibacterial properties of, 87
antifungal activity and, 46
antimicrobial activity and, 71, 248
bacterial cell wall and, 23
essential oils and, 39
impact of geographical position on, 41
inhibitory effects of, 25
insecticidal effects of, 137–138
isoprenes and, 40
oxygen and, 40, 248
reduction of mitochondrial content, 51
Tetraclinisarticulata essential oil, 7
thuja essential oil, 278
thyme essential oil
antibacterial properties of, 71
antifungal activity and, 52–53
antimicrobial activity and, 71
antioxidant activity of, 7, 103, 107
antiviral activity of, 128
carvacrol and, 17, 52–53
combined with chitosan, 94
cosmetics industry and, 7
food industry applications, 2
food preservation and, 2, 13
free radicals and, 6–7
genotoxicity and, 270
nanoemulsions and, 217
release of cellular components, 45
thymol and, 52
thymol
active food packaging and, 255
antibacterial properties of, 23, 71
antifungal activity and, 52–54
antimicrobial activity and, 71, 87
antioxidant activity of, 107
cosmetics industry applications, 6
cytotoxicity and, 270
effect on bacterial toxins, 29
food preservation and, 239
inhibitory effects of, 92
livestock and poultry breeding industry applications, 5
permeability of *E. coli* cell membrane and, 44
pesticides industry applications, 6
time, 40–43, 71, 89–90
time sterilization method, 19
tomato yellow leaf curl virus (TYLCV), 124
toxicity
cell toxicity test cases of EOs, **273–275**
cytotoxicity and, 270, 272, 275
of EOs-NE in food systems, 220
essential oils and, 228, 257, 263, 269–270
evaluation methods of, **271**
genotoxicity and, 270
mutagenicity and, 270
phototoxicity and, 275
Trachyspermum ammi Sprague essential oil, 40
traditional medicine, 121
trans-cinnamaldehyde, 23
transmission electron microscopy (TEM), 45
triterpenes, 138
turmeric essential oil (TEO), 1–2
TVB-N (total volatile basic nitrogen), 3
Tyndall effect, 213

U

unenveloped viruses, 122
US Food and Drug Administration (FDA), 46, 122, 263

V

vapor phase of EOs
agar plug method and, 75
antifungal activity and, 76
antimicrobial activity and, 69–70, 74–76, **77–78**
autoclaved polycarbonate vial, 76
boxes of different materials and, 75
food packaging and, 69, 76, 78
inverted Petri dish method and, 75
potential applications of, 76
Vibrio, 4
Vibrio cholerae, 29
Vibrio fischeri, 27
Vibrio parahaemolyticus, 3, 71, 87, 93–94
virus infection, *see also* antiviral activity of EOs
characteristics of, 121
enveloped/unenveloped viruses, 122, 128
essential oils and, 122–124, 127–129
immunity and, 129
inhibition of plant viruses, 124

as molecular infection, 126–127
overview of, 122
SARS-CoV-2 and, 121–122, 124, 126, *127*
virus plaque reduction method, 127
volatile compounds
detection methods of, 72, **73**
drying conditions and, 42
essential oils and, 69
inverted Petri dish method and, 75
vapor phase of EOs and, 74, 78
volatile oils, 1, 13, 95, 140

W

wall materials, 148–149
weather conditions, 41
whey protein
active food packaging and, 2
edible coatings and, 2, 13, 178–179, 184, 193, 252
microencapsulation technology and, 162
nanoemulsions and, 217
whey protein concentrate (WPC), 193

Y

Yersinia enterocolitica, 93

Z

Zanthoxylum bungeanum essential oil, 95
Zataria multiflora essential oil, 7, 178
Zingiberaceae essential oil, 107
Zingiber officinale essential oil, 269
Ziziphora essential oil, 176